Sustainable Development Goals Series

The **Sustainable Development Goals Series** is Springer Nature's inaugural cross-imprint book series that addresses and supports the United Nations' seventeen Sustainable Development Goals. The series fosters comprehensive research focused on these global targets and endeavours to address some of society's greatest grand challenges. The SDGs are inherently multidisciplinary, and they bring people working across different fields together and working towards a common goal. In this spirit, the Sustainable Development Goals series is the first at Springer Nature to publish books under both the Springer and Palgrave Macmillan imprints, bringing the strengths of our imprints together.

The Sustainable Development Goals Series is organized into eighteen subseries: one subseries based around each of the seventeen respective Sustainable Development Goals, and an eighteenth subseries, "Connecting the Goals", which serves as a home for volumes addressing multiple goals or studying the SDGs as a whole. Each subseries is guided by an expert Subseries Advisor with years or decades of experience studying and addressing core components of their respective Goal.

The SDG Series has a remit as broad as the SDGs themselves, and contributions are welcome from scientists, academics, policymakers, and researchers working in fields related to any of the seventeen goals. If you are interested in contributing a monograph or curated volume to the series, please contact the Publishers: Zachary Romano [Springer; zachary.romano@springer.com] and Rachael Ballard [Palgrave Macmillan; rachael.ballard@palgrave.com].

Nse Udohaya

Impact Investing and Financial Inclusion

Examining the Innovations that Empower the Underserved

palgrave
macmillan

Nse Udohaya
School of Business and Management
University of Dundee
Edinburgh, UK

ISSN 2523-3084 ISSN 2523-3092 (electronic)
Sustainable Development Goals Series
ISBN 978-3-031-87840-4 ISBN 978-3-031-87841-1 (eBook)
https://doi.org/10.1007/978-3-031-87841-1

Cover illustration: Nicolay Pandev gettyimages

This Palgrave Macmillan imprint is published by the registered company Springer Nature Switzerland AG
The registered company address is: Gewerbestrasse 11, 6330 Cham, Switzerland

If disposing of this product, please recycle the paper.

Preface

In a world where economic systems are increasingly being called upon to address pressing social and environmental challenges, the concept of impact investing is a transformative force. This book situates impact investing within the broader discourse on financial inclusion, exploring how these interconnected fields are shaping the future of economic empowerment and sustainable development. The inspiration behind this book stems from the increasing realisation that financial inclusion is not just an economic imperative but a fundamental driver of social transformation. Traditional investment models have long been governed by profit maximisation, often at the expense of broader societal well-being. However, a paradigm shift is underway, demonstrating that finance can serve as an instrument of positive change that balances profit with purpose. The principles of impact investing represent this shift, offering investors opportunities to generate financial returns while addressing some of the world's most pressing issues, including poverty, climate change, gender inequality, and access to essential services like healthcare and education.

This book charts the historical evolution of impact investing, tracing its roots from the ethical investment practices of religious communities to the development of Environmental, Social, and Governance (ESG) criteria and the sophisticated impact measurement frameworks that now guide global financial markets. It examines how financial innovations are shaping the future of inclusive economic growth. Moreover, the book provides real-world case studies illustrating the practical applications of impact investing, from microfinance initiatives in underserved regions to large-scale investments in clean energy and social enterprises. Beyond celebrating the successes of impact investing, this book also critically examines the challenges and limitations of this emerging field. Issues such as greenwashing, impact measurement inconsistencies, and the balance between financial returns and social impact are explored, ensuring a comprehensive understanding of the opportunities and risks inherent in impact investing. Addressing these concerns aims to contribute to the ongoing discourse on how to refine and scale impact investing effectively.

Written for academics, policymakers, investors, business leaders, and students, this book is a valuable resource for those seeking to navigate and contribute to the expanding world of impact finance. Synthesising research, case studies, and practical insights provide a thought-provoking guide that informs decision-making and inspires actionable solutions. As we stand at a critical juncture in global financial evolution, the question is not whether impact investing can succeed but how we can ensure its effectiveness, integrity, and scalability. The intersection of finance and social progress presents a profound opportunity, and I hope this book contributes to the momentum towards a more inclusive and sustainable global economy.

Scotland, UK Nse Udohaya

Acknowledgements First and foremost, I give all glory, honour, and gratitude to God Almighty, whose boundless grace, wisdom, and strength have guided me through this journey. Without His unwavering presence, this work would not have been possible. I extend my deepest appreciation to my mother, Dr. Uduak Udohaya, whose love, sacrifices, and steadfast belief in my potential have been a source of strength and inspiration throughout my life. Her unwavering support has shaped my resilience and determination. To my family, Ofon, Aniema and Idorenyin, your encouragement, patience, and love have been my anchor. Your belief in my dreams and continuous support have been invaluable in this academic and professional journey. A special and heartfelt thank you to Dr. Suzanne Fifield, whose guidance and mentorship throughout my PhD journey transformed me in profound ways. Her wisdom, patience, and relentless encouragement played a crucial role in my growth as a scholar and thinker. Her insights have left an indelible mark on my work and my approach to knowledge.

I also extend my sincere gratitude to my colleagues, friends, and all those who contributed to the development of this book. Their constructive feedback, discussions, and shared experiences have enriched this work in ways that words alone cannot express. This book is the culmination of collective efforts, support, and unwavering faith. I am deeply grateful to all who have played a role in this journey, Thank you.

Competing Interests The author has no competing interests to declare that are relevant to the content of this manuscript.

Contents

PART I

Understanding Impact Investing

CHAPTER 1

Introduction

1.1 The Historical Context

The evolution of impact investing reveals how human values, ethics, and social justice have progressively shaped financial systems to tackle pressing global challenges. The historical context shows a dynamic interplay between societal needs and financial innovation, illustrating how finance can transition from being solely profit-oriented to becoming a powerful tool for achieving meaningful, measurable social and environmental outcomes. This chapter traces the root of impact investing from the ethical investing practices of religious communities to its modern-day manifestation as a transformative force in global finance, providing a foundation that informs and inspires its future. By examining the origins, key developments, and ideological shifts, this section lays the groundwork for understanding why impact investing is more than a financial strategy—it is a movement towards a more equitable, sustainable, and inclusive world. This perspective enriches the reader's comprehension of the subsequent chapters and positions them to appreciate the transformative potential of impact investing in addressing the world's most urgent needs. This chapter uncovers the motivations, milestones, and ideologies that have informed the field, clarifying why this historical lens is indispensable for comprehending impact investing's significance today.

N. Udohaya, *Impact Investing and Financial Inclusion*, Sustainable Development Goals Series,
https://doi.org/10.1007/978-3-031-87841-1_1

1.1.1 The Origins: Ethical and Religious Foundations

The philosophical underpinnings of impact investing can be traced back to the ethical and religious traditions of the eighteenth and nineteenth centuries, where aligning investments with moral values first took root. Religious groups, particularly the Quakers and Methodists, were among the earliest proponents of what would later evolve into socially responsible investing (SRI) (Puaschunder, 2014). For instance, the Quakers adhered to a strict code of ethics that prohibited investment in industries they deemed morally objectionable, such as the slave trade. Similarly, Methodists avoided investments in sectors like tobacco, alcohol, and gambling, which they believed to be incompatible with their faith-based values. Schueth (2003) documents how the Jewish law laid down directives on investing ethically in biblical times. John Wesley, an English theologian and evangelist who founded Methodism in the mid-1700s, gave teachings from the Bible on the ethical use of money (Schueth, 2003). The Methodists, as well as the Quakers[1]—who had never condoned investing in war or slavery, began managing their investments using social screens. The religious groups sought to distance themselves from activities they viewed as objectionable, avoiding companies involved in alcohol, tobacco, gambling, and pornography. It is assumed that these religious groups brought the notion of socially conscious investing to the New World[2] (Bruyn, 1991; Schueth, 2003).

These early practices were not merely about avoiding sin but were grounded in a broader vision of stewardship and social justice. The Quakers' refusal to profit from slavery was a profound moral stance that laid the groundwork for future movements that sought to use capital for social change. However, it is important to recognise that these early forms of ethical investing were primarily exclusionary, focusing on what not to invest in rather than actively seeking to create a positive social impact. Following these earlier incidents of ethical investing, the Boston Ecclesiastical Group established the Pioneer Fund in 1928. It was the first public offering of a screened investment catering to investors seeking to screen "sin" industries (Sparkes, 2003). In the forty years that followed, the ethical investment approach remained a small phenomenon until the late 1960s.

1.1.2 *The Modern Socially Responsible Investment Movement: From Exclusion to Engagement*

The modern socially responsible investing movement began in the mid-twentieth century, influenced by the social and political upheavals of the 1960s and 1970s. The civil rights movement, the Vietnam War, and the rising awareness of environmental degradation catalysed a new wave of activism that extended into the financial markets.[3] Investors began to demand that their investments reflect their ethical and political beliefs. Students called for universities and other non-profit institutions to consider the social consequences of corporate activities from which the institutions derived endowment returns and demanded the institutions discontinue their engagements with corporations or organisations that were supporting or participating in immoral or unethical activities[4]—in particular companies that produced weaponry, military contractors, tobacco companies, and companies doing business in South Africa (Knoll, 2002; Simon et al., 1972). The drive for similar divestment grew from higher education institutions into other sectors. Millions of people, churches, universities, cities, and states have focused their investment strategies on putting pressure on perpetrators to dismantle oppressive systems and structures and on governments and corporations to act on social rights and environmental issues.[5] This drive led to the development of SRI as a formal investment strategy. In 1971, the Pax World Funds launched[6]—one of the seminal moments in the evolution of Socially Responsible Investment (SRI). It was the first mutual fund to explicitly incorporate social criteria into its investment decisions. The fund departed from the purely exclusionary approach of earlier ethical investing by adopting a more proactive stance—actively seeking to invest in companies aligned with the fund's social and environmental values. It began a significant trend in the investment world[7] (Gittell et al., 2012; Gutterman, 2020). During this period, shareholder activism became a powerful tool for promoting corporate responsibility. Investors began using their stakes in companies to push for changes in corporate behaviour, from divesting from apartheid South Africa to advocating for environmental sustainability. The creation of the Interfaith Center on Corporate Responsibility (ICCR) in 1971 further institutionalised this approach, providing a platform for interested investors to engage with corporations on issues ranging from labour rights to environmental stewardship (Ermann & Clements, 1984).

By 1977, Leon Sullivan, a minister and board member of General Motors, established the Sullivan Principles.[8] A set of ethical guidelines for American companies operating in apartheid-era South Africa, aiming to leverage economic influence to promote social justice and improve working conditions for non-white employees in South Africa (Larson, 2020). The principles included the non-segregation of facilities, equal and fair employment practices, equal pay for equal work, training programmes for non-whites, increasing the number of non-whites in management, improving quality of life outside the workplace, and working to eliminate laws impeding social justice. Despite their initial adoption by many corporations, the principles were criticised by anti-apartheid activists who argued they were insufficient to challenge the apartheid system fundamentally.[9] The principles showcased how ethical standards could guide corporate behaviour, even in challenging political environments—the drive to do good while making money continued.

In 1982, Joan Bavaria founded Franklin Research & Development, later renamed Trillium Asset Management—the first asset management firm dedicated entirely to SRI (Jones, 2016). Bavaria's approach involved avoiding investments in companies with poor social or environmental records and actively engaging with companies to improve their practices. Trillium Asset Management played a significant role in the growth and development of the SRI field, advocating for corporate responsibility and sustainable investment practices.[10] Its work demonstrated the viability and importance of integrating social responsibility into investment strategies. Two years later, in 1984, the Social Investment Forum (SIF), now known as the US Sustainable Investment Forum (US SIF), was established to advance the practice and growth of socially responsible investing. It provided resources and support to investors who wished to consider Environmental, Social, and Governance (ESG) criteria in their investment decisions. The Forum played a crucial role in promoting sustainable investment practices and advocating for better corporate responsibility and transparency through research, knowledge products, training, advancing public policy initiatives, public awareness of sustainable investing, and providing opportunities for the sustainable and impact investing community to convene at the national conference and local events. Other countries, including the United Kingdom and the Euro Zone, later established the UK Social Investment Forum, now the UK Sustainable Investment and Finance Association (UKSIF) and the EURO Sustainable Investment Forum (EUROSIF), respectively.

Gutterman (2020) notes that while forming these forums was crucial to the sustainable development movement, the Brundtland Report "Our Common Future", issued in 1987 by the World Commission on Environment and Development (WCED), laid the foundation for modern global development policies. The report stated that the enormous poverty of the South and the non-sustainable patterns of consumption and production in the North were the cause of critical global environmental problems. It called for a strategy to unite development and the environment. The term "sustainable development" was coined. The report conceptualised sustainable development as "development that meets the needs of the present without compromising the ability of future generations to meet their own needs" (World Commission on Environment and Development, 1987, p. 37).

Later, in the 1980s and through the 1990s, shareholder activism emerged and continued to grow as investors used their shares to influence corporate behaviour. With the Bhopal, Chornobyl and Exxon Valdez incidents, awareness of global warming and ozone depletion and the need for environmental responsibility began to rise (Pienitz & Vincent, 2000).[11] The concept of "double bottom line" investing gained prominence during this period, and the idea challenged the conventional wisdom that financial returns should be the sole objective of investment, positing instead that investments could and should generate both financial returns and positive social impact. The period also witnessed the rise of community investing, which sought to direct capital to underserved communities through vehicles such as community development banks and credit unions (Emerson & Twersky, 1996). These developments significantly broadened the scope of SRI, laying the groundwork for a more holistic approach to impact investing that would emerge in the following decades.

The creation of the MSCI KLD 400 Social Index, known initially as the Domini 400 Social Index, in 1990 in the United States for the first time, provided a tool for investors to gauge the performance of socially responsible investments relative to the broader market. The index tracks the performance of US companies that have met specific social and environmental standards, such as environmental sustainability, social responsibility, and ethical governance. As a result of tools like this, in the wake of the South African apartheid, companies who supported or did business in South Africa were screened out. Knoll (2002) documents that by 1993 almost $625 billion had been invested in ethical stocks while funds had been diverted away from such companies. It was not until late 1993 when Nelson Mandela urged the international community to end all sanctions against South Africa at the United Nations Special Committee on Apartheid, that most of the social investment community ceased screening out companies investing in South Africa. By 1995, only $162 billion was being screened, indicating the end of the South African boycott. This led to speculation regarding the demise of SRI. However, in the two years that followed, the assets screened for social responsibility tripled, reaching $529 billion in 1997 and nearly tripled again in 1999, hitting $1.497 trillion.

Critique of Socially Responsible Investment in the Late Twentieth Century

Many scholars and intellects heavily criticised the "social" criterion in investing. A famous example is Milton Friedman's seminal article for the New York Times in 1970—*The Social Responsibility of Business is to Increase its Profits*, where he argued that the primary responsibility of business is to maximise shareholder value. Friedman asserted, "There is one and only one social responsibility of business—to use its resources and engage in activities designed to increase its profits so long as it stays within the rules of the game" (Friedman, 1970, p. 17). According to Friedman, diverting focus from profit maximisation to social responsibilities undermines the fundamental principles of free-market capitalism and leads to inefficiencies. For him, investing had one goal—profit maximisation; whether that came at a cost to other systems was inconsequential, and there was no reason to take into consideration the interdependence of the economic, social, and ecological systems and the externalities of the financial market activities when aiming for financial gain. Drucker (1986) offered a more nuanced critique of SRI. Writing about the social responsibility of a business in his book *Innovation and Entrepreneurship: Practice and Principles* he argued that the primary social responsibility of business is to create wealth and jobs and that other institutions should pursue broader social goals. Drucker's pragmatic approach questioned the efficacy of SRI in addressing complex social issues and whether that was necessary since businesses, by their nature, already induce positive impacts on society. From a political economy standpoint, Tollison and Wagner (1988) also critiqued SRI as often amounting to moral posturing rather than economic efficiency and tends to lead investors away from the fundamental principles of free-market economics towards a politically charged and morally ambiguous agenda. This perspective highlighted the potential for SRI to advance specific political or social ideologies rather than focusing on financial returns and economic growth. However, from the tendency to undermine profit maximisation to businesses assuming responsibilities that are not theirs to bear and to it being a tool for virtue signally, these criticisms and others were not unfounded. Let us look at some of the events that shaped the criticism in Box 1.1.

Box 1.1: Events That Shaped the Criticism

(i) **Economic Crises and Market Volatility**

1973 Oil Crisis: The 1973 oil crisis, triggered by the OPEC oil embargo, led to a sharp increase in oil prices and significant economic turmoil. The event underscored the vulnerabilities in global supply chains and highlighted the interconnectedness of geopolitical events and economic stability. Critics argued that SRI, with its focus on ethical considerations, might overlook such geopolitical risks, leading to suboptimal investment decisions. The crisis prompted a re-evaluation of investment strategies, emphasising the need for a focus on economic fundamentals over ethical considerations.

1980s Savings and Loan Crisis: The Savings and Loan (S&L) crisis of the 1980s was another significant event that shaped the critique of SRI. It involved the collapse of numerous savings and loan associations in the United States, resulting in substantial financial losses. The crisis was partly attributed to deregulation, risky lending practices, and inadequate oversight. Critics of SRI argued that socially responsible investments might not be immune to such systemic risks and could potentially divert attention from the need for rigorous financial scrutiny and regulation.

(ii) **Corporate Scandals and Ethical Lapses**

Lockheed Bribery Scandal (1976): The Lockheed Bribery scandal, where Lockheed Corporation was found to have bribed foreign officials to secure contracts, was a major ethical lapse that influenced the critique of SRI. This scandal highlighted the potential for corruption and unethical behaviour within corporations, leading to scepticism about the sincerity of corporate commitments to social responsibility. Critics argued that SRI might serve as a façade for companies to engage in moral posturing while continuing unethical practices behind the scenes.

Nestlé Boycott (1977): The Nestlé boycott, initiated in 1977, protested the company's aggressive marketing of infant formula in developing countries, which was linked to increased infant mortality. This boycott underscored the power of consumer activism and the importance of ethical considerations in business practices. However, it also highlighted the complexities of assessing corporate behaviour, as Nestlé continued to face criticism despite its efforts to address the issue. Critics of SRI pointed to such incidents to argue that ethical investment criteria could be subjective and challenging to enforce consistently.

(iii) **Rise of Neoliberal Economic Policies**

Reaganomics (1980s): The economic policies of U.S. President Ronald Reagan, known as Reaganomics, emphasised deregulation, tax cuts, and a free-market approach. These policies reflected a broader shift towards neoliberalism, which prioritised market efficiency and shareholder value over social considerations. Critics of SRI argued that these policies demonstrated the effectiveness of free-market principles in driving economic growth and innovation, questioning the need for incorporating social responsibility into investment decisions.

Thatcherism (1980s): Similarly, the economic policies of British Prime Minister Margaret Thatcher, known as Thatcherism, promoted privatization, deregulation, and a reduction in state intervention. These policies were seen as successful in revitalising the British economy. Critics of SRI contended that such market-oriented approaches were more effective in achieving economic prosperity than socially responsible investments, which they viewed as potentially hindering economic efficiency and growth.

(iv) **Emergence of Environmental and Social Movements**

Earth Day (1970): The first Earth Day in 1970 marked the beginning of the modern environmental movement, raising awareness about environmental issues and the need for sustainable practices. While this movement supported the principles of SRI, it also attracted criticism from those who viewed environmental regulations as burdensome to businesses. Critics argued that SRI, influenced by such movements, might prioritise environmental goals at the expense of economic growth and competitiveness.

Anti-Apartheid Movement and Divestment Campaigns (1980s): The anti-apartheid movement in the 1980s, which included widespread divestment campaigns against companies operating in South Africa, was a significant event in the history of SRI. While these campaigns highlighted the power of ethical investing in driving social change, they also faced criticism for their economic impact. Opponents argued that divestment could harm portfolios' financial performance and that political activism should be kept separate from investment decisions.

The criticisms raised important questions about the role of businesses in society and the effectiveness of SRI. While the critiques highlighted potential challenges and inefficiencies associated with SRI, they also underscored the need for a balanced approach to integrating social responsibility into investment decisions.

1.1.3 Integration of Environmental, Social, and Governance (ESG) Criteria

The early 2000s witnessed a significant shift in the investment landscape with the growing integration of Environmental, Social, and Governance (ESG) factors into investment decision-making processes. This development emerged as a response to the increasing recognition that ESG issues could materially affect a company's financial performance and long-term sustainability.[12] High-profile corporate scandals, like Enron and WorldCom, highlighted the importance of strong governance practices and demonstrated that poor governance could lead to significant financial losses, prompting investors to scrutinise governance structures more closely (O'Brien, 2005). Social issues, such as labour practices and human rights, also gained prominence as globalisation expanded supply chains. Companies faced growing scrutiny regarding their treatment of workers, sourcing practices, and community impacts (Avery, 2001). Additionally, the increasing awareness of climate change and environmental degradation underscored the risks associated with environmental factors. Investors began to recognise that these factors could affect a company's brand value, customer loyalty, and operational continuity. This meant that overlooking ESG factors could lead to an incomplete assessment of a company's risk profile and growth potential. ESG integration emerged as a holistic approach to investment analysis, aiming to capture a broader range of risks and opportunities that traditional financial metrics might miss. This shift was driven by ethical considerations and the practical recognition that ESG factors could have a direct and measurable impact on financial returns. Regulatory frameworks and reporting standards also evolved to support ESG integration.

By 2004, the United Nations invited eighteen financial institutions from 9 countries with total Assets Under Management (AUM) of over 6 trillion USD to collaborate and develop a report that will provide guidelines and recommendations on how to better integrate ESG issues in asset management, securities brokerage services and associated research functions. The final report, *Who Cares Wins: Connecting Financial Markets to a Changing World,* was produced by the United Nations Global Compact (UNGC) and endorsed by 20 financial institutions. The following year, the UNEP Finance Initiative (2005) Freshfields Report was released, which gave the first evidence of the economic relevance of ESG issues and a discussion of the concern of fiduciary duty in the use of ESG information in investment decisions. The two reports launched the UN-backed Principles for Responsible Investment (PRI) in 2006. The PRI provides a framework for investors to consider long-term sustainability and ethical impacts. It currently has more than 5,391 signatories, including asset owners, investment managers, and service providers in nearly 100 countries, with a total of US$121.3 trillion in AUM. Their market power allows them to influence corporate practices in ESG domains through various governance mechanisms (Gond & Piani, 2012). Establishing the UN PRI provided investors with tools to assess and compare companies' ESG performance and enabled the institutionalisation of the approach.

1.2 Emergence of Impact Investing

The 1990s and early 2000s were a critical period in the evolution of impact investing, as the concept began to coalesce into a distinct field. While SRI had mainly focused on avoiding harm and promoting corporate responsibility, impact investing sought to take a more intentional and proactive approach to generating positive outcomes.[13] One of the pivotal moments in the formalisation of impact investing was the publication of *The Double Bottom Line: Investing in California's Emerging Markets* report by the California State Treasurer's Office in 2000. The report framed the double bottom line concept and advocated for public-sector leadership, particularly in underserved markets (Angelides, 2000). By late 2000, impact investing had emerged as a distinct field, marking a significant evolution in the approach to investing with a dual focus on generating financial returns and creating positive social or environmental impacts (O'donohoe et al., 2010). The investment approach was formalised when the Rockefeller Foundation formally coined impact investing in 2007. The formalisation played a pivotal role in shaping and supporting the development of the field. It became a unique investment approach, distinct from traditional philanthropy or SRI. By 2009, impact investing exploded worldwide and was adopted by governments, private equity firms, and not-for-profits. These developments solidified the foundation for impact investing. As a result, the Global Impact Investing Network (GIIN) was created in the same year to increase the scale and effectiveness of impact investing by building critical infrastructure, fostering a supportive ecosystem, and advocating for the importance of impact measurement and management. GIIN's role has been central in unifying various stakeholders, including investors, asset managers, and entrepreneurs, around the goal of generating measurable social and environmental impacts alongside financial returns (GIIN, 2024; O'donohoe et al., 2010).

GIIN launched the Impact Reporting and Investment Standards (IRIS) as part of its role. The IRIS+ became the industry's first widely accepted system for measuring, managing, and reporting on the social, environmental, and financial performance of impact investments (Jordan-Kirwan & Tengtio, 2020). IRIS+ provides a standardised framework that allows investors to compare the impact performance of different investments, thereby enhancing transparency and accountability.[14] Later in 2009, the US Government opened the Office of Social Innovation and Civic Participation to support impact investments, social innovation, and civic engagement. Also, the United Kingdom launched the Big Society Scheme to support the social sector and encourage impact investment. In August 2010, the first Social Impact Bond (SIB) was launched in London to reduce reoffending rates or recidivism among short-sentenced offenders in the United Kingdom while generating financial returns for impact investors. In the following years, the SIB became a popular financial mechanism for impact investment.[15] The period witnessed the launch of a number of impact investing funds and vehicles, which demonstrated that it was possible to achieve both financial performance and positive societal outcomes.[16] These early funds targeted various sectors, including renewable energy, affordable housing, microfinance, and healthcare, showcasing the breadth of opportunities within the impact investing landscape. These developments helped establish impact investing as a viable and attractive option for investors seeking to align their investments with their values. Intermediary organisations and industry standards of impact investment continued to develop.

1.3 Growth, Institutionalisation, and Standardisation

The 2010s saw rapid growth and increased recognition of impact investing as a mainstream strategy. Academic research played a significant role in validating the impact investing approach. For example, a study by Mudaliar and Bass (2017), published by GIIN analysed the financial performance of over 1300 impact investments and found that most met or exceeded expectations for both impact and financial performance. Similarly, research by Brest and Born (2013) examined the trade-offs between impact and financial returns and argued that impact investing could achieve competitive returns while generating significant positive externalities. Impact investing expanded significantly, with many institutional investors, family offices, and high-net-worth individuals entering the space. This influx of capital was driven by a combination of factors, including heightened awareness of global challenges such as climate change and inequality and a growing belief that financial markets could be leveraged to address these issues (Bugg-Levine & Emerson, 2011; Clark et al., 2014).

Additionally, regulatory developments in regions like Europe with the Sustainable Finance Disclosure Regulation (SFDR) pushed for greater transparency and accountability in impact investing. In March 2011, the GIIN launched ImpactBase, a global directory of impact investments.[17] In September of the same year, the B Lab[18] launched the Global Impact Investing Rating System (GIIRS)—a comprehensive and transparent system for assessing the social and environmental impact of developed and emerging market companies and funds. It became one of the prominent criteria for assessing the social impact for impact investors. These standards and the IRIS+ became a critical component of impact investing and marked significant developments in impact measurement. They provided a standardised framework for measuring and reporting the social and environmental performance of impact investments.

As it continued gaining traction among specialised investors and within the broader financial industry, major financial institutions such as banks, pension funds, and insurance companies started incorporating impact investing strategies into their portfolios, recognising the potential for these investments to deliver social and financial value. Impact investing benefited from the creation of dedicated funds and vehicles. The funds catered to various asset classes, including private equity, venture capital, fixed income, and real estate. Vehicles included green bonds, social impact bonds, and dedicated impact funds managed by leading investment firms. They provided investors with various options to achieve both financial returns and impact, depending on their risk tolerance, return expectations, and impact objectives. The field continued to evolve, with ongoing efforts to refine impact measurement practices, improve reporting standards, and scale the market. The growth was also supported by the UN's Sustainable Development Goals (SDGs), which provide a global framework for addressing critical challenges and offer a roadmap for aligning investments with sustainable development objectives.

1.4 The Mainstreaming of Impact Investing

The growth and the mainstreaming of impact investing occurred as investor demand increased. Impact-focused financial products proliferated, and the recognition of the role of private capital in addressing global challenges increased. The United Nations Sustainable Development Goals (SDGs) launched in 2015 provided a globally recognised framework for impact investors to align investment strategies with specific social and environmental objectives. Major financial institutions like Goldman Sachs, JPMorgan Chase, and BlackRock began to enter the impact investing space, signalling the acceptance of impact investment as a legitimate and potentially lucrative investment strategy. Also, the emergence of impact-focused venture capital funds highlighted the growing interest in financing early-stage companies with social missions. With the mainstreaming of impact investing, many investors continued to integrate ESG criteria into mainstream investment processes, where they needed more incentive to intentionally invest for social or environmental outcomes. While ESG and impact investing are distinct, ESG focuses on managing risks and opportunities related to non-financial factors. In contrast, impact investing emphasises intentionality and measurable impact; however, they are often complementary. The rise of ESG has helped to legitimise the idea that social and environmental factors are material to financial performance, creating a more conducive environment for impact investing to thrive.

A significant development in this period was the growing role of technology in impact investing. Advances in data analytics, artificial intelligence, and blockchain enabled more sophisticated impact measurement and reporting, enhancing transparency and accountability. In addition, other factors, like social and environmental crises and regulatory developments, supported or accelerated the movement of impact investing from the margins to the mainstream.

1.4.1 The Role of Technology in Mainstreaming Impact Investment

Advances in financial technology (FinTech), data analytics, blockchain, and digital platforms transformed how investors assessed, managed, and deployed capital, making impact investing more accessible, transparent, and efficient (Feyen et al., 2021). These technological innovations accelerated impact investing from a niche activity to a more mainstream practice. In particular, digital platforms democratised access to impact investment opportunities by allowing individuals and smaller investors to invest directly in projects previously available only to large institutions (Bajulaiye et al., 2020). Crowdfunding platforms and online marketplaces emerged, reducing barriers to entry and expanding the pool of available capital. These platforms also facilitated cross-border investments, enabling investors to support developing and emerging markets (Maula & Lukkarinen, 2022) projects. Also, transparency and accountability were made possible by blockchain technology through the provision of an immutable ledger. Blockchain ensured the accuracy of impact data and enabled real-time tracking of fund usage. That level of transparency helped build investor confidence and attracted more mainstream capital. Blockchain-based smart contracts also automated impact reporting, enhancing the efficiency and reliability of impact measurement. Reliable impact measurement was critical for impact investing to gain mainstream acceptance. Advances in data analytics and big data enabled the gathering and analysis of large volumes of impact data, significantly improving measurement quality and consistency. Machine learning and artificial intelligence (AI) helped investors identify trends, predict outcomes, and make informed decisions, enhancing credibility and attracting more capital to impact investments.

Furthermore, automation of investment processes, such as due diligence and reporting, reduced costs and sped up capital deployment, which was particularly important for impact investments with smaller deal sizes and complex metrics. Robo-advisors and AI-driven platforms offered personalised investment recommendations, making impact investing more accessible to retail investors and reducing the need for specialised knowledge. By enhancing accessibility, improving transparency, enabling effective impact measurement, and increasing efficiency, technology allowed more investors to participate and more capital to flow into impactful projects. Continued technological innovation will ensure that impact investing reaches its full potential and becomes a central part of the global investment landscape.

1.4.2 *The Impact of the Global Pandemic and Climate Change Awareness on Impact Investing*

The COVID-19 pandemic acted as a catalyst for accelerating impact investing. The global health crisis highlighted social, environmental, and economic interconnectedness. It exposed vulnerabilities in global supply chains, healthcare systems, and social safety nets, as well as systemic inequalities and the vulnerabilities of marginalised communities. It underscored the need for investments prioritising resilience, sustainability, and social welfare. As economies shut down and healthcare systems struggled, investors increasingly realised the limitations of traditional investment approaches that prioritised short-term gains over long-term resilience (Pavicevic et al., 2021), as they began to prioritise investments that contribute to building more resilient and sustainable economies.[19] The heightened awareness of global vulnerabilities and the interconnectedness of social, economic, and environmental systems profoundly shaped the investment landscape, including impact investing. This realisation led to a surge in interest in impact investing, which aimed at generating positive social and environmental outcomes alongside financial returns. Impact investors prioritised projects that addressed the inequalities by improving access to healthcare, education, and economic opportunities for underserved populations. During and after the pandemic, many investors sought opportunities to contribute to sectors that support societal well-being, such as healthcare, education, and affordable housing, resulting in increased capital flow into impact-focused projects (Craven et al., 2020).

In addition to the pandemic, environmental crises, such as wildfires, extreme weather events, and biodiversity loss, drove the growth of impact investing. The increasing frequency and intensity of these crises brought climate change and environmental sustainability to the forefront of investors' minds. Investments in clean energy, circular economy initiatives, and climate resilience projects became central to the impact investing agenda as investors recognised the need for sustainable solutions to ensure long-term economic stability. It catalysed innovation in financing models and instruments and allowed impact investors to channel funds into projects that support pandemic recovery, such as small business financing, affordable housing, and public health initiatives. The growth of these innovative financing models expanded the reach and scale of impact investing, enabling it to address urgent global challenges more effectively.

While the global pandemic and environmental crises accelerated the growth of impact investing, on the other hand, the economic uncertainty caused increased risk aversion among some investors, particularly in the early stages of the crisis. Many impact projects, especially those in developing and emerging economies, faced funding shortfalls as capital flows were redirected towards more established markets or lower risk assets (Kasyanenko et al., 2023). However, these challenges may have also presented opportunities for impact investing to demonstrate its resilience. The ability of impact-focused projects to deliver both financial returns and positive social outcomes during times of crisis has helped build investor confidence in the sector. The crises have also underscored the importance of long-term, sustainable investment strategies, encouraging more investors to incorporate impact considerations into their portfolios.

1.4.3 Regulatory Developments and Their Role in the Growth of Impact Investing

One of the key challenges in impact investing was the need for standardised metrics and reporting frameworks. Regulatory bodies and international organisations seek to establish consistent standards to measure and report social and environmental impact. Initiatives such as the European Union's Sustainable Finance Disclosure Regulation (SFDR) and the Task Force on Climate-related Financial Disclosures (TCFD) were created to encourage transparency and accountability in impact investing. The SFDR, introduced in 2021, requires asset managers and financial institutions to disclose how sustainability risks are integrated into their investment processes. This regulation has helped create a common framework for assessing and reporting the sustainability impact of investments, making it easier for investors to compare opportunities and make informed decisions. The TCFD provides guidelines for companies and investors to disclose climate-related risks and opportunities, promoting greater transparency and helping investors assess the long-term impact of their investments. These initiatives became instrumental in building trust among investors and driving more capital towards impact-focused projects.

Additionally, some governments used regulatory incentives to promote impact investing. They introduced various tax incentives, grants, and subsidies to encourage investments in sectors that generate social and environmental benefits. For instance, many countries offer tax breaks for investments in renewable energy projects, affordable housing, and other sectors contributing to Sustainable Development Goals (SDGs).[20] These incentives made impact investing more attractive by improving the risk-return profile of impact-focused projects. Governments and regulatory bodies also established frameworks for issuing green and social bonds, providing guidelines on their use and ensuring that the proceeds are directed towards projects that deliver measurable social or environmental impact. The growth of the green bond market, for example, has been driven by regulatory initiatives that provide clarity and credibility (Tolliver et al., 2020), making it easier for investors to participate in impact investing.

Another significant regulatory development that contributed to the growth of impact investing was the introduction of mandatory Environmental, Social, and Governance (ESG) integration for financial institutions. Regulators in several jurisdictions, including the European Union and the United Kingdom, introduced rules requiring asset managers and institutional investors to incorporate ESG factors into their investment decision-making processes. By making ESG integration mandatory, regulators ensured that impact considerations are embedded in mainstream investment strategies, encouraging more investors to allocate capital towards projects that generate positive outcomes. The EU's Green Taxonomy is a notable example of how regulation can drive impact investing. The taxonomy provides a classification system for environmentally sustainable economic activities, offering clear guidance for investors on what constitutes a sustainable investment. By defining the criteria for sustainable activities, the Green Taxonomy helped reduce greenwashing and provided a framework for investors to align their portfolios with environmental objectives.

Furthermore, regulatory developments also influenced the fiduciary duties of institutional investors, encouraging a shift towards long-term, sustainable investment strategies. In many jurisdictions, regulations clarified that fiduciary duty includes considering the long-term social and environmental impact of investments. For example, the UK Pension Schemes Act 2021 requires pension funds to consider climate-related risks and opportunities as part of their fiduciary duties. Similarly, the US Department of Labor's 2021 guidance on the Employee Retirement Income Security Act (ERISA) allows fiduciaries to consider ESG factors when making investment decisions, emphasising the importance of sustainability in meeting long-term financial goals. This encouraged institutional investors, such as pension funds and insurance companies, to incorporate impact considerations into their investment strategies, thereby increasing the flow of capital into impact investing. By promoting long-term, regulators have helped shift the focus from short-term financial gains to sustainable value creation. This shift has been crucial in driving the growth of impact investing, as it aligns the interests of investors with broader societal goals and encourages investments that contribute to positive social and environmental outcomes. Mainstreaming and growth of impact investing help to address pressing global challenges and build a more sustainable future.

By the 2020s, impact investing had become integrated into mainstream financial practices. The growing adoption by major financial institutions, advanced technology, heightened awareness of global challenges like climate change and the COVID-19 pandemic, and regulatory developments all contributed to the rapid expansion of impact investing of social and environmental impact are no longer seen as secondary to financial returns but are instead viewed as integral to achieving sustainable and resilient long-term growth.

1.5 Impact Investing and the SDGs

The Sustainable Development Goals (SDGs) represent one of the most ambitious and transformative efforts in global governance, embodying a collective vision for a sustainable and equitable future. Adopted in 2015 as part of the United Nations' 2030 Agenda for Sustainable Development, the SDGs serve as a comprehensive framework of 17 interconnected goals and 169 targets. They are designed to address the most pressing social, economic, and environmental challenges of our time, underpinned by the principle of "leaving no one behind" (UNSCEB, 2017). The SDGs emerged after the Millennium Development Goals (MDGs), an eight-goal framework that marked the international community's first coordinated effort to reduce extreme poverty and improve human welfare. While the MDGs achieved notable successes, including halving global poverty rates by 2015,[21] they were critiqued for their narrow focus on developing countries and their limited engagement with systemic global challenges such as inequality, climate change, and unsustainable consumption patterns.[22] In response, the United Nations initiated an inclusive global consultative process, engaging over eight million participants from governments, civil society, academia, and the private sector. This inclusive process culminated in the 2030 Agenda, formally adopted by all UN member states in September 2015, which integrates economic, social, and environmental dimensions of sustainable development into a cohesive, universally applicable framework (United Nations, 2015). Unlike its predecessor, each goal is interconnected for the SDGs, recognising that progress in one domain often catalyses advancements in others. For instance, addressing poverty (SDG 1) necessitates investments in quality education (SDG 4), gender equality (SDG 5), and access to clean water and sanitation (SDG 6). Similarly, combating climate change (SDG 13) requires sustainable practices in energy (SDG 7), urbanisation (SDG 11), and consumption patterns (SDG 12). The SDGs transcend traditional development frameworks by incorporating universally relevant goals. While issues like poverty and Hunger remain central, newer priorities such as innovation (SDG 9), reduced inequalities (SDG 10), and life below water (SDG 14) signal a broader understanding of development as a shared global responsibility. Since their adoption, the SDGs have catalysed notable progress. Countries have mobilised resources and reoriented policies towards achieving sustainable outcomes. The expansion of renewable energy sources, particularly solar and wind power, and progress in reducing child mortality are emblematic of successes in certain domains. However, the journey has been far from linear. Structural challenges, including entrenched inequalities and fragile governance systems, remain formidable barriers (Bhandari, 2024).

Furthermore, global crises such as the COVID-19 pandemic, escalating geopolitical tensions, and the climate emergency reversed gains in critical areas, particularly food security and poverty alleviation (Foreign Commonwealth & Development Affairs, 2023). The recent Sustainable Development Goals Report shows that progress on many SDGs, including zero Hunger (SDG 2) and reduced inequalities (SDG 10), is off track, underscoring the urgency of reinvigorated commitments. Specifically, 48 per cent of the target assessed exhibit moderate to severe deviations from the desired trajectory, 30 per cent show marginal progress and 18 per cent moderate progress. Alarmingly, 18 per cent reveal stagnation and 17 per cent regress below the 2015 baseline levels[23] (United Nations, 2024). The SDGs are not simply a roadmap but a rallying call for collective action. They signify a paradigm shift in how humanity envisions progress, emphasising inclusivity, sustainability, and partnership at unprecedented scales. Their success hinges on a multi-stakeholder approach that leverages the unique strengths of governments, international organisations, private enterprises, civil society, and individuals. Financial innovation, policy coherence, and grassroots activism are pivotal in transforming these global ambitions into tangible realities. The interaction between the Sustainable Development Goals (SDGs) and impact investing represents a profound convergence of aligned objectives. While the SDGs provide a comprehensive framework to address global priorities, impact investing offers the capital, creativity, and entrepreneurial energy necessary to operationalise these priorities. This convergence is rooted in a shared vision for social, economic, and environmental well-being, and it highlights the transformative potential of channelling private investment towards public good.

(a) **A Shared Vision of Systemic Change**

A shared vision of systemic, long-term change drives the convergence of the SDGs and impact investing. Both frameworks prioritise sustainable solutions beyond addressing symptoms to tackling the root causes of social, economic, and environmental challenges. With their ambitious scope, the SDGs call for integrated responses to complex global issues, from eradicating poverty and Hunger to promoting gender equality, climate action, and peace. Similarly, impact investing aims to create lasting social and environmental value by providing the capital necessary to drive innovation and empower solutions that address systemic inequities. Impact investing's emphasis on intentionality complements the SDGs' holistic framework. This alignment allows impact investors to systematically target specific SDGs, ensuring that investments are strategically placed to maximise impact. For instance, unlike traditional philanthropy, which often focuses on isolated projects, impact investing seeks to address the interdependencies between the goals, fostering scalable solutions that can sustain long-term development.

(b) **Addressing the SDG Financing Gap**

One of the critical challenges to achieving the SDGs is the significant financing gap. Estimates suggest an additional \$2.5 trillion annually is required to meet the SDGs in developing countries alone. Traditional funding sources—including government budgets, official development assistance (ODA), and philanthropic donations—are inadequate to address this gap comprehensively. Given the scale of the challenge, the participation of private capital is indispensable. Impact investing emerged as a powerful tool to bridge this financing gap by mobilising private capital towards sustainable development. The unique advantage of impact investing lies in its ability to attract diverse investors—from high-net-worth individuals and family offices to institutional investors such as pension funds and insurance companies—who are interested in creating both financial returns and societal impact. By integrating impact considerations into investment decisions, impact investing channels much-needed resources towards sectors critical for SDG achievement.

(c) **A New Paradigm for Corporate Responsibility**

The convergence of the SDGs and impact investing redefines corporate responsibility by integrating social and environmental goals into business strategy. Companies increasingly recognise that their long-term success is intrinsically linked to the well-being of the communities and environments in which they operate. Rather than viewing sustainability as an add-on or peripheral concern, corporations embed impact considerations into their core decision-making processes. Impact investing provides a pathway for businesses to directly contribute to the SDGs by financing projects that create measurable social and environmental benefits. For example, corporate venture funds are used to invest in clean energy startups, social enterprises, and other initiatives aligning with SDG priorities. This approach enhances a company's social licence to operate and positions it to capture emerging opportunities in the transition to a more sustainable economy. In addition, companies are leveraging new financial instruments, such as sustainability-linked bonds and social impact bonds, to demonstrate their commitment to the SDGs. These instruments explicitly link financial returns to the achievement of specific impact outcomes, thereby aligning investor interests with sustainable development objectives. By adopting these innovative tools, corporations are moving beyond traditional Corporate Social Responsibility (CSR) programmes towards a more integrated and impactful approach that drives financial performance and societal progress.

(d) **Interconnected and Measurable Goals**

A key feature of the SDGs is their interconnected nature, which requires integrated and holistic approaches to problem-solving. This interconnectedness mirrors the philosophy of impact investing, which often focuses on multiple dimensions of impact simultaneously. Impact investing's multidimensional approach allows investors to generate outcomes that contribute to several SDGs simultaneously. For example, investing in renewable energy not only advances SDG 7 (Affordable and Clean Energy) but also contributes to SDG 13 (Climate Action), SDG 9 (Industry, Innovation, and Infrastructure), and SDG 8 (Decent Work and Economic Growth). The focus on measurable impact is another critical area. The SDGs are defined by specific targets and indicators, providing a comprehensive framework for measuring progress towards each goal. Similarly, impact investors use metrics and reporting standards—such as the Global Impact Investing Network's (GIIN) IRIS+ system—to track and communicate their investments' social and environmental impact. By utilising these measurement tools, impact investors can demonstrate their contributions to the SDGs transparently and with accountability, attracting more capital to the impact investing ecosystem.

1.5.1 Contribution to the SDGs

Impact investing has made substantial contributions across various sectors that are instrumental to achieving the SDGs. Some primary sectors have experienced systemic change driven by impact investments.

Sustainable Agriculture and Food Security (SDG 2): Agriculture is vital for food security, livelihoods, and rural development, particularly in developing countries. Impact investing has contributed to SDG 2 (Zero Hunger) by funding projects that promote sustainable agricultural practices, improve productivity, and enhance the resilience of food systems. Investments in agroecology, climate-smart farming, and food supply chain innovations have enabled smallholder farmers to increase their yields, access new markets, and adapt to the impacts of climate change (Ma & Rahut, 2024). Impact investors have supported various initiatives to transform agriculture into a more sustainable and resilient sector. Investments in agroecology, climate-smart farming practices, and supply chain innovations have empowered smallholder farmers to improve their yields and adapt to the impacts of climate change. For example, Acumen has invested in agricultural enterprises in Africa and South Asia, helping small-scale farmers improve productivity through access to modern tools, improved seeds, and better irrigation systems[24] (Acumen, 2024). Such investments have also facilitated better market access for farmers, enabling them to achieve fairer prices for their produce and improve their livelihoods. Another example is the impact investment by the AgDevCo initiative, which focuses on developing sustainable agribusinesses in Sub-Saharan Africa. AgDevCo has provided catalytic funding to agricultural enterprises that serve local and regional markets, helping them expand operations, improve efficiencies, and support smallholder farmers through contract farming arrangements. By facilitating access to finance, technical assistance, and infrastructure, AgDevCo has contributed to enhancing the resilience of food systems while increasing incomes for smallholder farmers (AgDevCo, 2023). Agricultural technology (agritech) investments have also played a vital role in enhancing food security. For instance, impact investors have backed companies like Twiga Foods in Kenya, which uses technology to streamline the food supply chain and reduce post-harvest losses.[25] By connecting farmers directly with retailers and providing efficient logistics solutions, Twiga Foods ensures that fresh produce reaches urban markets at affordable prices, enhancing food security and supporting local farmers.

Climate resilience is a key focus of impact investments in agriculture, particularly in regions vulnerable to the effects of climate change. The One Acre Fund, an impact-driven social enterprise, has received significant investment to support smallholder farmers in East Africa.[26] Through training in climate-smart farming techniques, access to quality inputs, and financing, One Acre Fund helps farmers build resilience to unpredictable weather patterns and maintain food production even in challenging conditions. Such interventions are crucial in ensuring vulnerable communities have reliable access to nutritious food. Moreover, impact investing in agriculture has targeted gender inclusion, recognising women's critical role in food production and rural economies. Women comprise a substantial portion of the agricultural workforce in developing countries, yet they often lack access to land, finance, and training. Impact investors are working to bridge this gap by funding initiatives providing women farmers access to credit, agricultural inputs, and capacity-building programmes. For example, Root Capital has provided loans and technical assistance to women-led agricultural cooperatives in Latin America and Africa, empowering female farmers to boost productivity and contribute to household food security.[27] Investments in sustainable agriculture and food security impact investors are helping eradicate Hunger, fostering economic growth, reducing poverty, and promoting environmental sustainability. These investments contribute to building resilient food systems that can adapt to climate challenges and ensure a stable supply of nutritious food for all. SDG 2 is a key focus for impact investors aiming to create lasting, systemic change.

Healthcare and Well-Being (SDG 3): The healthcare sector is another area where impact investing has driven significant progress, particularly in underserved communities, by providing essential capital for healthcare infrastructure, medical technology, and affordable healthcare services. The healthcare sector has witnessed a significant increase in impact investment, with investments increasing from USD 4.349 billion in 2017 to USD 13.798 billion in 2022, currently standing at about USD 33 billion. These investments have been crucial in expanding access to quality healthcare, improving health outcomes, and addressing systemic healthcare inequalities. However, the current investment value still falls significantly short of the World Health Organisation's (WHO) estimated requirement of USD 371 billion annually to meet targets for Health and Well-Being for All in Lower Middle-Income Countries and to achieve

Universal Healthcare (UHC) for 52 per cent of the world's population[28] (World Health Organization, 2017). This gap highlights the need for increased private-sector engagement to complement government and philanthropic funding.

Impact investing in healthcare has supported initiatives that improve healthcare infrastructure, such as building hospitals and clinics in underserved areas, providing medical equipment, and improving healthcare delivery systems. Investments in renewable energy solutions for health facilities, such as solar power for clinics, have also played a crucial role in ensuring that essential services can be delivered even in areas without reliable electricity. For instance, organisations like the UBS Optimus Foundation have collaborated with partners to expand healthcare infrastructure in underprivileged regions, ensuring healthcare access for vulnerable populations. Medical technology and digital health solutions have also been a key focus for impact investors aiming to foster innovation and enhance healthcare accessibility. The COVID-19 pandemic was a pivotal moment that showcased the potential of impact investing in addressing urgent health needs. During the pandemic, impact investors mobilised capital to support healthcare solutions, including vaccine production and distribution, medical equipment, and digital health innovations like telehealth services. For example, the Global Health Investment Fund provided financing to support the production and distribution of COVID-19 vaccines in low-income countries. Additionally, impact investment firms like LeapFrog invested in digital health companies that expanded telehealth services to underserved populations, bridging the healthcare access gap during social distancing and healthcare disruptions.

Furthermore, impact investments have focused on healthcare initiatives that address pressing public health challenges in vulnerable communities. The Rockefeller Foundation invested in expanding COVID-19 testing capacity in Africa, aiming to mitigate the impact of the pandemic on vulnerable populations and support essential health infrastructure. Similarly, UBS Optimus Foundation collaborated with Gavi, the Vaccine Alliance, to facilitate vaccine delivery and expand healthcare access in underprivileged regions, demonstrating the power of partnerships in maximising impact. Impact investing in healthcare has also targeted social determinants of health, such as clean water, sanitation, and nutrition, to create a more holistic approach to health and well-being. By funding projects that improve access to clean water, provide nutritious food, and promote better living conditions, impact investors contribute to overall community health, further supporting SDG 3. These initiatives have led to improved health outcomes and increased economic and social inclusion for underserved populations, especially women and children, who are often disproportionately affected by health disparities. The contributions of impact investors are helping to address urgent global health challenges and promote well-being. However, a considerable gap exists between current investment levels and what is required to achieve universal healthcare. To bridge this gap, increased collaboration between governments, private investors, and philanthropic organisations is necessary to ensure that health and well-being are accessible to all.

Education (SDGs 4): Education is another sector that is driving sustainable development, and impact investing has played an essential role in its advancement by funding initiatives that provide access to quality education and vocational training, particularly in low-income communities. Impact investing in education has significantly increased access to quality education through funding for educational technology (edtech) platforms, vocational training, and infrastructure development. Investments in edtech platforms have expanded educational opportunities to millions of children and adults, especially in remote and rural areas. For example, Omidyar Network's investment in edtech startups in India and Africa has provided digital learning opportunities to students who otherwise would not have access to quality educational resources. Such investments have helped bridge the digital divide, making education more accessible to underserved communities and promoting lifelong learning (Dalmia & Kanodia, 2020). Edtech solutions like GUVI in India and Eneza Education in Kenya have revolutionised education by providing tailored learning experiences, affordable access, and innovative digital tools to support students. GUVI, incubated at IIT Madras, provides affordable and accessible coding courses in multiple Indian languages, making technical education available to students and working professionals. In contrast, Eneza Education provides mobile-based education to students in Kenya and other African countries, making learning materials accessible even in regions with limited internet connectivity.

Furthermore, impact investors have funded vocational training and skill development programmes, empowering youth and adults with the skills needed for employment and entrepreneurship. Such programmes have been crucial in addressing youth unemployment and promoting economic growth. For instance, Acumen, an impact investment fund, has supported several initiatives that provide vocational training in West Africa, equipping young people with technical skills that are in demand in the job market.[29] By empowering individuals with practical skills, impact investors are helping to reduce poverty and foster inclusive economic growth. In addition to edtech and vocational training, impact investments have focused on improving educational infrastructure, such as building schools, providing essential learning materials, and training educators (Convergence, 2022). Investments in community schools and low-cost private schools have helped increase enrolment rates and improve the quality of education in underserved regions. The COVID-19 pandemic highlighted the importance of resilient education systems (World Bank et al., 2022) and accelerated the adoption of digital learning solutions. During the pandemic, impact investors played a critical role in supporting edtech startups and remote learning initiatives that ensured continuity of education despite school closures. For example, the Global Schools Forum partnered with impact investors to support distance learning initiatives, providing digital devices and internet connectivity to support students' recovery post-COVID-19.[30]

While technology has made a wide range of informal learning opportunities accessible, Antoninis et al. (2023) note that technology use in education systems remains relatively limited in many countries, even in the most technologically advanced nations. The digital divide, particularly in low-income and rural areas, hinders access to quality education. Impact investing can help address these challenges by funding initiatives that ensure equitable access to digital infrastructure and support the development of localised education content tailored to the needs of specific communities. Education-focused impact investments are vital for breaking the intergenerational cycle of poverty, improving social mobility, and fostering inclusive growth. By funding initiatives that expand access to quality education, impact investors are not only contributing to SDG 4 but also empowering individuals and communities to achieve better socio-economic outcomes. The role of impact investing in education is crucial for building a more equitable world where everyone has the opportunity to learn, grow, and thrive.

Gender Equality (SDG 5): Gender equality is a fundamental human right and foundation for a peaceful, prosperous, and sustainable world. Impact investing has been crucial in advancing SDG 5 by supporting projects that economically empower women, provide access to essential resources, and address gender-based disparities across multiple sectors. Impact investing in gender equality has primarily focused on improving women's access to finance, promoting female entrepreneurship, fostering gender parity in education and employment, and supporting initiatives that reduce gender-based violence. Investments in these areas have helped create more inclusive economies where women have equal opportunities to thrive and contribute to social and economic development. One key area of focus for impact investors has been improving women's access to finance. Women often face significant barriers to capital and financial services, particularly in developing countries. To address these barriers, impact investors have supported programmes that provide women with the necessary capital to grow their businesses. For example, the International Finance Corporation (IFC) launched the Banking on Women programme, which has mobilised over USD 2 billion to support women entrepreneurs in developing countries. This initiative has enabled women to expand their businesses, create jobs, and contribute to local economic growth by providing access to credit and other financial services. In addition to access to finance, impact investing has also focused on promoting female entrepreneurship through initiatives that provide women with business education, mentorship, and networking opportunities. Goldman Sachs' "10,000 Women" initiative is an example. The programme provides women entrepreneurs with business education and access to capital, empowering them to grow their businesses and drive economic growth. Through initiatives like these, impact investors provide women with the tools they need to succeed and create a positive ripple effect that benefits entire communities.

Education and employment are also critical components of achieving gender equality. Impact investments have supported initiatives that promote gender parity in education, ensuring that girls have the same opportunities as boys to access quality education. For instance, projects like the Educate Girls Development Impact Bond in India have successfully increased school enrolment and improved learning outcomes for girls in underserved regions. By investing in education, impact investors are helping to break down the barriers that prevent girls from reaching their full potential and are contributing to a more equitable society. Employment and workforce participation are other important areas where impact investing has driven progress towards gender equality. Investments in companies that prioritise gender diversity and promote inclusive workplace policies have encouraged businesses to adopt practices that support women's advancement in the workforce. For example, impact investors have supported initiatives that provide training and upskilling opportunities for women in traditionally male-dominated sectors, such as technology and manufacturing, helping to close the gender gap in these industries.

Impact investing has also played a role in addressing gender-based violence and promoting women's safety. Initiatives that fund organisations working to prevent violence against women and provide support services for survivors have contributed to creating safer communities and empowering women to live free from violence. By addressing the systemic issues that perpetuate gender-based violence, impact investors are helping to create environments where women can thrive and fully participate in society. By focusing on gender equality, impact investors can help drive societal change that benefits women and entire communities, thereby contributing to sustainable development. Investments that empower women economically, provide access to education and healthcare, and promote gender parity in the workforce are instrumental in reducing gender-based inequalities and fostering inclusive development. The contributions of impact investing to SDG 5 pave the way for a more just and equitable world where everyone can reach their full potential.

Clean Water and Sanitation (SDG 6): Access to clean water and sanitation remains a significant challenge in many parts of the world, affecting public health, economic productivity, and overall well-being. Impact investing has been essential in advancing SDG 6 by funding initiatives that improve water infrastructure, develop innovative water purification technologies, and expand access to clean water and sanitation services, particularly in underserved regions. Impact investing in clean water and sanitation has focused on several key areas, including water infrastructure projects, affordable water purification technologies, decentralised water distribution, and sanitation infrastructure. These investments are crucial in enhancing health outcomes, improving the quality of life, and fostering economic and social inclusion for vulnerable populations. One prominent example of impact investing in water and sanitation is WaterEquity, an asset manager that has mobilised over USD 460 million to provide microloans for water and sanitation solutions to underserved populations. These microloans have enabled millions to access clean water and safe sanitation, significantly improving health outcomes and quality of life. By focusing on affordable and community-driven solutions, WaterEquity has empowered individuals and communities to take control of their water and sanitation needs. Aqua for All, a foundation focused on providing universal access to safe drinking water and sanitation services, has leveraged blended finance to attract private investments in water and sanitation projects across Africa and Asia. Focusing on affordable and sustainable clean water access solutions ensures that even the most marginalised communities can benefit from improved water infrastructure and services.

Innovative technologies have also been a key focus of impact investors working to address water and sanitation challenges. Solar-powered water purification systems and decentralised water distribution networks have received substantial backing from impact investors, enabling access to potable water in remote areas. Blue-Orchard, another impact investment firm, has invested in projects providing off-grid water solutions to rural communities, ensuring that people living in remote locations can access clean water. By supporting innovative and scalable solutions, impact investors are helping bridge the water access gap for underserved populations. Sanitation infrastructure is another critical area of focus for impact investors. Investments in the construction of sewage treatment facilities, eco-friendly toilets, and waste management systems have played a crucial role in reducing the spread of waterborne diseases, contributing significantly to public health outcomes. Access to proper sanitation improves health and enhances dignity and quality of life for millions of people. Impact investors have also supported initiatives that promote community-led sanitation programmes, empowering local communities to manage their sanitation needs and fostering long-term sustainability. The impact of these investments extends beyond immediate health benefits; they also enhance educational and economic opportunities, particularly for women and children. In many communities, women and children collect water, a time-consuming task that often prevents them from attending school or engaging in income-generating activities. By providing clean water and proper sanitation, impact investors are helping to alleviate this burden, enabling women and children to spend more time in school and participate in economic activities. This, in turn, fosters greater social and economic inclusion, helping to break the cycle of poverty.

Renewable Energy and Climate Action (SDG 7 and SDG 13): Transitioning to renewable energy is a cornerstone of efforts to combat climate change and ensure universal access to affordable energy. Investments in renewable energy projects, such as solar, wind, hydro, and geothermal, have been central to these efforts. International Energy Agency (2024) reports that global investments in renewable energy surpassed $2 trillion by 2023 and are projected to reach $3 trillion by the end of 2024. Impact investors have provided critical funding for these projects, particularly in regions with limited access to electricity. These investments have improved energy access, reduced greenhouse gas emissions, and contributed to climate adaptation by providing reliable and sustainable energy solutions to underserved communities. Funds like the Global Climate Partnership Fund (GCPF) have financed renewable energy and energy efficiency projects in emerging markets. By providing loans to local financial institutions, GCPF helps Small and Medium-sized Enterprises (SMEs) and households adopt renewable energy technologies, such as rooftop solar panels and energy-efficient appliances.[31] These projects contribute to climate mitigation by reducing carbon emissions and support economic growth by lowering energy costs for businesses and households. In addition to funding renewable energy projects, impact investors have focused on enhancing climate resilience through investments in climate-smart infrastructure and technologies. The Rockefeller Foundation, for example, has invested in distributed renewable energy systems to provide resilient power solutions for critical infrastructure, such as hospitals and schools, in regions vulnerable to climate-related disruptions.[32] These investments help ensure essential services remain operational during extreme weather events, contributing to SDG 7 and SDG 13.

The development of green bonds has further accelerated investments in renewable energy and climate action. Green bonds are financial instruments that raise capital for projects with environmental benefits, such as renewable energy installations and energy efficiency upgrades. The green bond market has grown exponentially, with annual issuances exceeding $500 billion in 2023. Impact investors are key in purchasing these bonds, providing much-needed capital for large-scale renewable energy and climate resilience projects. For instance, the African Development Bank (AfDB) has financed solar and wind energy projects across Africa by issuing green bonds, helping the continent transition to a low-carbon economy. Impact investments in renewable energy and climate action have also focused on promoting innovation in clean energy technologies. Companies like Breakthrough Energy Ventures have invested in early-stage companies developing cutting-edge renewable energy technologies, including advanced battery storage, hydrogen fuel, and carbon capture and storage (CCS) solutions.[33] These investments aim to accelerate the development and deployment of technologies essential for achieving deep decarbonisation and mitigating climate change. Impact investors are making significant contributions to SDG 7 and SDG 13. These investments help reduce greenhouse gas emissions, mitigate climate change, and provide affordable and reliable energy to underserved populations, fostering sustainable economic development. The role of impact investing in advancing renewable energy and climate action is crucial for building a more resilient and sustainable future, making SDG 7 and SDG 13 central targets for investors seeking to create lasting, systemic change.

Resilient Infrastructures and Innovations (SDG 9): The goal focuses on building resilient infrastructure, promoting inclusive and sustainable industrialisation, and fostering innovation. Impact investing continues to significantly advance this goal by channelling funds into projects that enhance infrastructure, stimulate industry, and support technological development in underserved regions.[34] Projects that align with this goal encompass a broad spectrum, including infrastructure development, industrialisation efforts, and technological innovations. Impact investors have supported initiatives that improve transportation systems, energy infrastructure, and communication networks, which are critical to economic growth and social development. For example, investments in renewable energy infrastructure, such as off-grid solar solutions, have enabled communities without access to the national power grid to benefit from reliable and clean energy sources. Companies like SunFunder have mobilised millions in impact capital to support solar power projects, directly contributing to sustainable industrialisation and energy access across Africa.[35] Moreover, impact investing has catalysed the growth of Small and Medium-sized Enterprises (SMEs) by providing essential capital for innovation and expansion—these enterprises often serve as the backbone of industrialisation, creating jobs and fostering economic development. Funds like the Small Enterprise Assistance Funds (SEAF) target SMEs in emerging markets, promoting inclusive growth by empowering local businesses and encouraging innovation.[36] Impact investors also foster innovation by backing projects that develop new technologies and innovations in manufacturing, agriculture, and digital services. Investments in tech-driven startups are helping to build digital infrastructure and promote innovation in areas like fintech and agritech, which are essential for expanding economic opportunities and modernising industries. For instance, investments in digital payment systems and mobile banking solutions improve financial inclusion, allowing businesses to thrive in regions that traditionally lack access to financial services.

The ease of securing funding for SDG 9 projects, however, varies. Many SDG 9 targets lag due to insufficient funding, and a considerable gap remains to be addressed. However, private finance, including corporate funding, has increased by 10 per cent annually over the past decade, driven by growing stakeholder and consumer expectations for Environmental, Social, and Governance (ESG) activities. This trend presents a significant opportunity to channel private resources towards SDG 9 initiatives. Impact investing, therefore, serves as a powerful mechanism for addressing the challenges posed by SDG 9. By funding infrastructure projects, supporting SMEs, and driving innovation, impact investors are helping to build resilient economies and create inclusive industrial growth that leaves no one behind. The contributions of impact investing to SDG 9 are paving the way for sustainable development that is both equitable and innovative.

1.5.2 *Challenges and Opportunities*

Impact investing has made significant strides in advancing the SDGs, yet several challenges must be addressed to unlock its potential fully. One of the major challenges is the lack of standardised impact measurement frameworks. Although frameworks like IRIS and the SDG Impact Standards have been developed, the impact investing sector still lacks universally accepted standards for measuring and reporting impact. This lack of standardisation creates several issues for investors. For example, different impact investing funds may use varying metrics to assess social and environmental impact, making it challenging to compare the effectiveness of investments across funds. Inconsistent measurement approaches also hinder investors' ability to aggregate data across portfolios, leading to difficulties in assessing overall performance. Furthermore, the absence of standardisation often results in a lack of transparency and accountability, as impact claims may not be backed by comparable evidence, reducing the confidence of both investors and beneficiaries in the reported outcomes. More reliable, comparable metrics are crucial to scaling impact investing effectively. Another challenge is the limited flow of investment capital into high-risk regions, particularly in developing countries and conflict-affected areas. These regions often face systemic barriers, such as political instability, weak governance, and underdeveloped infrastructure, which can deter private investors. Impact investing in these regions requires innovative financing mechanisms that de-risk investments and provide long-term support to ensure sustainability. Blended finance, which combines concessional public capital with private investment, has shown promise in mobilising funding for high-risk regions, but its scale remains insufficient to meet the massive needs of these areas. Increased collaboration among governments, Development Finance Institutions (DFIs), philanthropic organisations, and private investors is required to bridge the funding gap.

Capital absorption capacity is another significant challenge. Many local enterprises and initiatives that align with the SDGs cannot absorb large-scale capital due to limited operational scale or financial maturity. These enterprises often need technical assistance, capacity-building support, and patient capital to develop into investment-ready opportunities. Addressing this issue requires impact investors to adopt a holistic approach beyond just providing finance; it requires building local capacity, fostering entrepreneurship, and supporting the development of sustainable business models that can scale and attract larger investments. Despite these challenges, impact investing has substantial opportunities to drive systemic change and contribute meaningfully to the SDGs. The growing interest in sustainable finance and increased awareness of social and environmental issues have created a favourable environment for impact investing. Institutional investors, including pension funds and insurance companies, are increasingly integrating Environmental, Social, and Governance (ESG) criteria into their investment strategies, which aligns well with the objectives of impact investing. This trend opens up new avenues for mobilising large-scale capital to fund projects addressing the SDGs. Technological advancements also present significant opportunities for impact investing. Innovations in fintech, digital platforms, and data analytics have made identifying, assessing, and managing impact investments easier. Digital platforms have democratised access to impact investing opportunities, allowing smaller investors to participate alongside institutional investors. Data analytics and blockchain technology have improved transparency, making it easier to track the social and environmental impact of investments and enhance accountability. These technologies can help overcome some of the impact measurement and reporting challenges, providing investors with the information they need to make informed decisions.

Moreover, Public–Private Partnerships (PPPs) have emerged as a powerful tool for financing SDG-related projects. By combining the strengths of the public sector—such as regulatory support and concessional financing—with the efficiency and innovation of the private sector, PPPs can address critical infrastructure and development needs. For example, in the renewable energy sector, PPPs have been used to finance large-scale solar and wind projects in developing countries, helping to expand energy access and mitigate climate change. The SDGs have also fostered a greater sense of collaboration among stakeholders. Governments, non-profits, corporations, and investors increasingly work together to address complex challenges requiring coordinated solutions. This collaborative approach has led to the development of multi-stakeholder initiatives, such as the Global Impact Investing Network (GIIN) and the Blended Finance Taskforce, which aim to create enabling environments for impact investing to thrive. By pooling resources, sharing knowledge, and aligning efforts, stakeholders can accelerate progress towards achieving the SDGs and ensure that impact investing remains a key driver of sustainable development. Despite these challenges, the growing interest in sustainable finance and the increasing recognition of the SDGs as a framework for investment present significant opportunities. The SDGs have fostered collaboration among governments, non-profits, and the private sector, creating an enabling environment for impact investing to thrive. By pooling resources, sharing knowledge, and creating innovative solutions, these stakeholders can accelerate progress towards achieving the SDGs and ensure that impact investing remains a key driver of sustainable development.

Impact investing provides a compelling pathway to achieving the Sustainable Development Goals by leveraging private capital to address critical social, environmental, and economic challenges. By aligning investments with the SDGs, impact investors can drive systemic change across key sectors such as renewable energy, healthcare, education, and sustainable agriculture. While challenges remain, the growing momentum behind impact investing and sustainable finance offers immense potential to contribute to a more equitable and sustainable future. As the world strives to meet the SDGs by 2030, impact investing will be increasingly crucial in mobilising the resources and driving the innovation needed to create lasting positive change.

1.6 Conclusions

The historical journey of impact investing reveals profound insights into the evolving role of finance as an agent of transformative social change. Rather than merely recounting its evolution, this chapter invites us to critically examine the broader implications of how financial mechanisms have been repurposed to tackle pressing global challenges. The narrative of impact investing is not just about the milestones reached but about the paradigm shifts they represent—a movement from viewing capital purely as a vehicle for profit to understanding it as a catalyst for systemic change. One of the key revelations is the recognition that finance, traditionally an instrument of wealth accumulation, can be harnessed to serve the public good without sacrificing economic value. The progression from ethical investing, focusing on moral avoidance, to impact investing's proactive pursuit of measurable benefits underscores a shift in the ethos of capital allocation. This shift reflects a growing consensus that financial returns and social impact are not mutually exclusive but can be mutually reinforcing. The chapter also underscores the critical role of diverse stakeholders in shaping the trajectory of impact investing. The increasing participation of institutional investors, high-net-worth individuals, Development Finance Institutions, and government bodies is indicative of a broader shift in the financial ecosystem. This inclusivity has expanded the pool of available capital and democratised the notion of who gets to influence the future of sustainable development. The evolution of tools like IRIS+ and the rise of social bonds signify a move towards greater accountability, providing the necessary infrastructure for impact to be tracked, measured, and scaled effectively. Another important insight drawn from this historical analysis is the power of intentionality in investment strategies. Unlike traditional investing, where social outcomes may be incidental, impact investing places intentionality at its core. This approach challenges entrenched notions of value creation by directly embedding social and environmental objectives into financial decision-making. The chapter reveals that this intentionality is a moral imperative and a practical strategy for mitigating risks and enhancing long-term portfolio resilience.

The conclusion of this chapter highlights that the maturation of impact investing is not merely about the proliferation of financial instruments or the increase in capital flows. It is fundamentally about redefining the purpose of finance. The real success of impact investing lies in its potential to inspire a fundamental shift in how markets operate—towards inclusivity, sustainability, and long-term value creation. The challenge ahead is maintaining momentum by refining impact measurement, ensuring transparency, and striking a delicate balance between financial performance and the pursuit of public good. These elements will determine whether impact investing can fulfil its promise of contributing to a more equitable and sustainable world.

Notes

1. An influential proto-evangelical Christian movement.
2. Schueth (2003: 189) writes that social investing, socially responsible investing, ethical investing, socially aware investing, socially conscious investing, green investing, value-based investing, and mission-based or mission-related investing are often used interchangeably and all refer to the same general process. Having defined ethical investing as the exercise of ethical and social

criteria in the selection and management of investment portfolios, generally consisting of company shares, Sparkes (2003) argued that ethical investing more appropriately describe how the non-profitmaking bodies such as churches, charities, and environmental groups (rather than companies) invested their funds.

3. The Civil Rights Movement in the United States fought for racial equality and justice, challenging systemic racism and discrimination, and sparked widespread protests, demonstrations, and civil disobedience, leading to significant legislative changes and social progress. The Vietnam War was a prolonged and divisive war, primarily fought in Vietnam, which fueled anti-war protests and a growing disillusionment with government policies in the United States and other involved countries. This led to increased activism and a demand for greater transparency and accountability from corporations and governments. And a growing concern over environmental degradation and pollution led to the emergence of the global environmental movement. Activists raised awareness about issues like climate change, deforestation, and water pollution, prompting calls for sustainable practices and corporate responsibility. This marked the early stages of the sustainable and impact investing movement, which continues to grow and evolve today.
4. These concerns became prominent during student-led conferences at Yale in 1968 and 1969, where participants challenged the ethics of university investments linked to companies involved in apartheid South Africa and the Vietnam War.
5. Schueth (2003) writes about socially concerned investors focusing their investment strategies to effect change in South Africa apartheid, and on the provision of healthy working conditions in factories around the world.
6. The founders, Luther Tyson and Jack Corbett, were United Methodist ministers who sought to create an investment vehicle that did not support the Vietnam War or other unethical activities.
7. Gittell et al. (2012) provides a detail discussion on the fund.
8. The Sullivan Principles were formally announced on March 1, 1977. In 1999, six years after the Apartheid, it was relaunched, alongside the then United Nations Secretary General Kofi Annan as the Global Sullivan Principles.
9. Critics contended that the principles allowed corporations to continue their operations without making significant political changes, thus enabling the apartheid regime to maintain its power. See Larson (2020).

10. Jones (2016) explores Joan Bavaria's journey to becoming a campaigner for corporate sustainability and her work through Trillium Asset Management and Ceres.
11. The Bhopal Disaster in 1984 was a catastrophic gas leak at a Union Carbide pesticide plant in Bhopal, India, resulted in the release of toxic methyl isocyanate gas, leading to thousands of deaths and long-term health problems for hundreds of thousands more—see Lapierre and Moro (2009) for detailed account. Chernobyl Disaster in 1986 was a nuclear accident at the Chernobyl Nuclear Power Plant in Ukraine caused widespread radioactive contamination across Europe. The disaster had severe environmental and health consequences, including increased cancer rates and birth defects—see Plokhy (2018) for detailed account. Exxon Valdez Oil Spill in 1989 where its oil tanker ran aground in Prince William Sound, Alaska, spilling millions of gallons of crude oil into the pristine waters. The spill caused significant environmental damage to marine life, coastal ecosystems, and local communities—Ott (2008).
12. See Hart and Ahuja (1996), Margolis and Walsh (2003), Orlitzky et al. (2003), Preston and O'Bannon (1997), Waddock and Graves (1997).
13. The key distinction lay in the focus on measurable impact—investors were not only concerned with the ethical implications of their investments but also with the tangible social and environmental benefits they could create.
14. It combines impact investing's most widely used impact performance metrics with research, evidence, and practical implementation guidance into a single, curated system. See https://iris.thegiin.org/about/.
15. Palandjian (2017) discusses the success of the first Social Impact Bond.
16. Acumen Fund in 2001, Elevar Equity in 2008, and Triodos Sustainable Bond Fund in 2006. These funds were among the first to explicitly align financial returns with measurable social impact, paving the way for the broader impact investing movement.
17. ImpactBase has listed 5220 accredited impact investor subscribers and 440 active impact investment funds.
18. A non-profit organisation that created the B Corporation certification for for profit organisations.

19. Bell (2021) in his report for EY, discusses how the post-pandemic investment landscape is set to place greater value on Environment, Social, and Governance (ESG) disclosures.
20. See Thomson (2023) for some incentives to accelerate the renewable energy transition.
21. The number of people living in extreme poverty halved, from 1.9 billion in 1990 to 836 million, primary school net enrolment rate in the developing regions grew to 91 per cent, while dropout rate fell by almost half by 2015. Also there was notable progress in gender equality in education, governance, and in the workplace globally by the end of the strategic goal. Child mortality dropped by more than half by the end of 2015 (Nations, 2015).
22. See Fehling et al. (2013) for a discussion on some of the limitation of MDGs.
23. According to the report, 135 out of the 169 were assessed, 34 targets lacked sufficient trend data.
24. Recently announcing a five-year $300 M agricultural adaptation commitment that will help close agrifood adaptation finance gap aimed at reaching forty million smallholder farmers and feeding one billion people by 2030.
25. See https://empowerafrica.com/kenyan-agritech-startup-twiga-secures-funding-amidst-workforce-restructuring/.
26. In July 2024, One Acre Fund secured $1.4 Million investment from Impact Bridge Asset Management to Support African Farmers.
27. See Root Capital (2024) for most recent annual report.
28. WHO estimates an ambitious scenario, and finds that to achieve the SDG health targets, new investments increasing over time from an initial US$ 134 billion annually to $371 billion or $58 per person are needed by year 2030.
29. See https://acumen.org/regions/west-africa/.
30. https://globalschoolsforum.org/impact-scale-labs.
31. See https://gcpf.lu/how-we-work/.
32. See The Rockerfeller Foundation (2015).
33. See https://www.breakthroughenergy.org/.
34. See report Hand et al. (2023).
35. See https://www.sunfunder.com/about.
36. See https://www.seaf.com/.

References

Acumen. (2024). *Acumen announces $300M agricultural adaptation commitment.* Retrieved November 28, 2024, from https://acumen.org/news/acumen-announces-300m-agricultural-adaptation-commitment/

AgDevCo. (2023). *Annual report 2023.*

Angelides, P. (2000). *The double bottom line: Investing in California's emerging markets.* California State Treasurer's Office.

Antoninis, M., Alcott, B., Al Hadheri, S., April, D., Fouad Barakat, B., Barrios Rivera, M., Baskakova, Y., Barry, M., Bekkouche, Y., & Caro Vasquez, D. (2023). *Global education monitoring report 2023: Technology in education: A tool on whose terms?* UNESCO.

Avery, C. (2001). Business and human rights in a time of change. In *Liability of multinational corporations under international law* (pp. 17–73). Brill Nijhoff.

Bajulaiye, O., Fenwick, M., Skultetyova, I., & Vermeulen, E. P. (2020). *Digital transformation in the hedge fund and private equity industry* (Working Paper 2020-1). Lex Research Topics in Corporate Law & Economics.

Bell, M. (2021). *Why ESG performance is growing in importance for investors.* Ernst & Young.

Bhandari, M. P. (2024, July 15). What is next for the sustainable development goals, what are the challenges concerning SDG 10—Reduced inequalities? *Sustainable Earth Reviews, 7*(1), 23. https://doi.org/10.1186/s42055-024-00093-8

Brest, P., & Born, K. (2013). Unpacking the impact in impact investing. *Stanford Social Innovation Review, 11*(4), 22–31.

Bruyn, S. (1991). *The field of social investment.* Cambridge University Press.

Bugg-Levine, A., & Emerson, J. (2011). *Impact investing: Transforming how we make money while making a difference.* Wiley. https://books.google.co.uk/books?id=LhxIwPYYhfEC

Clark, C., Emerson, J., & Thornley, B. (2014). *The impact investor: Lessons in leadership and strategy for collaborative capitalism.* Wiley.

Convergence. (2022). *Blended finance for health and education* (Data Brief, Issue).

Craven, M., Liu, L., Mysore, M., & Wilson, M. (2020). *COVID-19: Implications for business* (p. 8). McKinsey & Company.

Dalmia, N., & Kanodia, S. (2020). *EdTech in India: A turning point.* Omidyar Network India. Retrieved July 12, 2024 from, https://www.omidyarnetwork.in/insights/edtech-in-india-redseer-report-2019-20

Drucker, P. F. (1986). *Innovation and entre-preneurship: Practice and principles.* Harper & Row.

Emerson, J., & Twersky, F. (1996). *New social entrepreneurs: The success, challenge and lessons of non-profit enterprise creation.* Roberts Foundation Homeless Economic Development Fund.

Ermann, M. D., & Clements, W. H. (1984). The interfaith center on corporate responsibility and its campaign against marketing infant formula in the third world. *Social Problems, 32*(2), 185–196. https://doi.org/10.2307/800787

Fehling, M., Nelson, B. D., & Venkatapuram, S. (2013). Limitations of the millennium development goals: A literature review. *Global Public Health, 8*(10), 1109–1122.

Feyen, E., Frost, J., Gambacorta, L., Natarajan, H., & Saal, M. (2021). *Fintech and the digital transformation of financial services: Implications for market structure and public policy*. BIS Papers.

Foreign Commonwealth & Development Affairs. (2023). *International development in a contested world: Ending extreme poverty and tackling climate change.*

Friedman, M. (1970, September 13). A Friedman Doctrine: The social responsability of business. *The New York Times*. https://www.nytimes.com/1970/09/13/archives/a-friedman-doctrine-the-social-responsibility-of-business-is-to.html

GIIN. (2024). *About GIIN*. Retrieved May 12, 2024, from, https://thegiin.org/about/#:~:text=Since%202009%2C%20the%20Global%20Impact,impact%20investing%20around%20the%20world.

Gittell, R., Magnusson, M., & Merenda, M. (2012). *The sustainable business case book*. Saylor Foundation.

Gond, J.-P., & Piani, V. (2012, March 01). Enabling Institutional investors' collective action: The role of the principles for responsible investment initiative. *Business & Society, 52*(1), 64–104. https://doi.org/10.1177/0007650312460012

Gutterman, A. S. (2020). *What is impact investing*. SSRN 3823837.

Hand, D., Sunderji, S., & Pardo, N. (2023). *Impact investing allocations, activity & performance*. Market GIINSight.

Hart, S. L., & Ahuja, G. (1996). Does it pay to be green? An empirical examination of the relationship between emission reduction and firm performance. *Business Strategy and the Environment, 5*(1), 30–37.

International Energy Agency. (2024). *World energy investment report*. https://www.iea.org/reports/world-energy-investment-2024

Jones, G. G. (2016). *Joan Bavaria and multi-dimensional capitalism*. Harvard Business Review Press (China Case Studies).

Jordan-Kirwan, S., & Tengtio, K. (2020). *How to measure the gender impact of investments: Using the 2X challenge indicators in alignment with IRIS+.*

Kasyanenko, S., Kenworthy, P. G., Ruch, F. U., Vashakmadze, E. T., Vorisek, D. L., & Wheeler, C. M. (2023). *Investment in emerging market and developing economy regions: Trends, prospects, and policy options*. World Bank Group.

Knoll, M. S. (2002). Ethical screening in modern financial markets: The conflicting claims underlying socially responsible investment. *The Business Lawyer, 57*(2), 681–726.

Lapierre, D., & Moro, J. (2009). *Five past midnight in Bhopal: The epic story of the world's deadliest industrial disaster*. Hachette UK.

Larson, Z. (2020). The Sullivan principles: South Africa, apartheid, and globalization. *Diplomatic History, 44*(3), 479–503. https://doi.org/10.1093/dh/dhaa002

Ma, W., & Rahut, D. B. (2024, April 29). Climate-smart agriculture: adoption, impacts, and implications for sustainable development. *Mitigation and Adaptation Strategies for Global Change, 29*(5), 44. https://doi.org/10.1007/s11027-024-10139-z

Margolis, J. D., & Walsh, J. P. (2003). Misery loves companies: Rethinking social initiatives by business. *Administrative Science Quarterly, 48*(2), 268–305.

Maula, M. V., & Lukkarinen, A. (2022). Attention across borders: Investor attention as a driver of cross-border equity crowdfunding investments. *Strategic Entrepreneurship Journal, 16*(4), 699–734.

Mudaliar, A., & Bass, R. (2017). GIIN perspectives: Evidence on the financial performance of impact investments. *Global impact investing network.*. Available at: https://thegiin.org/research/publication/financial-performance

Nations, U. (2015). *The millennium development goals report 2015.*

O'Brien, J. (2005). *Governing the corporation: Regulation and corporate governance in an age of scandal and global markets*. Wiley. https://books.google.co.uk/books?id=o5VEMbZQj94C

O'donohoe, N., Leijonhufvud, C., Saltuk, Y., Bugg-Levine, A., & Brandenburg, M. (2010). Impact investments: An emerging asset class. *JP Morgan, 6*, 3–91.

Orlitzky, M., Schmidt, F. L., & Rynes, S. L. (2003). Corporate social and financial performance: A meta-analysis. *Organization Studies, 24*(3), 403–441.

Ott, R. (2008). *Not one drop: Betrayal and courage in the wake of the Exxon Valdez oil spill*. Chelsea Green Publishing.

Palandjian, T. (2017). *World's first social impact bond shows positive results*. https://socialfinance.org/insight/worlds-first-social-impact-bond-shows-positive-results/

Pavicevic, O., Stepanovic, I., & Ilijic, L. (2021). Resilience in the times of Covid-19: Rethinking the neoliberal paradigm and creating new strategies for battling the global change. *Društvene i humanističke studije: časopis Filozofskog fakulteta u Tuzlil, 6*(3[16]), 165–182.

Pienitz, R., & Vincent, W. F. (2000). Effect of climate change relative to ozone depletion on UV exposure in subarctic lakes. *Nature, 404*(6777), 484–487.

Plokhy, S. (2018). *Chernobyl: The history of a nuclear catastrophe*. Basic Books.

Preston, L. E., & O'Bannon, D. P. (1997). The corporate social-financial performance relationship: A typology and analysis. *Business & Society, 36*(4), 419–429.

Puaschunder, J. M. (2014). Ethical and socially responsible investing. In *Investor behavior* (pp. 513–532). https://doi.org/10.1002/9781118813454.ch28

Root Capital. (2024). *Women in agriculture annual report October 2024*. Annual Reports. https://rootcapital.org/wai-report-2024/

Schueth, S. (2003). Socially responsible investing in the United States. *Journal of Business Ethics, 43*(3), 189–194. http://www.jstor.org/stable/25074988

Simon, J. G., Powers, C. W., & Gunnemann, J. P. (1972). *The ethical investor: Universities and corporate responsibility*. Yale University Press.

Sparkes, R. (2003). *Socially responsible investment: A global revolution*. Wiley.

The Rockerfeller Foundation. (2015). *City resilience framework*. https://www.rockefellerfoundation.org/wp-content/uploads/City-Resilience-Framework-2015.pdf

Thomson, E. (2023, July 16). Half-price energy bills if you live near a wind farm. Here's some incentives to accelerate the renewable energy transition. *Energy Transition—World Economic Forum*. https://www.weforum.org/stories/2023/05/renewable-energy-incentives-households-countries/

Tollison, R. D., & Wagner, R. E. (1988). *Social cost, rent seeking, and smoking: A public choice perspective*. Lexinton Books.

Tolliver, C., Keeley, A. R., & Managi, S. (2020, January 20). Drivers of green bond market growth: The importance of Nationally Determined Contributions to the Paris Agreement and implications for sustainability. *Journal of Cleaner Production, 244*, 118643. https://doi.org/10.1016/j.jclepro.2019.118643

UNEP Finance Initiative. (2005). *A legal framework for the integration of environmental, social and governance issues into institutional investment*.

United Nations. (2015). *Summit charts new era of sustainable development*. Retrieved September 18, from, https://www.un.org/sustainabledevelopment/blog/2015/09/summit-charts-new-era-of-sustainable-development-world-leaders-to-gavel-universal-agenda-to-transform-our-world-for-people-and-planet/

United Nations. (2024). *The sustainable development goals report 2024* (9210582594).

UNSCEB. (2017). *Leaving no one behind: Equality and non-discrimination at the heart of sustainable development*.

Waddock, S. A., & Graves, S. B. (1997). The corporate social performance–financial performance link. *Strategic Management Journal, 18*(4), 303–319.

World Bank, UNESCO, UNICEF, FCDO, USAID, & Foundation, B. a. M. G. (2022). *The state of global learning poverty*.

World Commission on Environment and Development. (1987). *The Brundtland report: 'Our common future'*. Medicine and War.
World Health Organization. (2017). *WHO estimates cost of reaching global health targets by 2030*.

CHAPTER 2

The Concept Impact Investing

2.1 Introduction

Impact investing has emerged as a transformative approach to finance, blending financial returns with measurable social and environmental benefits. As this field evolves, it is crucial to examine its foundational principles, the structures through which it is implemented, the legal and regulatory frameworks governing its practice, and the challenges it faces, including the risks of failure. At its core, impact investing is governed by key principles that distinguish it from other responsible investment strategies. Intentionality is paramount, ensuring that positive social or environmental outcomes are deliberately pursued rather than occurring as incidental byproducts. This principle requires investors to embed impact objectives at every stage of the investment lifecycle—from strategy development and decision-making to active management and performance evaluation. Equally significant is the expectation of financial returns, ranging from concessionary (below-market) to risk-adjusted market rates. Maintaining this dual focus enables impact investing to attract diverse stakeholders, including private investors, philanthropic organisations, and Development Finance Institutions. Another fundamental component is Impact Measurement and Management (IMM), which systematically tracks, assesses and optimises investment outcomes. Effective IMM frameworks ensure transparency, accountability, and alignment with global

N. Udohaya, *Impact Investing and Financial Inclusion*, Sustainable Development Goals Series,
https://doi.org/10.1007/978-3-031-87841-1_2

impact objectives. Finally, Contribution Beyond Capital highlights the role of investors as active partners who provide strategic guidance, capacity-building, and market access rather than merely supplying financial resources.

The practical application of impact investing is shaped by various investment structures that balance risk, return, and impact and financial tools, direct capital towards environmental and climate-related initiatives, and reinforce accountability through impact-linked incentives. As impact investing gains traction, the legal and regulatory landscape is essential in safeguarding its integrity. Regulatory frameworks vary across jurisdictions, influencing how impact funds operate, report outcomes, and interact with traditional financial markets. Standardised disclosure requirements, investor protection mechanisms, and incentives for sustainable finance are critical in ensuring that impact investing remains credible and effective. However, challenges arise when impact investing fails to deliver on its promises. The sector faces risks such as mission drift, where enterprises prioritise financial gains over impact objectives, and impact washing, where investments are branded as impactful without demonstrating measurable outcomes. A lack of standardisation in impact measurement frameworks further complicates transparency and comparability across investments.

Recognising and addressing failures in impact investing is imperative for sustaining the credibility of the field. Instances where investments do not achieve their intended impact, whether due to misaligned incentives, inadequate measurement frameworks, or unforeseen external factors, provide valuable lessons for investors and policymakers. Learning from these setbacks allows the impact investing ecosystem to evolve, adopt more rigorous methodologies, and foster resilience in future investments.

2.2 Defining Impact Investing

As impact investing evolved, its definition has become progressively refined. In the early days of the concept, some texts highlighted the ambiguity surrounding the definitions of impact investing.[1] Trelstad (2016) pointed out that the ambiguity and broad scope of the earlier definitions, which encompassed a diverse range of asset classes, themes, and return orientations, underscored the potential scale of impact investing. He added that although these definitions were problematic, they provided a helpful starting point for understanding impact investing. The Monitor

Institute (2009) supported this argument by stating that a strict clarification of the definition of impact investing, aimed at protecting its integrity from manipulation or dilution, sometimes had the unintended effect of alienating certain constituencies. They suggested a more diplomatic and nuanced approach to its definition. It was not until the late 2010s that the definitions became more specific and, in some cases, more complex.[2] Hockerts et al. (2022) argued that impact investing should indeed be considered an essentially contested concept, positing that it is challenging to define it categorically in an unequivocal manner. Nonetheless, one of the most widely cited definitions comes from the Global Impact Investing Network (2023), which defines impact investing as "investments made with the intention to generate positive, measurable social and environmental impact alongside a financial return". Other groups specify details that align impact investing with their particular focus. For example, focusing on social outcomes, the Organisation for Economic Co-operation and Development defines impact investment as "the provision of finance to address social needs with the explicit expectation of a measurable social, as well as financial, return" (OECD, 2019, p. 28). The European Investment Bank situates impact investing within the framework of additionality, climate action, and social cohesion in the EU context, defining it as an investment approach that assesses the value a business creates in the form of solutions to society and attempts to translate this solution into value for investors. Despite the variations and nuances in how different corporations, scholars, and policymakers define impact investing, it can be readily derived that it is an approach to investment that seeks to create positive environmental or social change in addition to financial gains. This distinguishes impact investing from other investment approaches, such as socially responsible investing (SRI), which primarily seeks to avoid harm by screening out harmful investments, and traditional investing, which focuses solely on financial returns. However, impact investing does not stop with intention; it goes on to establish and state specific social and environmental objectives, set performance metrics, monitor and manage performance, and report results (GIIN, 2023).

2.3 Core Principles of Impact Investing

Unlike traditional investing, which focuses solely on financial returns, or philanthropy, which prioritises social good without an expectation of financial return, impact investing seeks to achieve both. This dual focus is governed by core principles defining how impact investments are structured, implemented, and evaluated. These principles are crucial in ensuring that the intended positive outcomes are genuinely achieved, and they help differentiate impact investing from other forms of responsible or ethical investment. The following section discussed the core principles of impact investing, how they are incorporated into investments, possible challenges, and how they can be mitigated.

2.3.1 *Intentionality*

Intentionality constitutes the foundational principle of impact investing, representing a deliberate and proactive commitment to achieving positive social or environmental outcomes alongside financial returns (GIIN, 2019). Unlike traditional investments, where positive societal or environmental impacts may emerge as incidental byproducts, impact investing places these objectives at the forefront of its mission. Also, investments creating social or environmental impact coincidentally (as an unintended side effect) cannot be considered impact investing, whatever the actual social impact is ex-post (Hockerts et al., 2022). For intentionality to have true significance, it must be embedded in every aspect of the investment process.

Intentionality in Investment Strategy: Integrating intentionality into the investment strategy from the inception is pivotal for achieving meaningful and sustainable impact. Intentionality requires investors to explicitly define the social or environmental issues they aim to address and to embed these objectives systematically throughout the investment lifecycle. Articulating clear and well-defined impact goals is imperative for guiding decision-making across all stages of the investment process. For instance, an impact investment fund targeting the expansion of educational opportunities in low-income countries must identify investees whose core mission and operations are intrinsically aligned with this objective. Such alignment must transcend superficial commitments and inform all business activities, mission statements, and stakeholder engagement strategies, ensuring that the investee's operations are deeply integrated with their impact goals. Establishing well-defined, measurable impact objectives enables investors to align their investment processes better to achieve tangible outcomes. Bouri et al. (2018) discuss the importance of intentionality in guiding investments towards meaningful impact, emphasising that clearly defining objectives and expectations can help align investment activities with intended outcomes.

Moreover, embedding intentionality in all aspects of the investment helps to ensure that resources are effectively utilised to foster meaningful change rather than merely generating profit. Intentionality, therefore, serves as the compass for impact investors, guiding them to ensure that each investment is purpose-driven. This approach enhances accountability, helps mitigate the risk of "impact washing", and fosters deeper alignment between financial sustainability and positive societal outcomes (Findlay & Moran, 2019). A genuine commitment to intentionality distinguishes impact investing from other responsible investment approaches, ensuring that investments yield lasting and meaningful social and environmental benefits (GIIN, 2021).

Intentionality in Investment Decision-Making: Intentionality plays a crucial role in the investment decision-making phase, compelling investors to systematically evaluate social and environmental impacts alongside traditional financial performance metrics. During this phase, investors must engage deeply with the mission and values of prospective investees, assessing how effectively their activities align with the intended impact goals. During due diligence, intentionality requires a rigorous evaluation of both the mission of the investee organisation and the feasibility of achieving the desired impact outcomes. This involves going beyond financial viability to ensure that the investee's business model and operational strategy are designed to generate significant and measurable impact. For instance, Acumen Fund, a pioneer in impact investing, emphasises targeting business models that directly address poverty alleviation. Acumen's approach to intentionality is reflected in its strategic focus on sectors such as agriculture, clean energy, and education—sectors with the potential to create a transformative impact for underserved populations in developing regions (Impact Alliance, 2020). Embedding intentionality into the due diligence process allows investors to ensure that their investments align with financial and impact objectives from the outset. Clark et al. (2014) discuss that integrating intentionality during due diligence can help ensure that investment decisions remain aligned with both financial goals and impact objectives, reducing the risk of mission drift. Such a disciplined approach enables investors to select investees whose missions and business practices align closely with their overarching impact strategy, thereby enhancing the likelihood of achieving meaningful and sustainable outcomes. Intentionality, therefore, is not merely a guiding principle but a practical tool for shaping the selection of investments. It transforms due diligence into a comprehensive evaluation of an investee's capacity to deliver on intended social and environmental outcomes, ensuring that each investment decision is purpose-driven and aligned with broader impact objectives.

Intentionality in Active Management and Monitoring: The concept of intentionality is integral not only at the inception of an investment but also throughout its active management and monitoring phases. This ongoing commitment ensures that investments consistently align with their intended social or environmental objectives, alongside financial goals. Active management necessitates the establishment of specific, quantifiable metrics to assess the success of impact-related activities. These metrics should be defined during the due diligence process and revisited regularly to evaluate progress. Transparency in communicating these metrics and outcomes to stakeholders is also essential for accountability. The International Finance Corporation (IFC) emphasises that impact objectives should inform strategic decisions from initial screening through to exit, advocating for adaptive management to achieve intended outcomes (IFC, 2019). For example, an investor supporting an organisation developing affordable housing must monitor metrics such as the affordability of units, resident demographics, and broader community benefits over time. This oversight ensures that the project continues to fulfil its original intent of providing affordable living spaces to underserved populations. The GIIN underscores that intentionality in active management is crucial for mitigating mission drift and ensuring that investments remain aligned with their social and environmental objectives (GIIN, 2021). This perspective highlights the importance of adaptive management, enabling investors to respond effectively to challenges and recalibrate strategies as needed. Embedding intentionality into every investment process enables investors to enhance impact and financial performance, ensuring that investments remain purpose-driven and deliver measurable, meaningful outcomes over time. Intentionality in active management and monitoring is vital for ensuring that investments remain purpose-driven and deliver measurable, meaningful outcomes. This approach not only upholds the integrity of impact investing but also fosters trust among stakeholders and contributes to the sustainable growth of the impact investing ecosystem.

Stakeholder Engagement and Intentionality: Embedding intentionality within stakeholder engagement is fundamental to the success of impact investing. Intentionality ensures that every interaction with investees, beneficiaries, communities, and other partners with the overarching goals of the investment. Effective stakeholder engagement connects investment objectives with the lived realities of target communities. This begins with clearly defining impact goals and collaborating with stakeholders to address specific challenges and opportunities. The literature on impact investing underscores the importance of intentionality in stakeholder engagements to enhance the credibility and effectiveness of impact investments. Eccles et al. (2013) emphasise that transparent communication and active collaboration build trust and ensure alignment between community needs and investment objectives. Stakeholder feedback becomes a valuable tool for refining strategies and adapting to changing conditions, ensuring investments remain relevant and impactful. An investor in healthcare initiatives aimed at underserved regions would engage with healthcare providers, patients, and local organisations to identify barriers to access and co-create solutions that address these challenges effectively. Such engagements ensure that investments deliver tangible benefits while remaining adaptable to evolving community needs. Intentionality as a guiding principle in impact investing ensures that stakeholder engagements are purposeful and aligned with investment goals. This approach strengthens trust, enhances accountability, and delivers outcomes that are both meaningful and enduring.

Challenges to Maintaining Intentionality

Maintaining intentionality can be particularly challenging, especially as impact investing becomes more mainstream and larger pools of capital seek financial and social returns. Some sceptics claim that while impact investors express their willingness to support the creation of social and environmental value, they often employ the logic upheld by the mainstream investment approach (Bengo et al., 2021). The dynamic nature of markets, evolving societal needs, and the inherent challenges in ensuring impact necessitate continuous diligence from investors. A nuanced understanding of factors that may erode intentionality is essential to ensure that investments remain aligned with their intended social and environmental goals. Ensuring the alignment is an ongoing commitment that underscores the importance of intentionality as both a guiding principle and a practical challenge within the impact investing landscape.

Balancing Impact and Financial Returns: The growing popularity of impact investing has brought significant opportunities and challenges, particularly in maintaining intentionality. As mainstream institutional investors increasingly participate, their expectations for market-rate financial returns can inadvertently dilute the focus on achieving meaningful impact. This dynamic creates a fundamental challenge for maintaining intentionality. Impact investors aiming for market-rate returns often face pressures to prioritise financial metrics over impact goals (Urban & Mapula-e Lehasa, 2024). This shift risks undermining the integrity of the investment's mission, as short-term financial gains can overshadow the long-term systemic changes that impactful investments strive to achieve. Navigating these competing priorities is crucial for preserving intentionality. The misalignment between the timelines of financial returns and impact outcomes is the issue here. Financial returns are often expected within relatively short periods, whereas social and environmental impacts may take years, or even decades, to materialise fully. This misalignment can lead investors to prioritise ventures with immediate financial profitability, potentially at the expense of depth and sustainability of impact. Transparency about trade-offs is critical to maintaining intentionality. Investors must openly articulate the compromises they are willing to make between financial returns and impact outcomes. Without this clarity, intentionality cannot be sustained. It thrives when investors actively acknowledge and navigate these trade-offs, building trust and credibility with stakeholders (Clark et al., 2014).

Risk of Impact Washing: The proliferation of impact washing poses a critical threat to the integrity of impact investing. Impact washing occurs when investments are branded as impactful without demonstrating substantive or measurable social and environmental outcomes. This phenomenon undermines the authenticity of the sector and erodes trust among stakeholders, particularly when there is a lack of transparency and accountability. Investors and fund managers may intentionally or inadvertently exaggerate the social and environmental benefits of their investments to attract capital, leading to reputational risks and reduced credibility of the sector. The competitive pressure to demonstrate both financial returns and impact can incentivise superficial adherence to impact principles rather than a genuine commitment to achieving measurable outcomes. Impact washing dilutes the legitimacy of intentionality and creates a crowded marketplace where authentic impact-focused investments struggle to differentiate themselves. The lack of robust verification mechanisms exacerbates this issue, making it difficult for stakeholders to distinguish between genuine impact investments and those merely adopting the label for financial gain. As noted by Nicholls (2021), this environment risks alienating serious investors and diminishing the overall value proposition of impact investing. The challenge is further compounded by the increasing complexity of global capital markets, where multiple layers of intermediaries and stakeholders operate with varying degrees of alignment to impact goals. Without a shared understanding or stringent validation processes, the sector remains vulnerable to actors who prioritise profit over purpose, thereby weakening the foundational principles of intentionality.

Fragmented Impact Measurement Standards: Maintaining intentionality also requires effective measurement and communication of impact outcomes. Castellas et al. (2018) highlight the risk that impact logic could be overshadowed by investment logic if the disparity in rigour between financial performance measurement and impact performance measurement persists. They suggest that if financial performance is measured with more precision than impact performance, there is a risk that the focus on achieving positive impacts (social and environmental) could be diminished or overshadowed by the focus on financial returns, potentially leading to a situation where impact objectives are neglected or sidelined in favour of financial goals. The lack of universally accepted impact measurement standards poses a challenge to the credibility of impact investing. Different funds and investors may use varying metrics, making comparing impact performance across investments difficult. This lack of standardisation impedes transparency and poses challenges to demonstrating the tangible impact that investments create. While frameworks such as the Global Impact Investing Network (GIIN)'s IRIS+ system have made strides towards standardisation, they are not without limitations. Its adoption is not universal, and many investors continue to rely on disparate or bespoke measurement frameworks, leading to inconsistencies that undermine transparency and comparability across the sector (GIIN, 2021). IRIS+ addresses some of these issues by offering a standardised framework; its effectiveness is hindered by incomplete adoption and the complexity of applying uniform metrics across diverse investments. Stakeholders may still encounter subjective interpretations or "impact inflation", where outcomes are overstated due to a lack of rigorous verification processes. According to Bengo et al. (2021), the development and broader adoption of robust, universally recognised frameworks remain essential to enhancing the credibility of impact investing. Such frameworks would help align diverse stakeholders and ensure intentionality remains central to the investment process.

Financial Intermediaries and Complexity of Deal Structures: Financial intermediation such as Development Finance Institutions, private equity funds, and other stakeholders is essential to the impact investing ecosystem, connecting investors with opportunities that align capital with social and environmental goals. However, the involvement of these intermediaries can introduce layers of complexity that challenge the maintenance of intentionality. Misaligned incentives among intermediaries and stakeholders can dilute impact objectives, making it difficult to sustain a consistent focus on social and environmental outcomes. The multi-layered nature of financial intermediation may lead to competing priorities among stakeholders or varying levels of prioritisation (Casasnovas & Jones, 2022), resulting in compromises on impact goals, particularly when financial returns are emphasised over social or environmental outcomes. The intricacies of deal structures further exacerbate the challenge. Complex arrangements involving multiple investors, intermediaries, and project partners can obscure accountability and reduce transparency. Each layer of intermediation introduces the potential for mission drift, where the original intent of the investment is diluted through competing priorities and incentives. This fragmentation calls for the need for clear, enforceable impact performance expectations that are aligned across all stakeholders (Bengo et al., 2021). The variability of impact measurement frameworks employed by intermediaries and stakeholders can also pose an issue. Without standardised tools across board to assess and report impact outcomes, intermediaries may adopt inconsistent or non-transparent practices, complicating efforts to maintain intentionality throughout the investment lifecycle. Establishing robust contractual agreements and embedding accountability mechanisms at every stage can help address these gaps, ensuring that all parties remain committed to the intended social and environmental objectives. While financial intermediation is indispensable to the scalability of impact investing, its complexity poses significant challenges to intentionality. Addressing these challenges requires coordinated efforts to align stakeholder incentives, enhance transparency, and standardise impact measurement practices, ensuring that the integrity of impact objectives is preserved throughout the process.

Mission Drift in Scaling Operations: Scaling operations presents a significant challenge for impact-focused enterprises as they navigate the balance between growth and maintaining their core mission. While scaling can amplify positive impact, it often necessitates securing additional funding or forming strategic partnerships that may not fully align with the enterprise's original mission. This misalignment introduces the risk of mission drift, where social or environmental objectives are deprioritised in favour of financial or operational goals. Ebrahim and Rangan (2014) highlight that as social enterprises grow, the pressures to reduce costs, improve efficiency, and achieve profitability often intensify. These pressures can shift organisational priorities, leading to compromises on the original impact objectives. This challenge is particularly acute for early-stage companies as their growth trajectories are heavily influenced by external capital and the strategic interests of new stakeholders. Without robust governance and accountability structures, these enterprises risk diluting their foundational commitments to social and environmental impact. Maintaining intentionality in this context requires enterprises to develop clear, impact-centred scaling strategies and to seek out investors and partners who share their commitment to the intended outcomes. Mission drift during scaling is a pervasive challenge for impact-focused enterprises. Addressing this issue demands intentional alignment of growth strategies with core impact objectives, alongside strong governance mechanisms to safeguard the enterprise's mission and ensure its long-term integrity.

Mitigating the Challenges

Addressing the challenges to maintaining intentionality in impact investing requires a seamless interplay of strategies that align transparency, accountability, and engagement with practical tools and frameworks. At the heart of these efforts is the need for investors to demonstrate consistent commitment to the social and environmental missions that distinguish impact investing. Globally recognised frameworks, such as the International Finance Corporation (IFC) Operating Principles for Impact Management and the Global Impact Investing Network (GIIN)'s

IRIS+ system, are pivotal in setting standardised methodologies. These tools enhance comparability and provide stakeholders with the confidence to trust that intentionality is being upheld. Central to maintaining this trust is the establishment of robust accountability mechanisms. Investors must embed enforceable impact performance expectations into their contracts, ensuring alignment across all stages of the investment process (Geczy et al., 2021). Regular monitoring and transparent reporting are essential, providing opportunities for course correction while reinforcing stakeholder confidence (OECD, 2017). These practices safeguard the original mission of the investment and signal a measurable commitment to outcomes, anchoring intentionality within every phase of the lifecycle.

Stakeholder engagement is an indispensable element of this process. Consistent and meaningful dialogue with beneficiaries and communities ensures investments remain relevant and impactful. Integrating stakeholder feedback into decision-making enables investors to avoid misalignment and respond effectively to evolving societal needs. Such engagement also fosters trust and collaboration, creating a participatory ecosystem where those most affected by investments have a voice in shaping their outcomes. This approach is particularly critical in dynamic contexts, where responsiveness to shifting priorities ensures that intentionality is not compromised by changing external conditions. Equally important is the relationship between investors and their investees. Open and transparent communication is essential to keeping social and environmental objectives at the forefront. Beyond capital provision, investors can offer technical assistance and capacity-building resources, ensuring enterprises have the tools to scale responsibly and sustain their missions. This alignment is crucial in mitigating mission drift, a common risk as enterprises grow and face new pressures, particularly from profit-driven stakeholders. Clear goal-setting and regular check-ins help maintain a shared focus on impact. The role of industry associations and certifications further reinforces these efforts. Third-party validations from organisations like GIIN or certifications such as B Corp provide benchmarks for credibility, distinguishing genuinely committed impact investors from those who may engage in superficial practices. These standards also encourage collaboration among stakeholders and provide a platform for sharing best practices, strengthening the overall impact investing ecosystem.

Balancing financial returns with impact objectives remains a core challenge, one that requires thoughtful strategies to align these dual goals. Transparency about trade-offs and setting realistic financial expectations are key to navigating this tension. Addressing the tension head-on enables investors to ensure that the pursuit of profit does not eclipse their mission-driven objectives. Flexibility and adaptability are equally critical in the face of ever-changing societal needs. The dynamic nature of global challenges necessitates continuous learning, feedback loops, and adjustments to strategies. Adaptive management ensures that investments remain relevant and effective over time, allowing investors to anticipate and respond to emerging issues without losing sight of their core missions. This approach transforms adaptability from a reactive measure to a proactive component of intentionality.

The ability to mitigate these challenges lies in the deliberate integration of these practices into a cohesive strategy (Lydenberg et al., 2017). Fostering transparency, building accountability, and engaging meaningfully with stakeholders enable impact investors to navigate complex obstacles such as mission drift and impact washing. The unwavering commitment to intentionality not only differentiates impact investing from traditional models but also positions it as a vital tool for addressing pressing global challenges. Through thoughtful alignment of purpose and practice, the investment can deliver lasting and measurable change while upholding its foundational principles.

2.3.2 Financial Return Expectation

Impact investing operates on the principle that it is possible to achieve both social or environmental impact and financial return. Unlike philanthropy, which does not expect a financial return, impact investors seek to generate a return on their capital. The financial return expectation can vary widely, from below-market returns (concessionary) to risk-adjusted market rates. This flexibility allows impact investing to attract a diverse range of investors. The financial return expectation is also integral to ensuring the scalability and sustainability of impact investments. Demonstrating that both impact and profit can be achieved encourages more capital to flow into sectors that address global challenges through impact investing. The dual value proposition helps attract mainstream investors who may be hesitant to participate in purely philanthropic activities. The financial return expectation varies significantly, ranging from below-market concessionary returns to market-rate returns that are comparable to traditional investments. This flexibility enables impact investing to attract a broad spectrum of investors, each with different levels of risk tolerance and return expectations.

Concessionary Returns: Concessionary returns prioritise social or environmental benefits over financial returns. It addresses critical funding needs that traditional capital often overlooks. This range of return is marked by a willingness to accept below-market returns, reflecting a commitment to achieving meaningful impact in areas that may not yet be economically viable or that carry higher risk (GIIN, 2023). Investors in this space, typically foundations, family offices, high-net-worth individuals, and certain mission-driven institutions, see value beyond immediate financial gains. They focus on transformative change in sectors such as healthcare, education, affordable housing, and renewable energy, particularly within underserved and emerging markets. The decision to accept below-market returns often stems from a broader philanthropic mission. Unlike traditional impact investors who aim for market-rate returns, concessionary investors measure success in terms of lives improved, ecosystems preserved, or communities uplifted. Concessionary funding seeks to address "market failures" by deploying capital where private investment may not venture due to financial constraints or perceived risks, ultimately aiming to catalyse long-term social change. The investment provides an essential proof of concept, demonstrating viability in high-risk areas and paving the way for other capital sources to follow with less caution. This de-risking of projects and absorption of financial risk upfront makes such projects more appealing for later-stage or larger-scale investors, increasing the scalability and sustainability of high-impact projects over time.

Concessionary investments face inherent challenges due to their trade-off between financial and social returns. Investors in this space must carefully weigh the social impact against the lower returns they accept. The effectiveness of these investments depends on setting clear and achievable impact goals, robust measurement frameworks, and strategic sectoral focus. However, as these investments become more common, some argue for an evolution in how concessionary returns are perceived—not merely as sacrifices but as intentional choices that generate value in ways traditional investments cannot. Filling critical funding gaps and supporting high-impact initiatives makes concessionary impact investors crucial in advancing systemic change and creating long-term social and environmental benefits that ripple through communities and economies.

As the field of impact investing matures, concessionary investments are gaining recognition as a strategic tool for sustainable development. Their role in leveraging additional capital and demonstrating impact in challenging sectors positions them as vital components of the impact investing landscape. With growing interest from mission-driven institutions, concessionary impact capital is likely to expand, unlocking new opportunities in high-impact but underfunded areas.

Risk-Adjusted Market Returns: The pursuit of risk-adjusted market returns within impact investing embodies a core belief, which is that investments that address social or environmental challenges can also generate financial returns comparable to those of traditional investments.[3] This perspective is particularly significant for institutional investors who must fulfil both fiduciary responsibilities and, increasingly, sustainability goals. These investors operate on the conviction that social impact and financial performance are compatible and, when strategically managed, mutually reinforcing. For investors seeking risk-adjusted market returns, the returns often act as a validation of the impact investment's quality and resilience. Profitable returns suggest financial stability and the robustness and scalability of the impact goals embedded within the project or company. As several studies have shown that impact investments can, indeed, yield competitive returns, particularly in renewable energy, clean tech, affordable housing, microfinance and sustainable agriculture.[4] These sectors are projected to grow as the world confronts climate change, inequality, and resource scarcity, businesses that offer solutions to these challenges are well-positioned for growth and resilience in an evolving market, reinforcing the economic value of impact-driven investments.

Achieving risk-adjusted returns while targeting impact requires advanced risk management, which is often heavily data-driven (Adeoye et al., 2024). Metrics such as Environmental, Social, and Governance (ESG) scores, alongside more detailed impact measurements, serve as tools for both assessing potential returns and managing risks. Investors look closely at these metrics to evaluate the alignment between financial performance and impact objectives, using data to uncover areas where high-impact investments may offer competitive returns or reveal unforeseen risks. Moreover, data-driven impact measurement is becoming increasingly sophisticated, helping investors to identify and assess potential risks associated with social and environmental factors. For instance, companies with strong ESG profiles tend to have lower volatility and a reduced risk of regulatory, reputational, or operational challenges (Chen et al., 2023)—factors that can enhance long-term stability and returns. As the growth of impact investing continues, the development of consistent, robust metrics and transparent reporting standards will be critical for helping institutional investors navigate risk-adjusted impact returns. Enhanced regulatory frameworks, such as the EU's Sustainable Finance Disclosure Regulation (SFDR), are beginning to shape the impact investing landscape, compelling funds to disclose how sustainability risks are managed. As transparency increases, it may become easier to harmonise social impact goals with financial returns, ultimately lowering perceived risks and encouraging more institutional capital to flow towards impact investments. The pursuit of risk-adjusted market returns in impact investing is reshaping how profitability and purpose intersect.

2.3.3 Impact Measurement and Management

Another defining principle of impact investing is the commitment to impact measurement and management (IMM). It is the process that encompasses the systematic tracking, assessment, and optimisation of social and environmental outcomes associated with investments. IMM involves actively using this data to improve strategies, mitigate risks, and align investments with their intended impact goals (Gianoncelli & Gaggiotti, 2021). In doing so, it provides investors with the information needed to ensure their investments achieve measurable and meaningful positive outcomes. Investors are thus responsible for monitoring and managing the social or environmental impact they pledge to create. As impact investing attracts more capital, there is an increasing need to validate claims about social and environmental benefits. The imperative of IMM stems from the understanding that positive impact should be more than aspirational. Transparency and investment monitoring of projects are crucial. IMM frameworks serve as the backbone of accountability within this field, offering tools and methodologies that allow investors to substantiate their impact goals. Yelnikova and Kwilinski (2020) found that lack of measurement, impact management, and a uniform system of benchmarks are fundamental limitations which hinder the scaling of impact investment.

Like the other core principles in impact investing, IMM distinguishes impact investing from other forms of investing, such as ESG integration or SRI, by focusing on measurable change (Hummels & de Leede, 2014). It transforms impact investing into a results-driven discipline that prioritises accountability and alignment with stated objectives. Using structured measurement systems, IMM enables comparability, transparency, and rigour across investments, helping investors and stakeholders verify outcomes and assess their contributions towards broader global objectives. The GIIN highlights IMM as a best practice in impact investing and emphasises its role in transforming impact intentions into tangible metrics.

The Role of IMM in Impact Investing

Ensuring Accountability: IMM serves as the bedrock of accountability within impact investing. Establishing clear linkages between defined objectives and measurable outcomes ensures that IMM keeps investments aligned with their goals. This level of transparency reinforces the integrity of the investment process and builds confidence among stakeholders, including investors, beneficiaries, and regulatory bodies. IMM mitigates impact washing risk by providing a rigorous framework for tracking and reporting progress. Through IMM, impact claims can be verified, quantified, and communicated to stakeholders, fostering trust and preventing reputational risks (Financial Conduct Authority, 2023). Holding investors accountable for both financial returns and social or environmental outcomes strengthens the credibility and sustainability of the impact investing ecosystem through IMM.

Informing Strategy Development: IMM goes beyond monitoring and evaluation. It actively shapes strategic decisions by offering data-driven insights into what works and what does not. The iterative nature of IMM frameworks allows investors to refine their strategies over time, adapting to changing circumstances and emerging opportunities. Evidence-based adjustments enable investors to deploy resources more effectively and enhance the impact of their initiatives (Gutterman, 2024). IMM ensures that strategy development is dynamic and responsive, aligning investments with evolving community needs and market conditions.

Enhancing Decision-Making: Effective decision-making in impact investing hinges on the availability of reliable, actionable information. IMM provides investors with the tools and methodologies to evaluate the potential and actual outcomes of their investments. Analysing quantitative and qualitative data allows IMM to enable investors to make informed decisions on resource allocation for maximum impact. Insights from assessments help direct resources towards interventions that provide the greatest value to farmers and investors (Frede, 2021). Reducing the risk of resource misallocation and focusing on initiatives with proven efficacy enable IMM to enhance the overall efficiency and effectiveness of the impact investing process.

Managing Risks: Risk management is a critical component of any investment strategy, and IMM plays a vital role in identifying and mitigating risks specific to impact investments. Unlike traditional investments, impact projects often operate in complex environments with multiple stakeholders and unforeseen challenges. It provides early warning systems to identify unintended consequences and potential risks, enabling proactive responses. By systematically addressing risks, IMM protects beneficiaries from harm and ensures the long-term viability of impact investments (Aggarwal & Elembilassery, 2018).

Key Frameworks in IMM

The effectiveness of the IMM depends on applying robust frameworks and standardised metrics that enable consistency, comparability, and actionable insights. These frameworks provide the necessary structure for measuring, managing, and communicating impact, reinforcing the integrity and credibility of the impact investing process. The proliferation of disparate individual standards in impact measurement presents significant challenges. The absence of a unified framework leads to inconsistencies in reporting, impeding the comparability of impact data across different organisations and sectors. This fragmentation hampers stakeholders' ability to assess and benchmark performance effectively, undermining impact assessments' credibility. Moreover, the lack of standardisation can increase reporting burdens for organisations, as they may need to adhere to multiple frameworks to satisfy diverse stakeholder requirements. This complexity strains resources and detracts from the focus on achieving and enhancing actual impact.

Implementing a standardised and localised framework that is adaptable is imperative for advancing. Such a framework would establish a common language and set of metrics, facilitating comparability and transparency across the impact investing landscape. Simultaneously, its adaptability would allow for customisation to address specific contextual nuances, ensuring relevance across various regions and sectors. This balance between standardisation and flexibility is crucial for accommodating the diverse nature of impact investments while maintaining rigorous and comparable impact assessments.

Impact Reporting and Investment Standards (IRIS+): IRIS+, developed by the GIIN, is a primary standardisation in IMM. The system offers a comprehensive catalogue of over 400 metrics tailored to diverse impact themes and sectors, ensuring its applicability across various investment contexts. IRIS+ integrates globally recognised standards, including alignment with the Sustainable Development Goals (SDGs), offering a cohesive structure for understanding and reporting impact (Reisman & Olazabal, 2021). Bridging the gap between universal applicability and specificity allows IRIS+ to facilitate consistency and comparability in impact data. One of its key strengths is its ability to address fragmentation in IMM. Harmonising disparate approaches into a unified framework allows IRIS+ to reduce the burden on organisations that might otherwise struggle with adhering to multiple incompatible systems. This standardisation enhances credibility while enabling stakeholders to benchmark outcomes effectively across projects, sectors, and geographies. This promotes a collective approach to addressing global challenges, fostering alignment among investors, beneficiaries, and other stakeholders (Sarmento & Herman, 2020).

The IRIS+'s focus on decision-useful data differentiates it from other frameworks. Precisely, it aligns metrics with the Impact Management Project (IMP)'s five dimensions of Impact, namely what, Who, How Much, Contribution, and Risk, to ensure that collected data is descriptive and actionable.[5] Reisman and Olazabal (2021) highlight the importance of the dimensions in supporting investors in making informed strategic adjustments and optimising their impact over time. Another critical feature of IRIS+ is its iterative design. The framework evolves in response to user feedback and emerging industry needs, ensuring its continued relevance. Recent updates have expanded its capabilities to capture cross-sectoral impacts and address systemic challenges such as climate resilience and social equity. This responsiveness positions IRIS+ as a dynamic tool capable of adapting to the complexities of modern impact investing. Offering a standardised yet flexible approach allows IRIS+ to effectively balance the need for comparability with the necessity of contextual sensitivity. Its integration with SDG-linked metrics and its emphasis on transparency and accountability make it a pivotal framework in advancing the field of impact investing. Furthermore, IRIS+ helps mitigate the resource burden associated with bespoke methods, streamlining the measurement process while maintaining rigour and credibility.

While IRIS+ presents the above-discussed advantages in impact measurement, it has some downsides. It can be complex to implement. The extensive catalogue can be overwhelming for new or smaller organisations. The framework's comprehensiveness may require significant time and expertise to navigate, leading to steep learning curves and potential underutilisation. To address this, IRIS+ can integrate tiered guidance, providing simplified entry points for novice users while allowing advanced users to access deeper functionalities. Tutorials, case studies, and tailored templates could ease the onboarding process for smaller organisations. Another challenge with IRIS+ is that the implementation can be resource-intensive. It often demands considerable financial and human resources. This can be a barrier for smaller enterprises or organisations in resource-constrained environments, limiting its accessibility and adoption. Such enterprises may struggle to allocate funds for the additional personnel or technology required for IRIS+ adoption. Introducing subsidised training programmes or partnerships with Development Finance Institutions could alleviate resource constraints. Streamlining metrics for specific sectors could reduce the burden of selecting and applying relevant indicators. Limited flexibility for context-specific needs also sometimes poses a challenge when adopting IRIS+. Despite its adaptability, IRIS+ may struggle to fully capture the unique nuances of localised or sector-specific contexts. This can result in gaps where unique cultural, geographical, or organisational factors are inadequately represented. For example, in rural agricultural initiatives, specific metrics might need to account for traditional ecological knowledge or community-defined sustainability indicators outside the standard IRIS+ catalogue. A dynamic customisation module within IRIS+ would allow organisations to adapt metrics to their specific needs while maintaining alignment with the broader framework. Encouraging collaboration with local stakeholders can further enhance contextual relevance.

Addressing the abovementioned limitations can enable IRIS+ to solidify its position as the most preferred framework for impact assessment. Its continued evolution should focus on reducing complexity, improving accessibility, and maintaining relevance across diverse contexts. Strategic partnerships with local entities, technology integration, and robust capacity-building initiatives will further strengthen its usability and impact. These enhancements will ensure IRIS+ remains a dynamic and indispensable tool in the ever-evolving landscape of impact investing. Nonetheless, IRIS+ is an invaluable component in the impact investor's toolkit. Its use requires a thoughtful approach. Understanding its strengths and limitations allows investors to leverage IRIS+ for meaningful impact measurement while recognising where additional frameworks or metrics may be needed to capture the full scope of their investments' social and environmental contributions.

The Sustainable Development Goals (SDGs): The United Nations' SDGs is another essential framework in impact investing. It is a comprehensive framework for aligning impact goals with global developmental and environmental priorities. The SDGs provide a language for articulating impact, ensuring that diverse stakeholders, including investors, governments, and civil society, can align their efforts towards shared objectives (Gasper, 2019). Linking investment activities to specific SDG targets allows IMM frameworks to enable investors to assess their contributions to global priorities while fostering accountability and transparency. The SDGs also facilitate cross-sectoral collaboration by providing a universally accepted set of goals that transcend geographic and cultural boundaries. This universality enhances the relevance of the SDGs as a benchmark for evaluating the outcomes of impact investments. The integration of SDG-linked metrics within frameworks like IRIS+ amplifies their utility. This alignment ensures that impact measurement is standardised and strategically aligned with broader societal imperatives. For example, tracking progress against SDG targets can help investors prioritise interventions that address systemic challenges, such as building resilient infrastructure or reducing inequalities. Additionally, the SDGs encourage a long-term perspective by emphasising sustainability and resilience, ensuring that impact investments deliver enduring value. This forward-looking approach helps create investments that not only address immediate needs but also contribute to building stronger and more adaptive systems that can withstand future challenges. Embedding these principles into investment strategies enables the SDGs to ensure that solutions remain durable and impactful across generations.

The SDGs also emphasise inclusivity in impact measurement, recognising that sustainable development cannot be achieved without addressing the needs of marginalised and vulnerable populations. Focusing on measurable outcomes that directly improve the lives of these groups enables the SDGs to reinforce the principle of leaving no one behind. This inclusivity is operationalised through specific targets aimed at reducing inequalities, enhancing gender equity, and ensuring equitable access to resources and opportunities. The alignment of these priorities with the broader goals of impact investing creates a powerful framework for achieving transformative and equitable change. Furthermore, the SDGs provide a unified global agenda that facilitates collaboration among diverse stakeholders. This shared vision fosters cross-sector partnerships, encouraging alignment among governments, businesses, and civil society. Such collaboration enhances the effectiveness of impact investments by ensuring that resources are directed towards globally recognised priorities. The SDGs become an indispensable tool for articulating, guiding, and achieving meaningful and inclusive impact at scale by embedding equity, sustainability, and global alignment into its framework.

Despite these strengths, several challenges are associated with using the SDGs in impact investing. One notable limitation is that the SDG framework lack specificity for certain sectors or niche areas of impact. The SDGs were designed as broad global targets intended for various stakeholders, including governments, NGOs, and corporations, rather than being tailored specifically for investors. This broadness creates challenges for investors working in specialised fields where impact indicators are highly sector-specific. For instance, an impact fund focused on innovative financial inclusion solutions may find that the SDG indicators are too general to capture the nuanced effects of their activities on low-income communities. In such cases, the SDG indicators might need to be supplemented with more specialised metrics to provide a detailed understanding of the impact. Another challenge with the SDG framework is the difficulty of direct attribution in impact measurement. Since the SDGs address global, systemic issues, attributing the impact of an individual investment directly to specific SDG targets can be complex. For example, investments aimed at reducing poverty (SDG 1) or addressing climate change (SDG 13) contribute to large-scale goals that involve countless other contributing factors, making it challenging to isolate and quantify an individual investor's impact. This complexity can lead to challenges in accountability and measurement, as investors may struggle to demonstrate a direct connection between their activities and progress towards specific SDG targets. In many cases, impact investors need to approach SDG alignment as part of a broader, collective effort rather than attempting to claim precise attribution, which may not fully satisfy stakeholders seeking clear, attributable impact data. The SDGs' broad applicability, credibility, and relevance to systemic global issues make them an indispensable framework within the impact investing ecosystem. Investors who approach the SDGs thoughtfully—complementing them with tailored metrics where needed, committing to transparency in reporting, and acknowledging the challenges of direct attribution—can leverage the framework to drive meaningful impact while contributing to a global agenda for sustainable development.

Global Impact Investing Rating System (GIIRS): The **GIIRS,** developed by B Lab, is a well-regarded framework for assessing the social and environmental performance of companies and investment funds. GIIRS operates as a rating system, providing a standardised scoring mechanism that evaluates an organisation's impact. This framework is particularly valuable for impact investors who seek reliable and comparable data on the impact outcomes of their investments, facilitating more informed decision-making and helping investors align their portfolios with mission-driven goals. It offers a structured approach to impact performance rating and serves as a critical tool in the impact investing landscape, enabling investors to assess financial performance and the broader social and environmental contributions of their investments. The GIIRS assigns ratings based on rigorous social and environmental criteria. This standardisation is particularly beneficial for investors who wish to compare companies or funds across similar impact dimensions. The GIIRS score provides a straightforward way for investors to gauge impact performance, creating a common benchmark that enhances transparency and accountability. The framework simplifies complex impact data into accessible ratings, supporting investors in their efforts to make meaningful comparisons across a range of investment opportunities. This is particularly useful for institutional investors who manage diversified portfolios and need a quick, reliable way to assess the relative impact performance of various assets.

Additionally, GIIRS has a strong reputation for credibility and robustness. The B Lab has a rigorous approach to evaluating corporate practices, which lends GIIRS a layer of trust. Investors can be confident that a GIIRS rating reflects a thorough and unbiased assessment of a company's social and environmental contributions. The rigorous methodology gives stakeholders confidence that the impact ratings accurately reflect the performance of the companies and funds they evaluate. The framework's focus on supporting mission alignment is another noteworthy strength. GIIRS ratings help investors find and support companies that align closely with their impact goals. Offering a comprehensive view of a company or fund's impact across various social and environmental indicators, GIIRS enables investors to align their investments with specific impact missions. For example, an investor focused on climate action can use GIIRS ratings to identify funds or companies with strong environmental scores, ensuring their capital is directed towards initiatives that genuinely prioritise climate-related impact. This alignment is important for impact effectiveness and supports investors' efforts to demonstrate commitment to specific goals, whether reducing inequality, fostering sustainable communities, or advancing clean energy.

Despite these strengths, GIIRS does face some limitations and challenges that investors should consider. One such limitation is that the GIIRS framework, though effective in providing standardised scores, may lack the level of granularity that some investors seek. While GIIRS ratings summarise impact performance in an accessible way, they may not always capture the nuanced outcomes of complex impact projects. For investors who require a deeper understanding of specific impact metrics or who wish to explore the detailed social or environmental contributions of a company, GIIRS scores may offer only a high-level view that does not fully satisfy more in-depth impact analysis needs. In these cases, additional tools or frameworks may be necessary to complement GIIRS and provide a richer picture of impact performance. Another challenge is that GIIRS ratings primarily focus on broad social and environmental categories, which may not always align perfectly with sector-specific needs. For instance, an investor focused on the intricacies of healthcare access might find that GIIRS' generalised ratings do not provide sufficient detail on health-specific indicators. The broader categories used in GIIRS are valuable for facilitating comparability but may not capture the full spectrum of impact outcomes relevant to more specialised or niche sectors. This limitation could lead investors to seek additional frameworks or data sources that offer more targeted metrics tailored to the unique impact goals within particular industries or fields. Data accessibility and collection requirements can also present challenges, particularly for smaller organisations. GIIRS ratings require comprehensive data across multiple social and environmental domains, which necessitates a well-organised infrastructure for data gathering and reporting. For smaller companies or funds with limited resources, meeting GIIRS' data demands can be challenging. This barrier may prevent some organisations, especially in early-stage or resource-constrained contexts, from pursuing a GIIRS rating, potentially reducing their visibility within the impact investing community. While the rigour of GIIRS is a strength for investors who prioritise accuracy, it can also inadvertently limit the inclusivity of the framework by making it difficult for smaller organisations to participate. Finally, while GIIRS' comparability across different companies and funds is highly valuable, it may not always provide sufficient context regarding an organisation's unique circumstances or challenges. Impact performance is often influenced by regional, industry-specific, or operational factors that a standardised rating system like GIIRS may not fully account for. For example, a renewable energy fund operating in a challenging regulatory environment might face specific constraints that impact its performance metrics. The standardised nature of GIIRS ratings could overlook these contextual factors, leading to potentially oversimplified conclusions. Therefore, investors must consider the broader context of each company or fund they assess through GIIRS and, when necessary, complement the ratings with a more contextualised analysis.

2.3.4 Contribution Beyond Capital

The principle of Contribution Beyond Capital is fundamental in impact investing. It emphasises that investors are not merely financiers but also active partners in the growth and impact of the enterprises they support. While contribution beyond capital is not often explicitly articulated in the literature, many have argued that it is the case that contribution beyond capital is a core principle in impact investing.[6] This principle acknowledges that while financial resources are necessary, they are often insufficient to foster sustained impact, particularly in challenging sectors or underserved markets. Investors deepen their commitment to the enterprise's success and mission alignment by actively engaging in capacity-building and strategic support. It supports the alignment of interests between investors and investees. The shared responsibility for financial returns and impact outcomes creates a more cohesive partnership where both parties work towards a shared vision of sustainable growth and positive change. Furthermore, the insights gained from working closely with investees can inform the investor's broader strategy, offering valuable learnings that can be applied to future investments or shared with the broader impact investing community. Providing non-financial support such as technical expertise, capacity-building, strategic partnerships, and market access enables impact investors to significantly amplify the impact potential of their investments, creating a more robust and enduring foundation for social and environmental change. Holtslag et al. (2021) observe that contribution beyond capital, which is central in impact investing is often not applicable in traditional investing.

Capacity-Building: A key aspect of this principle is its focus on capacity-building. Early-stage and emerging social enterprises often face critical gaps in skills, governance structures, and operational processes. These gaps can limit their ability to grow, scale, or even survive, especially in complex markets (O'Reilly et al., 2023). In supporting capacity-building, impact investors address these foundational challenges directly. For instance, by offering guidance on financial management, operational efficiency, and impact measurement, investors enhance an organisation's resilience and readiness for growth. The support ultimately builds a stronger organisational infrastructure, enabling the enterprise to deliver on its mission with greater effectiveness and reach.

Strategic Partnerships: Impact investors function as strategic partners through contribution beyond capital. This is particularly visible in the case of Development Finance Institutions (DFIs), which often provide advisory services and governance support. DFIs typically work alongside investees to refine business strategies, improve governance frameworks, and optimise operational efficiencies (Moretto & Scola, 2017). This active involvement goes beyond a hands-off investor relationship, positioning the impact investors as advisors who shape the overall direction of the enterprise. Additionally, this type of involvement from the investors helps to de-risk the investment, making the enterprise more appealing to future investors and setting a standard for operational rigour that strengthens the enterprise's long-term impact. Such governance and operational guidance are especially valuable in regions where regulatory frameworks and market conditions may be unstable or evolving, as they offer a buffer against external pressures that could threaten the enterprise's sustainability.[7]

Market Access: Another crucial aspect of contribution beyond capital is the provision of market access and network connections. Early-stage enterprises often struggle to penetrate markets or establish partnerships that can enhance their impact. Investors with industry knowledge, strategic networks, or connections to key stakeholders can open doors that would otherwise remain closed, providing investees with critical access to new markets and growth opportunities. For example, a healthcare-focused impact fund might introduce a social enterprise to partnerships with public health institutions, enabling the enterprise to expand its services and access new funding streams.[8] This network-building can be transformative, as it embeds the investee within a supportive ecosystem that enhances both its reach and credibility (Martin, 2013). However, contribution beyond capital requires a strategic commitment of time, expertise, and resources from the investor, and this additional commitment can increase transaction costs. It may limit the number of investments an impact investor can make effectively. It requires investors to be highly selective, investing only in enterprises where they can bring both capital and complementary expertise to the table. This hands-on approach also demands that investors maintain a balance, offering guidance without undermining the autonomy or innovation potential of the enterprise. Striking this balance is essential for fostering a collaborative partnership that supports, rather than overshadows, the investee's unique mission and strategies (Nachyła & Justo, 2024).

The impact of contribution beyond capital is especially critical in underserved markets or in sectors where institutional and market barriers may prevent enterprises from scaling. Ruch (2020) observed that in emerging and developing economies, entrepreneurs face systemic challenges, including limited infrastructure, regulatory complexities, and workforce skill gaps. Financial investment alone cannot address these issues. Localised knowledge, regulatory insights, and culturally informed guidance are crucial to empower enterprises to overcome these obstacles. Impact investors create a sustainable business model that is more adaptable to local conditions for long-term impact through the principle of contribution beyond capital. It is a strategic and catalytic component of impact investing that goes beyond financial transactions to foster genuine, scalable impact. By addressing critical skill gaps, enhancing operational capacity, facilitating market access, and strengthening governance, impact investors create conditions that enable enterprises to thrive and achieve greater resilience and scale.

2.4 Structures of Impact Investments

Impact investment structuring is a critical phase in the impact investment process, where financial tools and mechanisms are devised to ensure the achievement of financial returns and measurable social or environmental outcomes. Impact investors approach this with a unique focus on aligning capital deployment strategies with mission-driven goals. The structuring begins with a deep understanding of the investment context, recognising the unique challenges and opportunities within targeted sectors or geographies. It also involves the knowledge of suitable capital types and financial instruments and how they can be tailored to suit the specific needs of investees, ensuring that risk is appropriately managed and all stakeholders are aligned on expected outcomes. The thoughtful alignment of financial strategy with social and environmental purposes embeds the dual mandate into the very foundation of the investment framework, ensuring that the flow of financial resources is directed towards outcomes that create measurable, positive change while maintaining the financial sustainability necessary to attract a broad spectrum of investors.

Impact investments are uniquely varied in their structure. They span a diverse spectrum of models. This diversity reflects the complex and multifaceted nature of the issues impact investments aim to address. No single approach can adequately address the various challenges—namely, improving global health outcomes, expanding access to education, or mitigating climate change. Each model is, therefore, tailored to the specific objectives of the investment, the context in which it operates, and the stakeholders it seeks to engage. This tailoring is essential because the mechanisms through which capital is deployed directly influence the extent to which impact is achieved. Deliberately aligning the design of investment vehicles with their intended impact ensures that financial resources catalyse meaningful change in a sustainable, scalable, and equitable way. This deliberate alignment between intent and mechanism positions impact investing as a transformative tool for global development and sustainability. In essence, structuring impact investments is more than a financial exercise; it is an act of strategy, vision, and responsibility amplifying capital's power to drive systemic change. The following section examines some fund types and approaches and how investors deploy them to align intent with impact.

2.4.1 Aligning Intent with Impact

Blended Finance—De-Risking and Leveraging Capital

Blended finance effectively harnesses the unique strengths of public, philanthropic, and private capital to address some of the world's most pressing challenges. Its central premise lies in leveraging concessional funds—typically from public or philanthropic sources—to reduce risk and enhance the attractiveness of high-impact projects for private investors. Carefully structuring these investments allows blended finance to channel substantial capital into sectors or regions otherwise deemed too risky, underdeveloped, or unprofitable under conventional investment frameworks. At the heart of this mechanism is the principle of risk absorption. Public and private funding often takes the form of grants, guarantees, or first-loss capital, which cushions against potential financial losses. This mitigates the risk perception for private investors, who might be wary of committing resources to projects in fragile markets, such as emerging economies or sectors like renewable energy, healthcare, and social infrastructure. By assuming a disproportionate share of the risk, these concessional contributions instil confidence in private actors, enabling them to deploy capital into projects that align with their financial goals while contributing to broader societal and environmental objectives. More than a tool for de-risking, blended finance is a mechanism for aligning incentives across stakeholders with differing priorities. Public and philanthropic institutions are typically driven by mission-oriented goals, such as reducing poverty, improving health outcomes, or combating climate change. On the other hand, private investors are motivated by financial returns, though they are increasingly considering ESG factors. Combining these sources of capital allows blended finance structures to create a synergy where each stakeholder's objectives are met without compromising the overall viability of the project. For instance, public funding can establish a foundation by covering upfront costs or enabling feasibility studies, while private capital scales the project, bringing efficiency and innovation. Blended finance is employed in various projects, often tailored to address specific developmental or sustainability challenges. These projects can include renewable energy initiatives aimed at expanding access to clean power, infrastructure development in underserved regions, or agricultural investments to enhance food security. The choice of project type typically depends on the developmental priorities of the targeted region or country and the nature of the public good being promoted.

Nonetheless, the effectiveness of blended finance hinges on its structuring. It requires a delicate balance to ensure that concessional capital is used judiciously, catalysing rather than crowding out private investment. Poorly designed structures risk distorting markets, creating dependency on public funds, or failing to attract sufficient private-sector participation. To avoid these pitfalls, blended finance must incorporate clear metrics for success, robust accountability mechanisms, and transparent allocation of risks and returns among stakeholders. Also, the structuring process must consider the unique challenges of the targeted markets and sectors. In fragile economies, blended finance can play a transformative role by fostering local capacity, creating jobs, and driving economic resilience. Similarly, in experimental technologies—such as clean energy or innovative healthcare solutions—it can accelerate development and adoption, bridging the gap between early-stage innovation and commercial viability.

Social and Development Impact Bonds (SIBs and DIBs): Outcome-Driven Innovation

SIBs and DIBs embody a transformative approach to financing social and developmental initiatives, directly tying financial returns to the achievement of predefined outcomes. These innovative instruments depart from traditional funding models, focusing on accountability, risk-sharing, and outcome-driven innovation. Aligning financial incentives with tangible social and developmental results allows impact bonds to create a framework that prioritises effectiveness, efficiency, and long-term impact. SIBs leverage private investment to fund social programmes, with repayment and potential returns contingent on the achievement of predefined outcomes. Governments or public agencies typically commission SIBs, entering into contracts with service providers often NGOs or social enterprises to deliver interventions. Investors provide upfront capital and are repaid by the government or a philanthropic funder if the intervention achieves its agreed-upon targets, as verified by an independent evaluator. SIBs aim to shift social service funding from input-based to outcome-based approaches, promoting efficiency and accountability in public service delivery. DIBs function similarly to SIBs but are applied primarily in low- and middle-income countries (LMICs). Unlike SIBs, where governments repay investors, DIBs typically rely on third-party funders, such as development agencies, philanthropies, or multilateral institutions, to make the outcome payments. This distinction makes DIBs particularly suited for contexts with constrained public-sector funding or fiscal capacity. DIBs address development challenges in areas such as education, health, agriculture, and workforce development, aligning incentives across stakeholders to improve effectiveness in delivering social impact. The results-oriented approach of SIBs and DIBs ensures that resources are used efficiently and that interventions make a difference. This accountability mechanism has a profound ripple effect on how services are designed and delivered. Service providers, whether non-profits or private entities, are incentivised to innovate and optimise their methods to achieve the desired outcomes. The focus shifts from merely implementing activities to ensuring meaningful impact, fostering a culture of evidence-based practices and continuous improvement.

Another defining feature of impact bonds is their ability to transfer financial risk from governments or development agencies to private investors. In a typical structure, investors provide upfront funding to service providers to implement social or developmental programmes. Governments or philanthropic organisations commit to repaying the investors, but only if the agreed-upon outcomes are achieved. This arrangement reduces the financial burden on public agencies, particularly in cases where interventions do not deliver the expected results. For private investors, the financial risk is offset by the potential for returns if the outcomes are met. This dynamic creates a strong incentive for all parties to collaborate effectively. Service providers focus on delivering results, investors are vested in supporting high-performing programmes, and governments benefit from improved efficiency and accountability in public spending. Shifting the risk in this manner allows impact bonds to unlock new opportunities for governments and development agencies to experiment with innovative solutions that might otherwise be considered too risky. This is particularly important in addressing complex challenges, such as improving health outcomes in underserved populations or reducing recidivism among formerly incarcerated individuals, where traditional approaches have often fallen short.

Where traditional investment structures lack the flexibility to accommodate the longer time horizons and uncertain outcomes often associated with social and developmental initiatives, impact bonds are explicitly designed to incentivise sustained, outcome-oriented approaches. They create a contractual framework that commits all parties to a shared goal, aligning financial rewards with social and developmental success over the medium to long term. This long-term orientation increases the likelihood of achieving meaningful impact and builds a foundation for scaling successful interventions. Once a programme demonstrates its effectiveness under an impact bond structure, it can attract further investment and be replicated in other contexts, multiplying its impact. Social and Development Impact Bonds represent a paradigm shift in financing and implementing public and social initiatives. They challenge traditional funding models that often lack accountability, encourage inefficiency, and fail to align incentives with outcomes. Instead, they place results at the forefront, fostering a collaborative ecosystem where investors, service providers, and governments work together to achieve shared objectives. While SIBs and DIBs present challenges, such as the intricacy of designing robust metrics and the risk of prioritising easily attainable outcomes, they have emerged as powerful tools for addressing complex societal and developmental issues. These instruments incentivise efficiency and accountability by linking financial returns to tangible impact and encourage collaboration among diverse stakeholders. SIBs and DIBs exemplify the potential of innovative financing to reimagine how resources are mobilised and applied, fostering solutions that deliver deep and enduring social progress.

Bonds and Sustainability-Linked Instruments: Mobilising Climate Finance

Sustainability bonds (including green bonds, blue bonds and others) and sustainability-linked instruments have also emerged as pivotal mechanisms for mobilising capital to address urgent climate and sustainability challenges. These instruments offer a compelling pathway for investors seeking returns and impact by bridging financial performance with environmental and social objectives. Their structuring, which ensures transparency, accountability, and flexibility, is crucial in aligning the interests of issuers, investors, and the broader global sustainability agenda. At the heart of sustainability bonds lies the principle of transparency, ensuring that capital raised is directed exclusively towards projects that deliver measurable environmental benefits. Unlike conventional bonds, green bonds come with rigorous use-of-proceeds requirements, which mandate that funds be allocated to verifiable green initiatives. These projects typically include renewable energy installations, energy efficiency upgrades, sustainable transportation, sustainable agriculture, and water conservation systems, among others. The transparency embedded in green bonds is essential for maintaining investor trust. Investors are increasingly wary of greenwashing.[9] Green bonds are often accompanied by third-party certifications, impact reporting, and ongoing monitoring to mitigate this risk, providing investors with confidence that their capital is contributing to verifiable outcomes. This trust enhances the appeal of green bonds and reinforces their credibility as a tool for financing climate action at scale. Sustainability-Linked Loans (SLLs) and Bonds (SLBs) build upon the transparency of green bonds but introduce a unique element of flexibility, broadening their applicability across various sectors and geographies. Unlike green bonds, which are tied to specific projects, SLLs and SLBs are linked to an issuer's overall ESG performance. Companies commit to achieving predetermined sustainability targets, such as reducing carbon emissions, improving water use efficiency, enhancing workplace diversity, or increasing renewable energy usage. The flexibility allows sustainability-linked instruments to align with the unique circumstances and goals of the issuer while still driving meaningful impact. For instance, a manufacturing company might use proceeds for general corporate purposes while committing to measurable ESG improvements. The financial terms of these instruments, such as interest rates, are often linked to the issuer's success in meeting these targets, creating a powerful incentive for sustained progress. The adaptability of sustainability-linked instruments makes them particularly effective in sectors or markets where green bonds might not be feasible, such as industries with diverse or evolving sustainability priorities. They also enable issuers to embed sustainability into their broader operations, fostering a more holistic approach to impact beyond individual projects.

The significance of SLLs and SLBs instruments extends beyond their immediate financial and environmental benefits. They represent a paradigm shift in how capital markets engage with climate and sustainability issues, catalysing investment in areas critical to achieving global goals such as the Paris Agreement and the United Nations SDGs. Providing a transparent, accountable, and flexible framework for financing sustainability enables these instruments to attract traditional investors and open the door for new participants, including impact-oriented funds and ESG-focused institutional investors. Without these tailored structures, it would be challenging to mobilise the scale of capital needed to address the climate crisis and broader sustainability challenges. Green bonds and sustainability-linked instruments exemplify how innovative financial mechanisms can align market interests with global imperatives, driving transformative change at the intersection of finance and sustainability. Their continued evolution and adoption are essential for bridging the financing gap and accelerating progress towards a sustainable future.

Venture Philanthropy: Balancing Concessionality and Scalability

Venture philanthropy represents a transformative approach to financing early-stage social enterprises by merging the altruistic intent of philanthropy with the rigour and efficiency of investment methodologies (OECD netFWD, 2014). This hybrid model is designed to address the unique challenges social enterprises face in organisations that operate at the intersection of social impact and financial sustainability. Combining concessional funding with strategic support enables venture philanthropy structures to ensure that these enterprises have the resources and resilience to achieve their long-term objectives (John, 2006). Traditional venture capital approaches, emphasising rapid returns, often fail to accommodate the unique dynamics of social enterprises. Many social enterprises operate in markets where profitability takes longer, mainly if they prioritise affordability and inclusivity over immediate revenue generation. Therefore, venture capital's short-term focus can place undue pressure on these enterprises, leading to mission drift or premature scaling efforts that compromise their impact.

At its core, venture philanthropy leverages concessional funding—grants, low-interest loans, or equity with low-return expectations—to provide a safety net for unproven ventures. Early-stage social enterprises often operate in high-risk environments or markets where traditional capital is either unavailable or prohibitively expensive. They may also be testing innovative solutions to complex social or environmental challenges, making their financial viability uncertain. Concessional funding absorbs much of this risk, enabling social enterprises to experiment, iterate, and refine their models without the immediate pressure to generate substantial financial returns (Cuccia, 2023). This safety net is critical because it allows these organisations to focus on their mission-driven objectives, such as improving access to healthcare, education, or renewable energy in underserved communities, rather than solely prioritising profitability. Taking on this initial risk enables venture philanthropy to create an enabling environment for innovation and social impact.

While concessional funding provides the initial runway for social enterprises to operate, venture philanthropy goes beyond short-term support. Its structures are intentionally designed to bake scalability into the model, helping high-potential enterprises transition from grant dependence to commercial viability. This involves financial investment and the provision of non-financial support, such as mentorship, capacity-building, and access to networks. The focus on scalability ensures that social enterprises can grow their impact over time, reaching more beneficiaries and addressing systemic challenges at scale. Structuring mechanisms often include milestones or performance-based funding, encouraging enterprises to adopt sustainable practices and business models. This emphasis on long-term viability distinguishes venture philanthropy from traditional philanthropic approaches, which often prioritise immediate outcomes over enduring solutions (John, 2006).

However, a key challenge in venture philanthropy lies in balancing concessionality with market discipline (OECD netFWD, 2014). While concessional funding is essential for de-risking early-stage ventures, an over-reliance on such funding can create dependencies that undermine the enterprise's ability to operate independently. To address this, venture philanthropy structures are designed to gradually introduce elements of market discipline, preparing enterprises for eventual engagement with commercial investors. For example, an enterprise that begins with grants might transition to concessional loans as it demonstrates its ability to generate revenue. Over time, as the enterprise matures and its risk profile improves, it may attract equity investments from impact investors or even traditional venture capitalists. This approach ensures the financial viability of the social enterprise and is also positioned to scale their impact sustainably (Roth, 2021). Venture philanthropy's ultimate value lies in its ability to drive systemic change. Successful ventures not only scale their impact but also demonstrate the viability of new models, attracting additional investment and inspiring replication in other contexts. Also, venture philanthropy fosters collaboration between philanthropists, investors, and social entrepreneurs. This collaborative ecosystem accelerates innovation, reduces duplication of efforts, and enhances the overall effectiveness of impact-oriented initiatives. Combining philanthropic funding with business principles enables venture philanthropy to achieve measurable social impact while fostering sustainable solutions.

Venture philanthropy bridges the gap between philanthropy and investment. Providing concessional funding as a safety net and embedding scalability into its structures ensures that social enterprises have the resources and resilience to achieve their mission-driven objectives. This balanced approach, which combines financial support with strategic guidance, enables ventures to transition from early-stage experimentation to commercial viability, ultimately driving sustainable impact at scale. In doing so, venture philanthropy can transform individual enterprises and contribute to systemic solutions for some of the world's most pressing challenges.

Patient Capital: Bridging Long-Term Horizons
Patient capital is a form of investment that prioritises long-term value creation over immediate financial returns. It is characterised by its commitment to enduring timelines, acceptance of greater uncertainty, and willingness to support ventures that tackle complex societal challenges. Unlike traditional investments that emphasise rapid returns, patient capital aligns its objectives with the growth and development trajectory of the investee, often with a focus on social and environmental impact. The concept of patient capital emerges from the recognition of market failures where traditional capital inadequately addresses systemic social and environmental challenges. Rooted in the principles of impact investing and development finance, patient capital is increasingly viewed as a hybrid approach that combines the financial discipline of commercial investing with the mission-oriented goals of philanthropy. Scholars like Bugg-Levine and Emerson (2011) have highlighted patient capital's pivotal role in creating transformative change by enabling risk-tolerant, flexible financing for ventures in nascent markets or underserved communities. These ventures often operate at the intersection of market innovation and public good, requiring investments beyond conventional time horizons. Patient capital often originates from foundations, family offices, or Development Finance Institutions committed to advancing equitable development. Patient capital provides more than monetary resources by blending financial tools such as equity, debt, or guarantees with technical assistance and strategic support. It creates an enabling environment for social enterprises to experiment, adapt, and achieve sustained impact over time. Patient capital bridges long-term horizons by addressing critical gaps in the financial ecosystem that traditional capital often overlooks. Its core value lies in its ability to:

- **Align with Mission-Driven Objectives**: Patient capital is uniquely suited to ventures prioritising social impact and financial sustainability. These ventures often operate in challenging environments where immediate profitability is unattainable. For instance, studies by Brest & Born (2013) demonstrate how patient capital have catalysed advancements in healthcare delivery, particularly in low-income countries. The flexibility of patient capital allows organisations to focus on scaling their impact rather than delivering short-term financial returns to investors.

- **Support High-Risk, High-Impact Ventures**: Many transformative initiatives, such as rural electrification, affordable healthcare solutions, or sustainable agriculture, entail significant upfront costs and uncertain revenue streams. Traditional investors may shy away from these opportunities due to perceived risks or delayed returns. Patient capital fills this gap by providing the necessary resources to de-risk projects, enabling them to achieve scale and sustainability over time. Research by Koh et al. (2019) emphasises patient capital's role in fostering inclusive business models that integrate marginalised populations into economic value chains.
- **Foster Innovation and Experimentation**: Innovation often requires time, particularly in sectors addressing systemic challenges. Patient capital creates a conducive environment for experimentation, allowing entrepreneurs and organisations to test, iterate, and improve their solutions without the constraints of short-term performance metrics. For instance, clean energy technologies, which often require prolonged R&D cycles, have benefited significantly from patient capital, as shown in studies examining the transition to renewable energy in emerging markets (Bolton & Foxon, 2015).
- **Stabilise Impact-Focused Ventures**: Patient capital provides a stabilising force for ventures operating in volatile or underserved markets by offering long-term financial commitments. Enterprises delivering essential services in conflict-prone regions, for example, rely on patient capital to maintain operations during periods of uncertainty. This stability is critical for creating lasting social and economic impact, as evidenced by initiatives like the Acumen Fund's support for enterprises addressing poverty (Impact Alliance, 2020).
- **Catalyse Systemic Change**: Addressing deep-rooted societal challenges requires sustained effort and coordinated action. Patient capital plays a catalytic role by attracting additional investment, fostering partnerships, and demonstrating the viability of impact-driven business models. For example, patient capital invested in renewable energy microgrids can pave the way for broader adoption of decentralised energy solutions, transforming entire energy ecosystems. Such systemic changes align with findings in the literature on sustainable development finance (Addison & Roe, 2018).

Catalytic funding: Unlocking Private Investment

Catalytic funding recognises that many high-impact projects struggle to secure financing due to structural market gaps. It is a transformative financial mechanism in impact investing, designed to mobilise private capital for projects that face challenges in securing traditional financing. These challenges often stem from perceived risks, market inefficiencies, or a lack of track record in underserved regions or sectors. Catalytic funding addresses these barriers by strategically deploying public or philanthropic funds to redistribute risk, enhance market confidence, and attract private investment. It bridges the gap between philanthropic resources and the vast pools of private capital required to address pressing global challenges. Aligning public and philanthropic funds with private capital enables catalytic funding to create an enabling environment for high-impact projects to thrive.

Catalytic funding operates on the principle of leveraging resources. Every dollar of concessional or catalytic funding is strategically designed to unlock multiple dollars of private investment, amplifying its impact. This leveraging effect is achieved through first-loss capital, guarantees, and concessional loans, which reduce the perceived risk for private investors. For instance, first-loss capital—where public or philanthropic funds absorb the initial losses in a transaction—provides a safety net, making investments more attractive and accessible to risk-averse investors. Projects in emerging markets or underserved sectors are considered too risky for conventional investors. Catalytic funding de-risks the opportunity by taking on the riskiest portion of the investment, encouraging private participation. For example, in renewable energy projects, catalytic funds may cover the early-stage development costs or provide guarantees against potential losses, thus reducing the risk profile and increasing investor confidence. This funding mechanism also targets market inefficiencies. Information asymmetries and a lack of infrastructure deter private investment in many underserved regions. Catalytic funding can bridge these gaps by funding feasibility studies, providing technical assistance, or building local capacity. These efforts reduce uncertainty and create a more straightforward pathway for private capital to flow into high-impact projects. Additionally, catalytic funding is pivotal in building a track record for emerging sectors or nascent markets. Private investors often hesitate to enter unproven markets due to a lack of historical data or success stories. Catalytic investments demonstrate the viability and profitability of these markets, paving the way for subsequent rounds of private investment. For instance, a catalytic investment in an off-grid renewable energy project might prove the feasibility of such models, encouraging private investors to fund similar initiatives.

The leveraging effect of catalytic funding extends beyond individual transactions. Demonstrating success in neglected sectors or regions enables catalytic funding to create a ripple effect that fosters long-term market development. For example, an initial catalytic investment in clean energy infrastructure can establish a proof of concept, attracting additional private investment and driving the broader adoption of renewable energy solutions. Over time, this market maturation reduces dependence on concessional funding as private capital becomes increasingly willing to invest independently. This Demonstration and Crowd-In Effect is a pivotal aspect of catalytic funding that highlights its role in transforming high-risk, low-confidence markets into thriving ecosystems of private investment. In addition to its direct financial impacts, the crowd-in effect has significant social and environmental implications. As private capital flows into underserved regions or sectors, it drives job creation, economic development, and the adoption of sustainable practices. This ripple effect magnifies the impact of catalytic funding, creating systemic change that benefits communities and ecosystems. By bridging gaps, demonstrating success, and mobilising private capital, it drives long-term market development and builds resilient, inclusive economies that benefit society as a whole.

Structuring catalytic funding effectively is critical to its success. Catalytic funding mechanisms must be deployed with precision to ensure that public or philanthropic funds are used optimally, catalysing rather than crowding out private capital. Transparent criteria, accountability measures, and performance tracking are essential components of this structuring process, ensuring that these mechanisms achieve their intended impact without unintended distortions. The importance of catalytic funding lies in its ability to mobilise private capital and its role as an enabler of systemic change. By mitigating risks and demonstrating the viability of investments in underserved areas, these tools shift perceptions and redefine what is possible within impact investing. They challenge the conventional boundaries of financial viability, opening up new opportunities for innovation, entrepreneurship, and development in sectors and regions that have long been overlooked. In essence, catalytic funding is more than a financial mechanism—it is a strategic tool that aligns diverse interests, breaks down barriers, and unlocks the transformative potential of private capital. By amplifying the reach of public and philanthropic resources, catalytic funding ensures that the capital needed to address the world's most pressing challenges flows to where it is needed most. Its thoughtful application unlocks immediate opportunities and lays the groundwork for more inclusive and resilient financial systems.

Guarantee Mechanisms: Mitigating Risks
Guarantees are financial instruments designed to reduce or redistribute the risk associated with investments, loans, or other financial transactions. Providing a safety net ensures that if a borrower or investee defaults, the guarantor—typically a government agency, multilateral development bank, or philanthropic organisation—assumes part or all of the financial liability. This mechanism encourages lenders and investors to engage in projects or markets they might otherwise consider too risky. Guarantees are pivotal in mobilising private capital for high-impact projects in underserved regions or sectors. They do not provide direct funding; instead, they act as insurance, assuring investors or lenders that a portion of their financial exposure is protected. This assurance is critical when risks such as credit defaults, political instability, or revenue unpredictability are significant deterrents to investment. Assuming part of the risk, guarantees lower the threshold for private-sector participation, enabling projects that would otherwise remain unfunded. Beyond their financial implications, guarantees signal institutional support and confidence in a project or market. This symbolic value reduces perceived risks and creates an enabling environment for private investment. For example, a guarantee backing an infrastructure project in a politically volatile country reassures investors that they are shielded from certain non-commercial risks, such as government policy changes or currency inconvertibility. These instruments catalyse systemic change by aligning public or philanthropic guarantees with private capital, driving sustainable economic and social development.

Applications of Guarantees
Guarantees are versatile tools with a wide range of applications across different financial and impact-oriented contexts:

- **Partial Credit Guarantees**: These guarantees share the credit risk between the guarantor and the lender, covering a specific portion of potential losses. This partial risk coverage ensures that the lender remains incentivised to conduct due diligence while being shielded from the full brunt of potential defaults. For example, partial credit guarantees reduce risk perception in high-risk sectors such as smallholder agriculture or early-stage technology ventures, allowing financial institutions to extend credit with greater confidence. The shared-risk model encourages prudent lending while unlocking financing for otherwise overlooked projects.

- **Loan Portfolio Guarantees**: Rather than focusing on individual loans, these guarantees apply to a diversified portfolio, spreading risk across multiple borrowers or projects. This aggregation reduces lenders' overall risk profile and encourages underserved groups' inclusion. Loan portfolio guarantees are particularly impactful in fostering financial inclusion, enabling banks and Microfinance Institutions to support small enterprises or marginalised demographics. Pooling risk allows lenders to experiment with innovative financing approaches without jeopardising their financial stability.
- **Political Risk Guarantees**: Political risk guarantees specifically protect investors against non-commercial risks, such as expropriation, currency inconvertibility, and political violence. These guarantees are indispensable for cross-border investments in politically unstable regions. By de-risking the operating environment, they enable the implementation of infrastructure projects, resource extraction, or large-scale manufacturing in emerging markets. The presence of a political risk guarantee reassures stakeholders that their investments will be safeguarded against unpredictable geopolitical shifts, fostering long-term confidence and engagement.
- **Revenue Guarantees**: Revenue guarantees ensure that a project generates a baseline level of income, shielding investors from fluctuations in demand or price. This type of guarantee is particularly effective in Public–Private Partnerships and infrastructure projects where revenue streams can be uncertain. By stabilising cash flows, revenue guarantees create a predictable financial environment that attracts investors. For instance, a toll road project with a government-backed revenue guarantee would assure investors of a minimum income, regardless of traffic volumes, thereby mitigating financial risks.

Guarantees are indispensable tools for mitigating financial, political, and market risks that deter investment in high-impact projects. While their ability to redistribute risks and foster confidence is transformative, their strategic use requires careful consideration of long-term impacts and market dependencies. Guarantees encourage private-sector participation by lowering barriers to entry, particularly in sectors such as renewable energy, infrastructure, and healthcare. They also improve access to finance for underserved markets by offering more favourable terms, such as reduced interest rates and extended repayment periods. However, it is crucial to acknowledge the nuanced challenges and guarantees present. Over-reliance on guarantees could inadvertently distort markets by creating dependency or diminishing the need for private actors to build their internal risk assessment capacities. In the long term, guarantees should serve as catalysts, bridging the gap between market hesitation and active investment, but not as a perpetual crutch. Their ultimate success lies in building market confidence to the point where guarantees are no longer necessary. One of the most transformative aspects of guarantees is their leverage effect, where a relatively small amount of public or philanthropic capital mobilises a disproportionately larger pool of private investment. This ensures the efficient allocation of resources, maximising their impact. When applied judiciously, guarantees amplify the effectiveness of development finance, making them a cornerstone of innovative, high-impact financial mechanisms that enable sustainable and inclusive growth.

2.4.2 *Instruments in Impact Investment*

Various financial instruments are used in impact investing. These instruments address the diverse needs of impact projects, facilitating the flow of capital into projects and initiatives, in many cases, that are traditionally underserved by conventional finance. By leveraging a mix of these instruments, impact investors cater to different risk appetites and expected returns, making it possible to align diverse stakeholder interests with the pursuit of positive impact. This section will explore these instruments in depth, highlighting how they have contributed to the growth and maturation of the impact investing sector.

Equity Financing

In impact investing, an equity instrument represents an ownership stake in a company or organisation expected to generate both a financial return and a measurable positive social or environmental impact. Investors provide capital in exchange for ownership stakes in a company. This type of financing is particularly suited to early-stage or high-growth enterprises, enabling them to scale operations, innovate, or enter new markets without the burden of debt repayment. Equity financing in impact investing is usually instrumental in fostering innovation, supporting mission-driven businesses, and aligning investor returns with long-term social and environmental outcomes. They are typically used to fund initiatives that require substantial upfront capital or have long gestation periods before achieving profitability. These investments often involve higher risks but offer the potential for significant financial returns alongside measurable impact.

Equity instruments have played a critical role in the development of the impact investing industry by providing the necessary capital to support innovative solutions addressing social and environmental challenges. By offering different structures that cater to varying risk appetites and return expectations, equity financing has enabled the growth and sustainability of enterprises focused on delivering positive impact. This financing model has been particularly instrumental in scaling initiatives that require substantial upfront investment, such as renewable energy projects, healthcare access innovations, and financial inclusion technologies. The infusion of equity capital has allowed such enterprises to attract talent, invest in research and development, and ultimately expand their impact. Beyond providing capital, equity financing has also helped shape the industry by bringing strategic support and expertise to the enterprises they invest in, offering mentorship and guidance which support impact-driven companies to professionalise their operations, enhance governance, and create robust business models capable of delivering both social and financial returns. Furthermore, the development of patient capital and revenue-linked equity has expanded the tools available to impact investors, offering more flexible and aligned financing options that accommodate the unique challenges faced by social enterprises. The evolution of the models has also fostered collaboration between public, private, and philanthropic capital.

The GIIN's 2024 State of the Market report shows a steady growth of equity investment in impact investing. It remains a dominant asset class in impact investing, accounting for a significant proportion of Assets Under Management (AUM). Private equity, in particular, is a preferred vehicle for impact investors, with 73 per cent of investors utilising it. It accounts for 43 per cent of the total impact Assets Under Management (AUM), highlighting its central role in the sector. Developed market-focused investors allocate 42 per cent of their AUM to private equity, while emerging market-focused investors allocate an even higher proportion of 48 per cent. This reflects the versatility of equity financing in addressing diverse needs across regions. This dominance of private equity underscores its role in financing scalable social enterprises and enabling innovation in high-growth sectors such as technology, healthcare, renewable energy, and financial services, promoting global priorities like climate action, digital inclusion, health equity and financial inclusion (Hand et al., 2024).

It aligns well with impact investors seeking market-rate or higher returns while prioritising long-term impact; particularly those with a higher risk tolerance and a focus on fostering innovation and scalability. The higher allocation of equity investments in emerging markets highlights its role in addressing gaps in traditional financing and driving inclusive growth. It prominence underscores investors' commitment to long-term impact strategies, despite the associated risks. It remains a cornerstone of impact investing, driving innovation and addressing systemic challenges across markets and sectors. While challenges such as rising interest rates and inflation persist, reports reveal the continued satisfaction of impact investors' with their equity investments' financial performance as they adapt to these challenges. The following section discusses some of the common types of equity financing models currently adopted in impact investment. It examines the structure and operational model, the lifecycle, the challenges and opportunities to mitigate such challenges and leverage the instruments optimally.

(a) **Venture Capital (VC)**

Venture Capital (VC) equity is a form of private equity financing that investors provide to startups and small businesses that are believed to have long-term growth potential. Unlike traditional financing methods, VC equity is typically invested in companies that are at an early stage of development, often without a proven track record or consistent revenue streams. These companies are characterised by innovative business models, high-risk profiles, and the potential for significant returns if successful. VC equity is provided by venture capital firms, which pool capital from a variety of sources, including institutional investors, high-net-worth individuals, and family offices. These firms invest in high-potential startups in exchange for equity, or an ownership stake, in the company. This ownership gives venture capitalists a say in the strategic direction of the business, allowing them to influence key decisions that could affect growth and profitability. The goal of VC investments is to support early-stage companies through their critical phases of growth, helping them to scale operations, develop products, and ultimately achieve profitability. Venture capital investments are inherently high-risk, as the majority of startups do not achieve long-term success. However, the potential rewards can be substantial, as successful startups can deliver exponential returns to their investors. Venture capital plays a crucial role in fostering innovation by providing the necessary capital and expertise that enable entrepreneurs to bring disruptive technologies and business models to market. VC equity is an essential component of the entrepreneurial ecosystem, bridging the funding gap for startups that do not have access to traditional bank loans or public funding.

Emerging as a formalised financing mechanism in the mid-twentieth century, to support high-risk, innovative ventures, the modern structure of VC took shape in the United States post-World War II, with several key milestones marking its evolution. The establishment of firms like American Research and Development Corporation (ARDC) marked the birth of institutional VC. During this period, investments were predominantly directed at high-tech companies in defence and aerospace sectors. The rise of Silicon Valley heralded a golden age for VC, driven by the tech boom. During this era, firms like Sequoia Capital and Kleiner Perkins became synonymous with landmark successes, including investments in Apple and Google. The dot-com bubble reshaped VC strategies, leading to more stringent due diligence and a shift in focus. Subsequent waves of innovation in software, biotechnology, and fintech have broadened the VC landscape, making it more diverse and sophisticated.

Core Structure and Operational Model

VC equity operates through structured partnerships designed to align risk, reward, and accountability across stakeholders. The structure supports the managing of the high-risk nature of early-stage investments while ensuring that all parties benefit from successful ventures. There are typically the General Partners (GPs)—these are the professional investors responsible for fund management, source deals, conduct due diligence, and drive post-investment growth.[10] GPs play an active role in guiding startups towards growth and profitability. And there also the Limited Partners (LPs), which provide the capital but take on a passive role. They include pension funds, endowments, family offices, and sovereign wealth funds. Their primary interest is in sharing the profits after GPs' fees are accounted for. LPs contribute the financial resources necessary for GPs to make investments but do not participate in day-to-day decision-making.

Investment Lifecycle

The investment lifecycle in venture capital is a multi-phase process that ensures the effective allocation, management, and realisation of investments. Each phase is crucial in shaping the overall success of the fund and its portfolio companies.

Fundraising: The investment lifecycle begins with fundraising, where GPs secure commitments from LPs. Fundraising typically involves presenting a clear investment thesis, highlighting potential sectors or geographies of interest, and showcasing the GP team's expertise. For example, a climate-focused VC fund might target clean technology startups in Southeast Asia, positioning itself to address pressing global challenges while generating attractive returns for investors. Fundraising can be a lengthy process, often taking 12 to 18 months, during which GPs must build relationships with prospective LPs and convince them of the fund's value proposition.

Deployment: Once capital commitments are secured, GPs proceed with deploying the funds. During the deployment phase, capital is allocated to startups that align with the fund's investment thesis. Typically, 70 per cent–90 per cent of the portfolio is allocated to early-stage ventures, as these companies offer higher growth potential despite their inherent risks. Deployment involves rigorous due diligence, including evaluating business models, market opportunities, management teams, and financial projections. The goal is to identify startups that have the potential for scalability and significant value creation. GPs often invest in multiple funding rounds (seed, Series A, B, etc.) to support portfolio companies through different stages of growth.

Growth and Support: Following the deployment of capital, the growth and support phase is critical to ensuring the success of portfolio companies. GPs provide much more than just funding—they offer hands-on guidance, mentorship, strategic advice, and access to valuable networks. This phase involves working closely with startups to refine their business models, optimise operations, and navigate challenges. GPs may also help startups establish key partnerships, attract top talent, and expand into new markets. The support provided by GPs can significantly enhance a startup's ability to scale effectively, overcome obstacles, and achieve milestones. This active involvement is a defining feature of venture capital, setting it apart from other forms of passive investment.

Monitoring and Value Creation: Throughout the growth and support phase, GPs actively monitor the performance of portfolio companies to ensure they are on track to meet key performance indicators (KPIs). Regular board meetings, performance reviews, and strategic discussions are part of the monitoring process. GPs leverage their industry expertise to help startups pivot when necessary, refine their product offerings, or enter new markets. Value creation during this phase is not limited to financial growth but also includes fostering strategic positioning, enhancing brand reputation, and achieving product-market fit.

Exit: The final phase of the investment lifecycle is the exit, where GPs realise returns on their investments. Exits can take several forms, including initial public offerings (IPOs), mergers, acquisitions, or secondary sales. The timing and strategy of the exit are critical, as they determine the ultimate financial outcome for both GPs and LPs. Successful exits, such as Uber's IPO or Facebook's acquisition of Instagram, often generate significant returns that compensate for other less successful investments in the portfolio. Planning an exit requires careful consideration of market conditions, company performance, and potential acquirers. Exits are also crucial for demonstrating success to LPs, which helps secure commitments for future fundraising rounds.

Challenges in Leveraging Venture Capital (VC) Equity for Impact Investing

Venture capital (VC) equity, while instrumental in driving innovation and supporting high-growth startups, faces several inherent challenges that can impact the success of investments and the ability to generate returns. They include the following.

Short-Term Focus vs. Long-Term Impact Goals: Venture capital's emphasis on rapid scaling and financial returns within five to seven years is inherently at odds with the long-term nature of most social and environmental impact objectives. This disconnect discourages investment in initiatives that require sustained effort and patient capital to address systemic issues, such as climate resilience or educational reform. For instance, clean energy projects often have long development cycles and may not yield significant financial returns during a VC's expected timeframe. This short-term orientation can lead to premature scaling, sacrificing depth of impact for perceived profitability, or abandoning ventures that fail to show immediate growth. Such outcomes hinder the realisation of transformative solutions and perpetuate a market environment where short-term metrics dominate over long-term societal benefits.

High-Risk and Uncertain Returns: VC investments inherently target startups with unproven models, but the risks amplify in impact-driven sectors where traditional revenue streams may not exist. For example, startups addressing basic healthcare in low-income regions often serve demographics with limited purchasing power, creating challenges for monetisation. The high-risk, low-return perception of impact sectors discourages capital flow, leaving socially driven startups underfunded. This aversion to risk creates a vicious cycle, where lack of investment stifles innovation in critical areas, further solidifying the narrative that impact ventures are financially untenable. Consequently, the capital gap for startups seeking to address urgent global challenges persists, leaving critical needs unmet.

Impact Measurement and Verification: The lack of robust frameworks for measuring and verifying impact exacerbates the challenge of ensuring accountability in VC-backed impact ventures. Startups at early stages often lack the capacity or resources to develop rigorous impact metrics, and investors face difficulties in assessing the validity of reported outcomes. This lack of standardisation and transparency undermines investor confidence, making it harder for startups to attract further funding. The absence of reliable metrics also weakens the ability to compare investments across a portfolio, limiting the strategic allocation of resources. This creates a fragmented ecosystem where the credibility of impact claims is questioned, ultimately hindering the sector's growth and sustainability.

Sectoral Bias and Limited Diversity: VC investments predominantly target high-growth sectors like technology and fintech, where scalability and quick returns are more feasible. While renewable energy and public health have gained prominence, other critical sectors face significant barriers. These include *Agriculture*, typically characterised by fragmented supply chains, insufficient access to technology, and infrastructure deficits, particularly in emerging markets; *Water and Sanitation* projects, characterised by complex regulatory requirements, high capital intensity, and long payback periods deter investments, despite the sector's essential role in human health and development; *Circular Economy Initiatives,* which tend to struggle with market readiness for recycled materials, the high cost of technology development, and fragmented regulatory standards. These limit scalability and investment. *Biodiversity Conservation initiatives,* which may also struggle with monetisation models, require patient capital. Often facing challenges in demonstrating financial returns despite their critical ecological importance; *Affordable Housing projects,* which require long-term investments and often operate within constrained regulatory and financial frameworks that discourage VC engagement. Moreover, geographic biases persist, with investments heavily concentrated in developed markets, neglecting high-impact opportunities in emerging economies. These biases reduce the diversity of VC portfolios and limit the potential for transformative impact in underserved regions.

Lack of Expertise in Impact-Focused Investing: VC firms often lack the specialised knowledge necessary to evaluate and support impact-driven startups effectively. Traditional financial metrics dominate decision-making processes, leading to the undervaluation of ventures with unconventional models that prioritise social or environmental outcomes over immediate profitability. This gap in expertise results in missed opportunities to nurture impactful innovations and creates a misalignment between investor expectations and startup needs. Without a deeper understanding of the unique dynamics of impact sectors, VC firms struggle to provide the strategic guidance necessary for these startups to thrive, further limiting their potential to scale and sustain their impact.

Opportunities for Mitigation

Aligning Exit Strategies with Impact Goals: To reconcile the short-term focus of VC with long-term impact goals, investors can design exit strategies that prioritise sustained impact. Structuring exits through mission-aligned buyers, such as impact-oriented private equity firms or foundations, ensures continuity in achieving social or environmental objectives post-exit. Additionally, embedding impact milestones into funding agreements and linking financial incentives to these milestones fosters alignment between investor and entrepreneur priorities, reducing the risk of prematurely abandoning impactful initiatives.

Blended Finance Models to De-Risk Investments: Blended finance, which combines concessional capital from public or philanthropic sources with private VC funding, offers a viable solution to the high-risk nature of impact investments. Concessional capital can absorb early-stage losses or provide guarantees, making investing in high-risk, high-impact ventures more attractive for traditional VCs. This model not only reduces financial barriers but also encourages capital flow into sectors and geographies often perceived as too risky, fostering innovation where it is most needed.

Developing Standardised Impact Metrics: Developing flexible, sector-specific impact measurement tools tailored to the needs of early-stage ventures can bridge the gap in accountability and transparency. Frameworks such as the GIIN's IRIS+ system or the HIPSO, which offers a catalogue of sector-specific metrics, allow startups to measure impact in a way that aligns with their business models and scales with growth. Leveraging digital tools and analytics platforms enables real-time impact tracking, reducing the resource burden on startups. Incorporating iterative processes for refining metrics as startups evolve ensures that impact remains central to operations without stifling innovation or overburdening early-stage companies with rigid reporting requirements. These approaches foster investor confidence by providing meaningful insights while empowering startups to demonstrate tangible outcomes.

Expanding Sectoral and Geographic Focus: Overcoming sectoral and geographic biases requires adopting more inclusive investment strategies. Collaborating with local organisations and Development Finance Institutions (DFIs) can provide the contextual knowledge and networks needed to invest effectively in emerging markets. Dedicated funds should address persistent barriers in renewable energy, public health, agriculture, water and sanitation, circular economy initiatives, biodiversity conservation, and affordable housing. VCs can mitigate perceived risks by leveraging blended finance structures or government partnerships and unlock opportunities in these crucial but complex areas. Diversifying portfolios to include emerging economies and these vital sectors enhances impact and creates new avenues for financial returns in untapped markets.

Building Expertise in Impact-Driven Ventures: VC firms can enhance their capacity to evaluate and support impact-focused startups by integrating sectoral expertise into their operations. Hiring professionals with backgrounds in fields like healthcare, education, or sustainability ensures that investment decisions are well-informed and aligned with impact objectives. Establishing advisory boards with specialists and providing ongoing training for investment teams can further bridge the knowledge gap, enabling more effective engagement with socially driven entrepreneurs. This specialised approach enhances the ability to identify high-potential impact ventures and supports their growth with tailored guidance.

Leveraging venture capital in impact investing presents unique challenges, particularly in aligning short-term financial objectives with long-term impact goals, managing high risk, and addressing sectoral biases. However, these challenges can be mitigated through innovative strategies such as blended finance models, standardised impact measurement frameworks, and capacity-building initiatives. By adopting a more inclusive and patient approach, VC firms can unlock the potential of startups to drive meaningful social and environmental change while achieving financial success. This dual focus enhances VC's credibility in impact investing and contributes to a more sustainable and equitable entrepreneurial ecosystem.

(b) **Growth Equity**

Growth equity is a strategic financing instrument designed to support mature companies that seek to scale their operations, enter new markets, or enhance their impact initiatives. Positioned between venture capital and traditional private equity, growth equity targets established businesses with proven revenue models that require additional capital for expansion. This approach blends financial support with strategic guidance, enabling companies to pursue ambitious growth objectives without resorting to excessive debt or ceding majority ownership. Growth equity investments are particularly suited for businesses that are past the startup phase but not yet ready for the control-oriented structure of private equity.

Core Structure and Operational Model

Growth equity operates within a well-defined framework that prioritises collaboration between investors and companies to achieve sustainable growth. Unlike other equity models, growth equity emphasises strategic partnership rather than operational control, making it a flexible and adaptable tool for scaling businesses. Growth equity investors typically acquire minority stakes in portfolio companies, ranging from 10 per cent to 49 per cent. The size of the acquisition allows founders and management teams to retain decision-making authority while benefiting from the investor's expertise, networks, and financial resources. The capital injection is primarily directed towards initiatives such as geographic expansion, product development, operational optimisation, or acquisitions. The minority stake structure aligns the interests of both parties, fostering a collaborative rather than controlling dynamic.

The operational model of growth equity relies heavily on the expertise of the GPs, who identify investment opportunities, conduct thorough due diligence, and provide ongoing strategic support. GPs focus on enhancing value through board participation, advisory roles, and leveraging their industry networks. LPs, on the other hand, are primarily passive investors interested in financial returns. LPs are drawn to growth equity for its lower risk profile compared to venture capital, as the target companies already demonstrate revenue stability and growth potential. Due diligence in growth equity involves rigorous financial analysis, market research, and risk assessment to ensure alignment with the investment thesis. The structuring process often includes provisions for governance rights, protective measures, and board representation to safeguard the investor's interests. By concentrating investments in sectors where they possess deep expertise, growth equity firms maximise their ability to add strategic value and drive impactful growth.

Investment Lifecycle

The growth equity investment lifecycle encompasses several stages, each meticulously designed to guide companies through their scaling journey and realise maximum value at exit. These stages include capital deployment, the growth phase, and the exit phase.

Capital Deployment: The lifecycle begins with capital deployment, where growth equity investors identify and invest in companies that align with their strategic goals. Target companies are typically mature businesses with consistent revenue streams and a proven product-market fit. These companies seek growth capital to accelerate expansion, diversify products, or optimise operations. During this stage, GPs conduct extensive due diligence to evaluate the financial health, market position, and growth potential of prospective investments. Key focus areas include assessing management capabilities, market dynamics, competitive positioning, and scalability. The deal structuring process is equally critical, involving the negotiation of governance rights, board representation, and performance incentives. Investments are usually sector-focused, enabling investors to leverage specialised knowledge and industry networks. Capital deployment is characterised by the establishment of a strategic partnership between the investor and the company. This phase sets the stage for a collaborative approach to scaling, with the investor providing both financial resources and strategic guidance to unlock the company's potential.

Growth Phase: The growth phase is the core of the investment lifecycle, during which investors work closely with the company's management to implement a comprehensive growth strategy. This phase focuses on value creation through strategic support, operational enhancements, and market expansion. Growth equity investors often take on board roles or advisory positions to guide decision-making. They collaborate with management to refine business models, identify growth opportunities, and address operational inefficiencies. Key initiatives during this phase include expanding into new geographic markets, launching innovative product lines, and scaling production capacity. Operational optimisation is a central focus, with investors bringing in industry experts or consultants to streamline processes, improve margins, and enhance supply chain management. By introducing scalable technologies and best practices, investors help companies build a robust operational foundation capable of sustaining rapid growth. Talent acquisition is another critical component of the growth phase. Investors assist in recruiting senior executives and strengthening management teams to support the company's growth trajectory. Additionally, they leverage their extensive networks to connect portfolio companies with potential partners, customers, and suppliers, further accelerating growth. Throughout the growth phase, performance monitoring is conducted through regular board meetings, progress reviews, and the tracking of key performance indicators (KPIs). This ensures the company remains aligned with its growth objectives and allows for course corrections when necessary.

Exit Phase: The exit phase is the culmination of the growth equity lifecycle, where investors seek to realise investment returns. The choice of exit strategy depends on market conditions, company performance, and strategic objectives. Common exit options include Initial Public Offerings (IPOs), mergers and acquisitions (M&A), and minority buyouts. IPOs are pursued when a company has achieved significant scale and market leadership, attracting interest from public market investors. M&A exits involve selling the company to strategic acquirers who can integrate its capabilities into their operations. Minority buyouts, often by other investors, provide another viable exit route. Timing is crucial in the exit phase. Investors aim to capitalise on favourable market conditions, such as robust economic environments or high industry valuations. A typical holding period ranges from 5 to 7 years, during which the company is scaled and prepared for a successful exit. Investors also consider the interests of all stakeholders, including founders, employees, and co-investors, to ensure a smooth transition and sustained growth post-exit.

Challenges in Leveraging Growth Equity in Impact Investing and Their Mitigation

Alignment of Impact and Financial Goals: Balancing financial returns with social or environmental objectives is a persistent challenge in growth equity impact investing. The dual mandate often creates structural and operational tensions. For instance, where the LPs prioritise measurable financial outcomes, it may lead to a myopic focus on profitability at the expense of deeper societal benefits. Such prioritisation risks marginalising long-term impact projects, particularly those addressing systemic issues like climate change or income inequality, as these often require extended timelines and substantial upfront investment. The detrimental effect is twofold: the dilution of the transformative potential of impact investing and the erosion of investor credibility when impact goals are deprioritised. Addressing misalignment of priorities requires integrating robust accountability mechanisms, such as linking fund manager incentives to both impact and financial milestones, thereby ensuring dual objectives are not just aspirational but operational.

High Dependency on Management Teams: The success of growth equity investments is deeply intertwined with the competency and vision of portfolio company management teams. While allowing management to retain strategic control can foster innovation and ownership, it also heightens exposure to risks such as poor leadership, resistance to impact-driven change, or an overemphasis on short-term financial metrics. Such risks can undermine the strategic objectives of the investment, particularly in cases where the management lacks experience in scaling impact initiatives or integrating sustainable practices into core operations. These challenges become particularly acute in high-growth scenarios, where operational pressures often displace impact priorities. The dependency not only jeopardises the realisation of intended outcomes but also magnifies reputational risks for investors. To counter this, investors must adopt stringent due diligence frameworks that evaluate both leadership quality and their alignment with impact objectives. Post-investment, targeted support mechanisms such as executive coaching and advisory networks can mitigate these risks and reinforce the alignment of management with investor goals.

Sectoral and Regional Limitations: The sectoral and geographic focus of growth equity often creates blind spots in addressing urgent social and environmental challenges. Investments tend to gravitate towards sectors with predictable financial returns, such as technology or consumer goods, while high-impact sectors like affordable housing, public health, or sustainable agriculture remain underfunded. Similarly, geographic biases towards developed markets exclude emerging economies, where the need for transformative impact capital is most acute. These limitations not only constrain the diversity of impact outcomes but also perpetuate existing inequities by failing to address systemic global disparities. Moreover, the structural risks in underserved regions, such as political instability or inadequate infrastructure, discourage investments despite their transformative potential. The consequence is a missed opportunity to catalyse large-scale change in communities that stand to benefit most. Overcoming these limitations requires innovative risk-sharing mechanisms, such as blended finance models that de-risk investments in underserved sectors and regions, alongside strategic partnerships with local organisations to enhance contextual understanding and operational feasibility.

Capital Intensity and Risk Management: The capital-intensive nature of growth equity investments poses significant challenges, particularly in the context of impact investing. Large-scale funding requirements can deter smaller investors and concentrate decision-making power among a few dominant entities, potentially skewing priorities towards financial metrics. Furthermore, the long holding periods inherent in growth equity expose investors to macroeconomic risks such as inflation, currency devaluation, and geopolitical instability. These risks are compounded in emerging markets, where regulatory uncertainties and market volatility are more pronounced. The interplay of high capital demands and elevated risk levels often results in conservative investment strategies that prioritise risk mitigation over ambitious impact goals. This conservative approach constrains the transformative potential of growth equity, reducing its effectiveness as a tool for systemic change. To address this, investors must adopt advanced risk management frameworks that integrate scenario analysis, real-time data monitoring, and dynamic portfolio diversification. Leveraging concessional capital through Public–Private Partnerships can also offset risks while enabling bold investments in high-impact initiatives.

Impact Measurement and Reporting: The absence of standardised frameworks for measuring and reporting impact is a critical bottleneck in the scalability of growth equity as an impact investing tool. Existing metrics are often fragmented, focusing disproportionately on financial performance while relegating social and environmental outcomes to secondary considerations. This imbalance not only undermines the credibility of impact claims but also creates barriers to meaningful stakeholder engagement. Inconsistent measurement practices further complicate the benchmarking of impact performance across investments, reducing transparency and accountability. The implications are severe: Without robust and universally accepted metrics, the sector risks losing trust among investors, beneficiaries, and policymakers. To mitigate these challenges, developing comprehensive impact assessment tools that align with global standards, such as the UN's Sustainable Development Goals (SDGs), is imperative. Employing independent third-party audits and leveraging technology, such as blockchain for transparent impact tracking, can further enhance the reliability and comparability of impact data.

(c) **Private Equity (PE) in Impact Investing**

Private equity (PE) is a strategic investment approach that focuses on acquiring equity ownership in companies, often with the aim of restructuring, scaling, or strategically realigning the business to enhance its value. Typically targeting mature or underperforming companies, PE investors work to maximise financial returns by implementing operational improvements and strategic shifts. While conventionally profit-driven, PE in impact investing integrates financial goals to generate measurable social or environmental impact. This dual-purpose approach positions private equity as a powerful tool for addressing societal challenges while delivering robust financial performance.

Core Structure and Operational Model

Like the previously discussed forms of equity, PE also operates through a structured and disciplined model designed to create value for investors and portfolio companies. However, unlike VC and growth equity, the PE model is characterised by its focus on active ownership, operational enhancements, and strategic alignment to drive long-term success. PE funds are typically organised as limited partnerships. LPs provide the bulk of the capital and the GPs) manage the funds, source deals, and oversee portfolio management. The GPs contribute a smaller share of the capital, typically 1 per cent to 5 per cent, and are responsible for executing the fund's investment strategy. A hallmark of PE investments is the acquisition of controlling stakes in portfolio companies. This contrasts with growth equity investments, which often involve minority stakes. By acquiring control, PE firms can implement significant strategic and operational changes. A common acquisition method is the leveraged buyout (LBO), where a combination of equity and a substantial amount of debt is used to finance the purchase. This structure leverages the company's cash flows to service the acquisition debt, maximising the returns on equity invested. Once a company is acquired, PE firms engage deeply with its management team to enhance performance and unlock value. This engagement often includes board restructuring, cost optimisation, technology upgrades, and exploring new market opportunities. PE's active involvement ensures that portfolio companies are positioned for growth and profitability, benefiting both investors and the company's stakeholders.

Investment Lifecycle

The private equity investment process is meticulously structured, encompassing several distinct stages designed to optimise returns and manage risks.

Fundraising: The private equity lifecycle begins with the critical phase of fundraising. During this stage, GPs secure commitments from LPs as pension funds, sovereign wealth funds, and endowments. Fundraising is a time-intensive process, often taking 12 to 18 months. GPs present a detailed investment thesis outlining the sectors, regions, and strategies they intend to pursue. Additionally, they highlight their track record and capability to deliver competitive returns. The success of this phase determines the pool of capital available for future investments, ensuring that the fund is equipped to build a diversified portfolio.

Deal Sourcing and Acquisition: Once the fund is established, the focus shifts to identifying investment opportunities. Deal sourcing involves scouting for companies that align with the fund's strategic goals, including underperforming businesses with turnaround potential or mature firms requiring strategic realignment. During this phase, extensive due diligence is conducted to evaluate the target's financial health, operational capabilities, competitive positioning, and potential risks. Valuation and negotiation are critical components in ensuring the acquisition is financially viable. LBOs are commonly employed during acquisitions, combining equity investment with substantial debt to amplify returns. The structured approach ensures a disciplined entry into investments.

Value Creation: The value creation phase lies at the heart of the PE model. Post-acquisition, the private equity firm collaborates closely with the management of the portfolio company to develop and implement strategic initiatives. These initiatives may include operational restructuring, cost-efficiency measures, and entering new markets. For companies requiring a turnaround, the focus might involve leadership changes, revising business models, and addressing inefficiencies. For growth-oriented investments, strategies often include product innovation, technology adoption, and enhancing distribution networks. The overarching aim is to improve profitability, expand market share, and enhance the company's overall valuation.

Monitoring and Performance Improvement: During the holding period (that is when a private equity firm invests in a company and when they exit the investment), which typically ranges from three to seven years, PE firms actively monitor the progress of their portfolio companies. This involves setting measurable key performance indicators (KPIs) and conducting regular reviews to track financial and operational performance. The General Partners maintain a hands-on approach, offering strategic guidance, fostering leadership development, and ensuring that the value creation plan remains on course. This phase also allows for iterative adjustments to strategies, ensuring that emerging challenges are addressed and new opportunities are leveraged effectively.

Exit: The final stage of the investment lifecycle is the exit, where the private equity firm realises returns on its investment. Exit strategies are carefully planned, with options including sales to strategic buyers, secondary buyouts by other PE firms, or initial public offerings (IPOs). The choice of exit route depends on market conditions, the performance of the portfolio company, and the broader economic environment. Like in any equity financing, timing is critical. A well-timed exit can maximise returns. The success of this phase is measured not only by financial gains but also by the enduring improvements and growth achieved by the portfolio company.

Challenges in Leveraging Private Equity (PE) in Impact Investing and Their Mitigation

Misalignment Between Control and Impact Objectives: Private equity's hallmark feature of acquiring controlling stakes can create tension between financial priorities and the collaborative ethos required for meaningful impact. While control enables decisive operational changes, it often sidelines the perspectives of existing stakeholders such as employees, local communities, and non-profit partners. This misalignment can erode trust, disrupt organisational culture, and undermine long-term impact efforts. For example, restructuring to maximise profitability might result in layoffs or changes to supplier practices that harm community stability or sustainability initiatives. The result is a diminished capacity to achieve the social or environmental outcomes that underpin the investment's impact objectives.

Inflexibility of Exit Timelines: The structured timelines of PE investments—often dictated by the fund's lifecycle—create rigid constraints that conflict with the timeframes needed for sustainable impact. Impact initiatives addressing deep-rooted challenges such as poverty alleviation, educational reform, or climate resilience typically require longer horizons to materialise. Premature exits driven by financial considerations can interrupt these initiatives, leaving them incomplete or deprioritised by new ownership. Furthermore, the pressure to deliver financial returns within a fixed timeline often leads to decisions that prioritise short-term gains over systemic, transformative impact.

Dependency on Leveraged Buyouts (LBOs): The reliance on LBOs as a cornerstone of PE financing introduces significant risks when applied to impact-driven sectors. These deals often burden companies with substantial debt, limiting their financial flexibility to pursue socially or environmentally impactful initiatives. For instance, in sectors like renewable energy or affordable housing, cash flows may be unpredictable or slow to materialise, making debt servicing a priority over investments in innovation or sustainability. The excessive financial strain can hinder a company's ability to achieve its impact goals, potentially compromising both its mission and operational viability.

Overemphasis on Operational Efficiencies: Private equity's traditional focus on cost optimisation, while enhancing profitability, can directly conflict with impact objectives. Efforts to reduce costs—such as workforce reductions, outsourcing, or minimising supplier expenditures—may undermine commitments to job creation, fair labour practices, or sustainable procurement. Additionally, operational decisions aimed at immediate margin improvements can deprioritise investments in areas like renewable energy adoption or community engagement, constraining a company's potential to generate long-term societal benefits.

Challenges in Scaling Impact: PE's preference for mature, stable companies often results in a conservative approach prioritising incremental improvements over bold, scalable innovations. While this strategy mitigates risk, it also limits the potential for transformative impact in sectors like biodiversity conservation or circular economy ventures, which require visionary, high-impact solutions. Furthermore, PE's focus on established markets excludes opportunities in emerging regions, where investments could drive significant change but come with higher perceived risks. This conservatism perpetuates inequities, leaving critical gaps in underfunded regions and sectors unaddressed.

Opportunities for Mitigation

Redefining Control to Foster Stakeholder Collaboration: PE firms can adopt governance models that integrate stakeholder input into decision-making to address misalignment issues. Establishing advisory boards that include community leaders, environmental experts, and employees ensures that strategic decisions reflect a broader range of perspectives. This collaborative approach can enhance trust, align interests, and strengthen the company's long-term impact.

Implementing Flexible Exit Strategies: PE firms can mitigate the rigidity of exit timelines by adopting phased or hybrid exit models. For example, retaining a minority stake post-exit or transferring ownership to mission-aligned entities, such as cooperatives or foundations, ensures continuity in impact objectives. Alternatively, structuring exits around milestones tied to both financial and impact metrics can create a balanced approach that aligns with long-term goals.

Evolving Beyond Traditional LBO Structures: Reducing dependency on LBOs by incorporating alternative financing models can provide portfolio companies with greater flexibility. Blended finance—combining equity, concessional funding, and grants—can lower debt burdens while enabling investments in impactful initiatives. Additionally, impact-linked loans, where repayment terms are tied to achieving specific outcomes, can align financial structures with social and environmental objectives.

Balancing Efficiency with Impact Investments: PE firms can integrate impact objectives into their operational playbooks, ensuring that cost-cutting measures do not compromise social or environmental outcomes. For example, prioritising green technologies or investing in workforce development programmes can enhance both efficiency and impact. Embedding sustainability experts into operational teams can also guide decisions that balance profitability with broader societal goals.

Expanding Impact Horizons Through Strategic Risk-Taking: To unlock transformative impact, PE firms can broaden their investment criteria to include emerging sectors and regions. Establishing specialised impact funds focused on high-risk, high-impact opportunities—such as off-grid renewable energy, regenerative agriculture, or water access projects—can drive innovation while addressing critical development gaps. Partnering with Development Finance Institutions (DFIs) or leveraging risk-sharing mechanisms can further de-risk these ventures and encourage bold investments.

Standardising Sector-Specific Impact Metrics: Developing tailored impact measurement frameworks for PE investments can address the complexity of tracking diverse outcomes. Sector-specific metrics aligned with global standards, such as the Impact Weighted Accounts Initiative (IWAI) or the Global Impact Investing Network's IRIS+, can provide clarity and comparability. Integrating these metrics into digital dashboards enables real-time monitoring, fostering transparency and accountability across portfolio companies.

Private equity's unique characteristics—from its focus on control to its operational rigour—present both challenges and opportunities in the realm of impact investing. Addressing issues such as misaligned timelines, over-reliance on LBOs, and operational trade-offs enables PE firms to adapt their strategies to better align with impact objectives. Leveraging innovative financing models, fostering stakeholder collaboration, and expanding their horizons to include transformative opportunities can position PE as a powerful driver of societal change. These approaches ensure that private equity not only delivers robust financial returns but also creates enduring social and environmental value. Private equity offers significant potential for value creation, but it also presents several challenges that require careful navigation. Opportunities for innovation and improved outcomes counterbalance these challenges.

(d) **Social Venture Capital in Impact Investing**

Social Venture Capital (SVC) represents an innovative investment approach that combines the traditional practices of venture capital with a focused commitment to generating positive social or environmental impact. Unlike conventional venture capital, which prioritises financial returns, SVC balances profitability with addressing critical societal challenges such as poverty alleviation, healthcare access, environmental sustainability, and educational advancement. This dual objective positions SVC as a transformative tool for supporting enterprises that aim to create scalable and sustainable change.

Core Structure and Operational Model

Social Venture Capital is built on a model that blends the rigour of traditional venture capital with a profound emphasis on achieving measurable social and environmental outcomes. This hybrid approach allows SVC to meet the expectations of investors seeking financial returns and those prioritising impactful change. The capital structure of SVC typically involves raising funds from a diverse range of investors, including institutional investors, family offices, philanthropic organisations, and high-net-worth individuals. These LPs commit to funds managed by GPs, who are responsible for identifying, evaluating, and executing investments. Unlike conventional venture capital, where financial returns dominate decision-making, SVC funds are structured to prioritise alignment with social impact goals, often targeting thematic areas such as renewable energy, healthcare access, or financial inclusion. A critical component of the SVC model is its investment focus on early to mid-stage social enterprises. These businesses are evaluated not only for their financial viability but also for their ability to address pressing societal challenges. The assessment includes a thorough analysis of the enterprise's scalability and its potential for long-term impact. KPIs are often aligned with global frameworks like the Sustainable Development Goals (SDGs), and they guide the tracking of outcomes such as employment generation, carbon emission reductions, or access to education. SVC investments are characterised by their preference for minority stakes. This approach ensures that social entrepreneurs retain control over their mission-driven enterprises, while the SVC firm actively participates in strategic decision-making through governance rights like board seats or advisory roles. This balance empowers founders to innovate and scale without compromising their core values. Beyond financial capital, SVC firms play a pivotal role in providing strategic support to portfolio companies. This includes mentorship, capacity-building, and access to critical networks, such as partnerships with NGOs, governmental agencies, and other impact-driven organisations. These resources are instrumental in helping social enterprises navigate operational challenges, refine their impact strategies, and achieve sustainable growth. By leveraging their networks, SVC firms amplify the capacity of social enterprises to create meaningful change, ensuring that both financial and societal objectives are met.

Investment Lifecycle

The lifecycle of Social Venture Capital investments involves a series of structured stages, each designed to maximise both financial and social returns.

Fundraising: The process begins with raising capital from impact-oriented investors. This stage requires presenting a compelling investment thesis highlighting the fund's thematic focus, targeted social outcomes, and anticipated financial returns. Fundraising is particularly challenging, as it necessitates aligning investor expectations with the dual objectives of financial sustainability and social impact. Successful fundraising often attracts contributions from institutional investors, Development Finance Institutions (DFIs), and philanthropic entities committed to driving change.

Deal Sourcing and Due Diligence: Deal sourcing involves identifying social enterprises that align with the fund's mission and investment criteria. Due diligence is a rigorous process that evaluates the enterprise's financial health, market opportunity, scalability potential, and impact strategy. This dual-focus analysis ensures that investments are both financially viable and capable of achieving meaningful outcomes. Assessments often include on-the-ground evaluations to understand the enterprise's operational challenges and opportunities for growth.

Value Creation and Strategic Support: Once an investment is made, the focus shifts to value creation. SVC firms work closely with portfolio companies to refine business models, improve operational efficiency, and enhance impact delivery. Strategic support includes mentorship, connecting enterprises with potential partners, and navigating regulatory landscapes. SVC firms leverage their networks to help social enterprises build resilience, scale operations, and achieve long-term sustainability.

Monitoring and Impact Assessment: Throughout the investment period, SVC firms actively monitor financial and impact performance. Regular reporting ensures that enterprises remain accountable for achieving their objectives, while ongoing evaluations provide insights for optimising strategies. Metrics are aligned with standardised frameworks to ensure consistency and transparency, enabling investors to assess the fund's overall contribution to social change.

Exit Strategy: Exiting investments in SVC requires balancing financial returns with the continuity of the enterprise's mission. Common exit strategies include selling to strategic buyers, secondary sales to impact investors, or buybacks by the company's founders. The timing and structure of the exit are carefully managed to ensure that the enterprise continues to generate social impact post-exit, reflecting the fund's commitment to long-term change.

Challenges in Leveraging Social Venture Capital (SVC) in Impact Investing

Balancing Financial and Impact Objectives: Social Venture Capital's dual mandate—combining financial returns with measurable social or environmental outcomes—presents a uniquely complex challenge. Unlike traditional venture capital, SVC requires investors and entrepreneurs to align on dual priorities that often diverge in terms of timelines, resource allocation, and strategic goals. For example, an enterprise focused on renewable energy access in underserved regions may need extended investment periods to build infrastructure, educate local communities, and achieve scale. Such timelines conflict with the expectations of some investors who prioritise quicker exits or immediate profitability. Moreover, the diversity of impact goals—spanning poverty alleviation, education, healthcare, and environmental sustainability—introduces additional complexities in defining success. Different stakeholders may prioritise varying aspects of impact, leading to potential conflicts. For instance, some investors might emphasise financial inclusion metrics, such as the number of individuals gaining access to credit, while others may prioritise outcomes like the improvement in borrowers' quality of life. This tension is further compounded by the lack of standardised methodologies to quantify how financial trade-offs affect social outcomes. As a result, social enterprises may feel pressured to compromise on their mission to meet financial expectations, risking mission drift and reducing the credibility of SVC as a transformative impact tool. Over time, failure to effectively balance these objectives can alienate both impact-focused investors and socially conscious entrepreneurs, limiting the growth and potential of the SVC sector.

Attracting Impact-oriented Capital: Raising funds for SVC is uniquely challenging due to the need to appeal to investors with differing priorities. Institutional investors often perceive SVC investments as riskier or less lucrative compared to traditional asset classes. This perception stems from the longer timelines required for social enterprises to scale and generate returns, as well as uncertainties around impact measurement and verification. Additionally, the emerging nature of many impact sectors, such as rural healthcare or renewable energy for underserved areas, heightens perceived risk. In emerging markets, the challenges multiply due to regulatory instability, infrastructure deficits, and cultural nuances that require nuanced investment strategies. These factors discourage traditional investors, creating a funding gap that stifles the growth of enterprises addressing critical systemic issues. The difficulty in attracting aligned capital reduces the pool of resources available to scale impactful ventures, leaving transformational opportunities untapped.

Scaling Social Enterprises: Scaling social enterprises presents unique difficulties rooted in their dependence on contextual factors like local partnerships, cultural alignment, and community trust. While SVC provides critical funding and strategic guidance, replicating success across diverse geographies often requires significant adaptation. For instance, a healthcare initiative that thrives in one region due to established community networks may struggle to scale in a different cultural or regulatory environment. The scaling challenge is further exacerbated by resource constraints and limited operational expertise. Many social enterprises lack the infrastructure or leadership capacity to manage rapid growth effectively, leading to inefficiencies or compromises in impact delivery. Additionally, partnerships with larger corporations or governments, often necessary for scaling, may impose constraints that dilute the enterprise's mission. These hurdles limit the systemic change that SVC seeks to achieve, reducing the broader societal benefits of its investments.

Measuring and Verifying Impact: Impact measurement remains a persistent challenge in SVC due to the absence of universally accepted standards and the diverse nature of social outcomes. Social enterprises often operate in resource-constrained environments, making it difficult to allocate sufficient resources to robust monitoring systems. This results in incomplete or inconsistent data, which undermines accountability. For example, an education-focused enterprise might report enrolment numbers as a success metric without measuring qualitative outcomes like improvements in literacy rates. Similarly, an environmental initiative may highlight the number of trees planted without tracking their long-term survival or ecological impact. Such gaps in measurement reduce the credibility of SVC as a tool for achieving systemic change. Without reliable data, it becomes challenging to attract investors, scale operations, or refine strategies for maximum impact.

Mission Drift and Stakeholder Alignment: As social enterprises grow, they face increasing pressure to balance competing priorities from diverse stakeholders, including investors, employees, and community beneficiaries. This dynamic often leads to mission drift, where enterprises shift focus to more profitable markets or strategies at the expense of their original social or environmental objectives. For instance, a renewable energy enterprise initially committed to off-grid solutions for rural communities might pivot to urban markets with higher profitability, leaving its target beneficiaries underserved. Mission drift is compounded by the entry of new investors who may prioritise financial returns over impact goals. These tensions can create internal conflicts and erode trust among stakeholders, including the communities served. The long-term consequences of mission drift include reduced impact credibility, diminished stakeholder confidence, and the potential failure of SVC to deliver on its transformative promise.

Opportunities for Mitigation

Integrating Impact into Investment Structures: SVC firms can foster better alignment between financial and impact objectives by designing innovative investment structures. For instance, integrating impact-linked performance incentives for both investors and portfolio enterprises ensures that achieving social outcomes is directly tied to financial benefits. These mechanisms could include revenue-sharing models where a percentage of profits is reinvested in social programmes or deferred investor payouts contingent on achieving specific impact metrics. Such approaches create accountability and strengthen the intrinsic value of impact for all stakeholders.

Building Investor Awareness and Ecosystem Engagement: Educating investors about the unique value proposition of SVC requires a multi-pronged approach. This includes publishing comprehensive case studies that showcase successful dual-return investments, hosting forums that bring together investors and impact practitioners, and developing tailored communication strategies that highlight risk mitigation and scalable success stories. Partnerships with platforms such as the Global Impact Investing Network (GIIN) can help create awareness about innovative investment models, while collaborative networks can amplify the visibility of SVC funds and attract diverse capital sources.

Strengthening Capacity-building Support: Capacity-building efforts must be tailored to the evolving needs of social enterprises. SVC firms can establish on-demand advisory programmes that offer strategic advice on navigating operational hurdles, scaling effectively, and refining business models. Additionally, embedding capacity-building elements, such as access to regional experts or cross-sector partnerships, into the early stages of investment can provide long-term value. SVC firms equip enterprises with the tools to tackle specific challenges—such as regulatory navigation or talent acquisition—driving sustainable growth while preserving impact.

Developing Standardised and Practical Impact Metrics: To overcome inconsistencies in impact reporting, SVC firms should adopt sector-specific impact metrics that are practical, transparent, and easy to implement. Leveraging technology, such as AI-based analytics platforms, can enhance the precision and scalability of impact measurement systems. For example, real-time dashboards that capture qualitative and quantitative data can enable dynamic tracking of social outcomes. Standardisation efforts should be complemented by collaborative benchmarking initiatives with other SVC funds to establish industry norms and facilitate cross-comparison of impact performance.

Safeguarding Mission Integrity Through Governance: Embedding impact into governance structures is crucial for mitigating mission drift. SVC firms can establish independent oversight committees that monitor alignment with social objectives and flag potential deviations early. Additionally, incorporating mission-aligned shareholder agreements that define explicit impact priorities ensures that growth strategies remain mission-centric even as enterprises expand. Regular engagement with a diverse array of stakeholders—from community representatives to industry experts—can further reinforce transparency and accountability.

Social Venture Capital is not merely a financial tool but a dynamic force capable of driving systemic change across global challenges. Its capacity to merge profit with purpose positions it uniquely within the broader impact investing ecosystem. While the dual mandate introduces complexities, these challenges offer opportunities for innovation, collaboration, and long-term value creation. Embedding impact within investment structures, aligning capital flows with social priorities, and championing accountability through standardised metrics, enable SVC to redefine how investments generate both returns and societal benefits. Capacity-building initiatives tailored to the unique demands of social enterprises further ensure sustainable scaling without compromising mission integrity. Looking ahead, the future of SVC depends on a shared commitment to fostering trust among stakeholders, creating ecosystems that embrace both risk and reward and prioritising the long-term impact over immediate gains. In doing so, Social Venture Capital can realise its full potential as a cornerstone of sustainable development, empowering enterprises to deliver transformative solutions for the world's most pressing challenges.

(e) **Equity Crowdfunding for Impact Investing**

Equity crowdfunding is a modern financing method that allows startups and small businesses to raise capital by offering equity ownership to a wide range of investors through online platforms. This approach connects businesses directly with individual investors, enabling them to access funding that may not be available through traditional financial channels. Equity crowdfunding empowers a diverse group of investors to participate in the growth of early-stage companies, fostering innovation and entrepreneurship, by decentralising access to investment opportunities. Through this mechanism, investors gain equity stakes in businesses, aligning their interests with the company's success and creating a dynamic ecosystem for economic growth and shared value creation.

Core Structure and Operational Model

At its core, equity crowdfunding operates as a decentralised and accessible financial model that utilises online platforms to bridge the gap between startups seeking funding and investors looking for opportunities. This model is underpinned by several key components that define its operational framework and appeal. The platforms themselves, such as Seedrs,

Crowdcube, and Wefunder, act as intermediaries, providing the infrastructure for businesses to create and manage their campaigns. These platforms play a vital role in ensuring compliance with regulatory requirements, vetting the legitimacy of participating businesses, and facilitating seamless transactions. For businesses, the platforms offer tools to craft compelling campaigns that showcase their value propositions, including detailed business plans, financial projections, and growth strategies. By connecting businesses to a large, diverse pool of potential investors, these platforms lower the barriers to raising capital and expand access to funding.

Equity crowdfunding campaigns are structured to offer ownership stakes in the business in exchange for investment. Unlike traditional financing models dominated by large institutional investors, this approach brings in numerous smaller investors, each contributing modest amounts. This democratisation of capital not only diversifies the investor base but also allows retail investors to access early-stage ventures that were previously exclusive to institutional players. Investors typically receive equity shares that grant them rights such as dividends and voting power, aligning their interests with the growth and success of the business. The campaign process is pivotal to the operational model. Businesses set clear funding goals, valuations, and timelines, all of which are critical to attracting investor interest. Most campaigns operate on an "all-or-nothing" basis, meaning that funds are only collected if the campaign reaches its target. This structure mitigates risk for both parties—investors are assured of sufficient capital being raised to meet the company's objectives, and businesses avoid partial funding scenarios that could hinder execution.

Regulatory compliance and due diligence form another cornerstone of equity crowdfunding's operational model. Platforms must navigate jurisdiction-specific securities laws to ensure legal integrity while also protecting investors. Rigorous due diligence processes are essential to verify the authenticity and viability of businesses, safeguarding the credibility of the ecosystem. This balance of compliance, transparency, and accessibility makes equity crowdfunding an attractive and reliable mechanism for raising capital.

Investment Lifecycle

The lifecycle of equity crowdfunding is a structured process designed to ensure transparency, efficiency, and alignment between businesses and investors. This journey unfolds across several key stages.

Campaign Preparation: The process begins with meticulous preparation. Businesses collaborate with crowdfunding platforms to craft their campaigns, which include creating engaging marketing materials, financial forecasts, and detailed business plans. The company's funding goals and valuation must be carefully determined to appeal to potential investors while reflecting realistic growth projections. Effective preparation often involves developing multimedia content, such as videos and testimonials, to build credibility and attract interest.

Campaign Launch: Once ready, the campaign is launched on the platform. During this phase, businesses actively promote their campaigns through social media, email marketing, and personal networks. Early momentum is critical, as initial investments create a sense of urgency and encourage further participation. A successful launch requires a strong value proposition, clear communication, and consistent engagement with the target audience.

Monitoring and Engagement: Throughout the campaign, businesses must remain actively involved in monitoring progress and engaging with potential investors. Transparent communication is essential to build trust and address investor concerns. Platforms often provide interactive features, such as discussion forums and Q&A sections, allowing direct interaction between businesses and investors. Regular updates on campaign milestones help sustain investor interest and foster a sense of collaboration.

Campaign Closure and Fund Collection: If the campaign achieves its funding target within the specified timeframe, it is deemed successful, and funds are collected. Investors receive equity shares in exchange for their contributions, formalising their role as stakeholders. For campaigns operating on an "all-or-nothing" basis, failure to reach the funding goal results in no funds being collected, ensuring fairness for both parties. The collected funds enable businesses to execute their growth strategies, whether by scaling operations, expanding markets, or launching new products.

Post-Funding Activities and Investor Relations: The relationship with investors continues beyond the campaign's closure. Businesses must maintain robust investor relations by regularly updating financial performance, strategic developments, and key milestones. This ongoing communication helps build investor confidence and loyalty, turning them into potential brand advocates. For businesses with a large investor base, streamlined communication tools and structured reporting are critical to managing stakeholder expectations effectively.

Challenges in Leveraging Equity Crowdfunding in Impact Investing

Attracting and Retaining Impact-focused Investors: Equity crowdfunding's inclusivity is both its strength and a significant challenge. While it democratises access to investment opportunities, the diverse motivations of retail investors can dilute the focus on long-term impact. Many retail investors prioritise financial returns over social or environmental outcomes, which can shift business strategies away from transformative goals towards more commercially viable but less impactful initiatives. This misalignment undermines the very essence of impact investing by steering capital towards ventures that prioritise short-term gains. Additionally, the lack of sophisticated knowledge among many retail investors regarding impact metrics complicates efforts to sustain interest and commitment to socially driven outcomes, limiting the overall effectiveness of crowdfunding as a tool for impact investing.

Balancing Impact Objectives with Profitability: The competitive environment of equity crowdfunding platforms pushes businesses to present compelling financial projections, often at the expense of their social or environmental objectives. This pressure creates a paradox: To attract investors, enterprises may overemphasise short-term profitability, which can lead to decisions that deprioritise investments in sustainable practices or community-focused programmes. For instance, a business addressing affordable housing might delay or reduce its impact-oriented initiatives to meet the expected financial milestones of its investors. This tension erodes the credibility of impact investing as the dual promise of financial and societal returns becomes increasingly difficult to achieve in practice.

Ensuring Transparency and Accountability: Transparency is paramount in equity crowdfunding, yet the decentralised nature of the model exacerbates the risk of incomplete or misleading disclosures. Startups eager to secure funding may unintentionally or deliberately overstate their impact potential or financial projections. Without stringent oversight, these misrepresentations undermine investor trust and discourage future participation in impact-focused campaigns. Furthermore, the lack of standardised metrics for reporting impact outcomes creates inconsistencies that make it difficult for investors to evaluate and compare opportunities. This opacity not only hampers the credibility of individual campaigns but also weakens the broader perception of equity crowdfunding as a reliable mechanism for impact investing.

Managing Post-Funding Investor Relations: Equity crowdfunding's model often results in businesses having a large and diverse investor base, making post-funding communication complex and resource-intensive. This challenge is amplified for impact-focused enterprises, which must provide updates on both financial performance and social or environmental outcomes. Without effective systems to manage these communications, businesses risk alienating their investors, leading to dissatisfaction and potential reputational damage. Moreover, inadequate post-funding engagement can weaken investor confidence in the enterprise's ability to deliver on its impact promises, reducing their willingness to participate in future campaigns.

Navigating Regulatory and Compliance Challenges: The regulatory landscape for equity crowdfunding is fragmented and complex, varying significantly across jurisdictions. For impact-focused enterprises, navigating these regulatory requirements adds a significant administrative burden, diverting resources away from mission-critical activities. This complexity is particularly challenging for businesses operating in multiple regions, where varying compliance standards can lead to delays or restrict access to certain investor groups. Regulatory missteps not only expose enterprises to legal penalties but also risk eroding investor trust, further complicating their ability to raise capital and deliver impact.

Opportunities for Mitigation

Building a Community of Impact-Driven Investors: Platforms and businesses can create a thriving community of impact-focused investors by adopting proactive and innovative approaches. Hosting interactive webinars that unpack the mechanics of impact investing, producing transparent and engaging impact reports, and spotlighting successful case studies can attract investors genuinely committed to social and environmental outcomes. Establishing crowdfunding platforms tailored specifically to impact-driven ventures ensures alignment of investor motivations, fostering deeper engagement and long-term commitment.

Enhancing Campaign Design for Balanced Objectives: To effectively balance impact and profitability, businesses must craft campaigns that articulate clear, measurable goals for both. For instance, embedding specific impact milestones within financial projections ensures investors understand the enterprise's dual focus. Leveraging testimonials from beneficiaries, detailed use-of-funds explanations, and transparent risk assessments can further enhance credibility. By presenting a holistic narrative that connects financial returns to societal benefits, campaigns can attract investors who value comprehensive success metrics.

Implementing Robust Due Diligence and Verification: Equity crowdfunding platforms can bolster transparency and accountability by instituting stringent due diligence protocols. Third-party audits of financial and impact claims enhance trustworthiness, ensuring that campaigns accurately represent their potential. Additionally, platforms can standardise impact reporting frameworks, such as integrating IRIS+ or aligning with the Sustainable Development Goals (SDGs), to provide investors with consistent and comparable metrics. These measures reduce information asymmetry and position equity crowdfunding as a credible avenue for impact investing.

Streamlining Post-Funding Investor Relations: Communicating with a large and diverse investor base requires leveraging technology and structured processes. Interactive dashboards that offer real-time updates on both financial and impact metrics can simplify engagement, allowing investors to track progress effortlessly. Establishing investor advisory boards and hosting periodic virtual meetings can also foster deeper connections and provide a platform for addressing concerns. By nurturing transparent and meaningful relationships, businesses can cultivate loyalty and encourage reinvestment from their investor community.

Advocating for Harmonised Regulatory Frameworks: Harmonising regulatory standards for equity crowdfunding across jurisdictions is essential for fostering global participation and scalability. Platforms and industry stakeholders can collaborate with policymakers to develop streamlined compliance frameworks that prioritise investor protection while minimising administrative burdens. Establishing international certifications or accreditations for impact crowdfunding platforms can further enhance cross-border trust and accessibility. Providing compliance support services, such as legal consultations and automated documentation systems, empowers enterprises to focus on delivering impact while confidently navigating regulatory complexities.

Equity crowdfunding represents a transformative model for democratising capital and fostering inclusive growth within the impact investing landscape. While the challenges of aligning investor motivations, ensuring transparency, and navigating complex regulations persist, these barriers also present opportunities for innovation, collaboration, and systemic improvement. Businesses and platforms can elevate equity crowdfunding into a reliable mechanism for advancing Sustainable Development Goals by building dedicated communities of impact-driven investors, enhancing campaign narratives, implementing rigorous verification processes, and leveraging technology for streamlined post-funding engagement. Additionally, advocating for harmonised regulatory frameworks will unlock greater global participation, enabling enterprises to scale their impact across diverse markets. With thoughtful strategies and collective action among stakeholders, equity crowdfunding can evolve into a cornerstone of impact investing, driving measurable societal change, fostering innovation, and empowering investors to contribute to a more equitable and sustainable future.

Debt Financing

Debt financing is the process of raising capital through issuing or utilising debt instruments, which represent a contractual obligation to repay borrowed funds, typically with interest, according to predetermined terms. It allows impact-focused companies to access capital with predefined repayment obligations, making it suitable for organisations with predictable cash flows. Unlike equity financing, debt provides capital without diluting ownership, but it requires consistent repayments.

Role in the Development of the Impact Investing Industry

Debt financing has played a pivotal role in the development of the impact investing industry by providing capital that aligns with the needs of impact-driven enterprises. It enables such enterprises to pursue their social and environmental missions while maintaining control over their operations. Whether private debt instruments such as traditional loans, microfinance, and convertible debt, or public debt instruments such as bonds, debt instruments have fostered the growth of the impact investing industry, ensuring that social outcomes are integrated into the financial landscape. Private debt provides working capital, fund infrastructure projects, or support Small and Medium-sized Enterprises (SMEs). It is currently utilised by 48 per cent of impact investors. Despite its widespread use, it accounts for only 14 per cent of the total impact Assets Under Management (AUM). Emerging market-focused investors allocate a significant portion of their AUM (43 per cent) to private debt, showcasing its importance in regions where financial inclusion and access to capital are critical. Conversely, developed market-focused investors allocate only 4 per cent of their AUM to private debt. Development Finance Institutions (DFIs) heavily rely on private debt, with the instrument representing nearly half (47 per cent) of their total AUM. This reliance underscores the role of DFIs in addressing financing gaps in underserved markets. Public debts on the other hand, enable large-scale funding for social and environmental projects through government or corporate bonds. These instruments are often used to finance public infrastructure, healthcare, and education initiatives. About 13 per cent of investors allocate impact capital through public debt, which accounts for 12 per cent of the total AUM. Public debts instruments are largely leveraged by Institutional Asset Owners (IAOs) to balance risk and impact in their portfolios. It is a significant portion of AUM for IAOs, at 44 per cent. In terms of allocations by investor type, investors who aim for market rate allocate 11 per cent of their AUM to private debt. Below-market-rate investors allocate a much higher proportion—26 per cent—to private debt, aligning with their focus on social impact over financial returns. Similarly, impact-agnostic investors allocate only 6 per cent to private debt, while impact-only investors allocate 26 per cent of their AUM to private debt, highlighting its importance in mission-driven portfolios.

Debt instruments, particularly private debt, are versatile tools for addressing diverse financing needs across markets and sectors. Their prominence in emerging markets underscores its role in fostering financial inclusion and supporting underserved communities. Public debt enables impact investors to channel large-scale capital towards systemic challenges, particularly in developed markets with more mature bond markets. Through their varied applications and growing importance, debt instruments remain integral to advancing the objectives of impact investing across geographies and sectors. The following section explores the common types of debt instruments, their structure, lifecycle, challenges, and mitigations.

(a) **Loans for Impact Investing**

Loans are a traditional form of debt financing that has evolved into a vital tool for impact investing. In this context, loans are structured to provide capital to enterprises or projects that aim to generate measurable social and environmental benefits alongside financial returns. These loans cater to organisations that require funding but prefer to avoid diluting ownership or are not yet prepared to issue equity. Loans empower organisations to pursue sustainable growth and achieve meaningful outcomes, by tailoring terms to meet the unique needs of impact-focused enterprises.

Core Structure and Operational Model

Loans in impact investing are designed to seamlessly integrate financial sustainability with measurable social and environmental outcomes. They provide a versatile financing option for enterprises that may struggle to access traditional capital markets. Structured with flexibility and purpose, these loans cater to organisations focused on addressing societal challenges, ranging from poverty alleviation to environmental conservation. The core structure of impact-focused loans begins with their providers, which include Development Finance Institutions (DFIs), impact funds, philanthropic organisations, and commercial banks. These institutions aim to bridge the financing gap for borrowers such as social enterprises, non-profits, cooperatives, and SMEs. The borrowers, often operating in underserved markets or addressing high-risk challenges, typically lack the collateral or credit history required for conventional loans. The terms of impact loans are tailored to align with the unique needs of the borrower and the dual objectives of financial and social returns. Interest rates are often concessional, reducing the financial burden on borrowers and ensuring affordability. Repayment schedules are customised to match the borrower's cash flow patterns, with options such as grace periods or income-contingent repayments providing additional flexibility. This adaptability ensures that loan repayment obligations do not hinder the organisation's ability to achieve its mission. A notable feature of these loans is the incorporation of impact-linked terms. These terms align repayment conditions with the borrower's success in achieving predefined social or environmental milestones. For example, enterprises meeting goals like reducing carbon emissions or improving access to education might benefit from reduced interest rates. This structure not only incentivises borrowers to prioritise impactful outcomes but also ensures alignment between the lender's objectives and the enterprise's mission.

To mitigate risks associated with lending to high-risk borrowers, impact loans frequently include mechanisms such as credit guarantees, subordinated debt, or first-loss capital. These risk-sharing tools, often supported by DFIs or philanthropic entities, de-risk the lending process and attract private investment. These structures enable lenders to support projects that might otherwise be deemed too risky, by combining concessional and commercial capital through blended finance models. Overall, the operational model of impact loans emphasises collaboration between borrowers and lenders. Lenders play an active role not only in providing capital but also in offering technical assistance, advisory support, and ongoing monitoring. This hands-on approach ensures that borrowers remain on track to achieve both their financial and impact objectives, fostering a symbiotic relationship that drives meaningful change.

Investment Lifecycle

The lifecycle of loans in impact investing encompasses distinct stages, from origination and underwriting to monitoring, repayment, and impact evaluation.

Origination and Underwriting: The process begins with identifying borrowers that align with the impact objectives of the loan programme. Underwriting involves a comprehensive assessment of the borrower's financial health, cash flow stability, business model, and capacity to deliver measurable impact. This phase also evaluates the borrower's management team and their ability to address the targeted social or environmental challenges.

Loan Structuring and Disbursement: Following the underwriting process, the loan is structured to balance the borrower's needs with the lender's financial and impact goals. Terms such as interest rates, repayment schedules, and impact-linked conditions are established during this phase. Disbursement occurs either in a lump sum or in tranches tied to the borrower's achievement of predefined milestones. Tranche-based disbursement ensures accountability and aligns funding with progress.

Monitoring and Support: Lenders actively monitor the borrower's financial and impact performance throughout the loan term. Monitoring involves collecting data on financial metrics (e.g., revenue, repayment status) and impact metrics (e.g., emissions reductions, number of beneficiaries served). Many lenders provide technical assistance or advisory support to help borrowers optimise their operations and achieve their objectives. This phase fosters a collaborative relationship between lenders and borrowers, ensuring alignment with shared goals.

Repayment and Impact Evaluation: During the repayment phase, borrowers make regular payments based on the agreed-upon schedule. If the loan includes impact-linked terms, repayment conditions may be adjusted based on the borrower's progress in meeting impact targets. Achieving significant milestones might result in benefits such as lower interest rates or extended repayment periods. Impact evaluation continues throughout this phase, assessing the outcomes of the investment to ensure alignment with the intended goals.

Exit and Impact Reporting: The lifecycle concludes with the full repayment of the loan. At this stage, lenders conduct a final evaluation of both financial and impact performance, documenting the results to share with stakeholders. This transparency demonstrates the value of the investment and informs future decision-making, helping to attract additional capital to the impact investing ecosystem.

Challenges in Leveraging Loans for Impact Investing and Their Mitigations

Debt Aversion Among Impact-Focused Enterprises: Debt aversion poses a unique challenge to impact investing, as many mission-driven enterprises prioritise their social or environmental goals over financial risk-taking. This reluctance stems from the inflexibility of loan repayment obligations, which can strain limited and unpredictable cash flows, particularly for organisations reliant on grants or subsidised services. In sectors like healthcare and education, where revenue streams are tightly linked to affordability and accessibility, the fear of over-indebtedness can lead enterprises to forego critical funding opportunities. This aversion not only limits the scalability of impactful initiatives but also constrains the diversity of financing instruments available to support systemic change.

Limited Access for Underserved Sectors and Geographies: Loans often bypass underserved sectors and geographies due to high perceived risks and weak financial infrastructure. For example, smallholder farmers or rural cooperatives may lack formal credit histories or tangible collateral, making them unattractive to traditional lenders. The absence of local banking networks or the logistical challenges of reaching remote areas further exacerbate this exclusion. By sidelining high-impact sectors like small-scale agriculture, clean water access, and renewable energy for rural communities, the loan system perpetuates inequities and fails to mobilise capital where it is most needed to drive transformative change.

Inflexibility of Traditional Loan Structures: The rigidity of conventional loan structures is particularly detrimental to enterprises operating in volatile or cyclical markets. Fixed repayment schedules and uniform interest rates do not account for the seasonal or uneven cash flows typical of many impact-driven organisations. For instance, agricultural enterprises dependent on harvest cycles or social enterprises with project-based funding face significant challenges in meeting monthly repayment obligations. This misalignment not only increases the risk of default but also discourages lenders from engaging with sectors that require customised financial solutions, leaving critical gaps in the ecosystem.

High Transaction Costs for Small Loans: Small-scale loans, which are often critical for micro-enterprises or grassroots initiatives, incur disproportionately high transaction costs. The processes of underwriting, disbursement, and monitoring are resource-intensive, particularly for loans that require impact assessments alongside financial evaluations. For borrowers, these costs can make loans unaffordable, while for lenders, they reduce the economic viability of servicing small-scale clients. This dynamic creates a structural barrier to the deployment of capital at the grassroots level, where it is often most effective in fostering inclusive growth and innovation.

Challenges in Measuring and Aligning Impact with Financial Performance: The dual mandate of financial returns and measurable impact introduces complexities that hinder the scalability of impact loans. Defining, tracking, and reporting impact metrics require sophisticated systems that many borrowers lack. Lenders, on the other hand, face challenges in integrating these metrics into their risk and performance assessments. The lack of standardised frameworks further complicates comparisons across sectors and geographies. Without robust alignment between financial and impact performance, loans risk losing credibility as a tool for advancing sustainable development, deterring both borrowers and investors from fully engaging.

Opportunities for Mitigation

Designing Flexible and Adaptive Loan Structures: To address the inflexibility of traditional loans, lenders can introduce loan structures that accommodate the cash flow realities of impact-driven enterprises. Income-contingent repayment models, for instance, tie repayments to the borrower's earnings, reducing financial pressure during low-revenue periods. Seasonal repayment schedules tailored to industries like agriculture can address cash flow fluctuations, enabling borrowers to meet obligations without compromising operations. Additionally, milestone-based disbursements ensure accountability by linking funding to progress on predefined financial and impact goals, creating a mutually beneficial dynamic between lenders and borrowers.

Expanding Access Through Technology and Partnerships: Technology can revolutionise access to loans for underserved regions and sectors. Digital lending platforms reduce overhead costs and streamline application processes, making financing more accessible to remote or marginalised communities. Partnerships with local cooperatives or community organisations enhance credit assessments by leveraging localised knowledge and trust networks. For example, working with agricultural cooperatives can improve risk profiling for smallholder farmers while providing borrowers with financial literacy training, strengthening repayment capacities and boosting lender confidence.

Innovative Risk-Sharing Models: Risk-sharing mechanisms such as impact bonds or shared-risk loan pools can make lending to high-risk sectors more viable. For example, establishing regional risk-sharing funds—backed by Development Finance Institutions—can protect lenders against defaults while incentivising investment in sectors like renewable energy or sustainable agriculture. Insurance-linked loan products could mitigate risks by covering losses from external shocks, such as climate events or market downturns.

Streamlining Loan Processes with Data-Driven Innovations: Data analytics and AI can transform the underwriting process by leveraging alternative data sources, such as satellite imagery for agriculture or mobile usage patterns for informal businesses. These tools enable more accurate credit assessments and reduce the time and cost of loan approvals. Aggregating small loans into portfolios for securitisation further enhances scalability by creating investment opportunities for institutional investors. This approach channels larger pools of capital towards grassroots-level initiatives without compromising accessibility or impact.

Building Community-Centric Lending Models: Community-centric approaches that involve borrowers in loan design and management can enhance trust and compliance. For example, participatory financing models—where community members collectively determine loan terms and monitor usage—can create accountability and mutual support among borrowers. Combining this approach with localised financial literacy programmes ensures that loans are used effectively to generate long-term impact.

Investing in Scalable Impact Measurement Systems: Developing scalable impact measurement frameworks supported by digital platforms can enhance transparency and comparability. For instance, real-time dashboards that aggregate data on financial and impact performance allow both borrowers and lenders to track progress dynamically. Partnerships with tech companies can facilitate the creation of sector-specific metrics, making impact measurement more accessible and meaningful for smaller enterprises.

Loans in impact investing offer a powerful mechanism for driving transformative change, particularly in sectors and regions where access to capital is limited. However, their effectiveness is constrained by challenges such as debt aversion, inflexible structures, and high transaction costs. By adopting innovative financial designs, leveraging technology, and fostering collaborative partnerships, stakeholders can unlock the full potential of loans as a tool for impact. Through strategies such as blended finance, adaptive repayment models, and investments in impact measurement infrastructure, loans can bridge the gap between financial markets and underserved communities. These efforts not only enhance access to capital but also ensure that financing is aligned with long-term societal goals, positioning loans as a cornerstone of sustainable and inclusive development.

(b) **Bonds**

Sustainability-Linked Bonds (SLBs) encompass a broad category of fixed-income instruments, including green bonds, blue bonds, and other sustainability-linked instruments. These bonds are designed to align capital-raising efforts with Environmental, Social, and Governance (ESG) objectives. Unlike traditional bonds that are detached from specific impact goals, SLBs connect financial terms, such as coupon rates, to the achievement of predefined Sustainability Performance Targets (SPTs). This integration incentivises issuers to improve their ESG practices, making SLBs a versatile and powerful tool for driving systemic change and addressing global challenges such as climate change, social equity, and resource conservation.

Core Structure and Operational Model

Sustainability-Linked Bonds (SLBs) are built upon a flexible yet disciplined framework that aligns the issuer's financial obligations with sustainability objectives. At the heart of this model is the integration of financial terms with measurable ESG performance indicators, known as Sustainability Performance Targets (SPTs). These targets are carefully selected to reflect the issuer's sustainability priorities, often tailored to address sector-specific challenges such as carbon reduction, biodiversity conservation, or workforce inclusion. The structure begins with the identification of SPTs, which must be both ambitious and credible to drive significant impact. These targets often align with established global frameworks, such as the United Nations Sustainable Development Goals (SDGs) or the Paris Agreement. Once the SPTs are defined, they are embedded within the bond's terms. This mechanism introduces a dynamic element to the bond, linking financial incentives—such as interest rate adjustments—to the achievement of the predefined targets. Conversely, failing to meet the targets may result in financial penalties, such as increased interest rates, thereby holding issuers accountable for their commitments. Issuers of SLBs range from corporations and municipalities to financial institutions and supranational entities. The flexibility of SLBs allows these issuers to allocate funds broadly, addressing operational and strategic needs while maintaining a focus on sustainability. The allocation of funds is not tied to specific projects, granting issuers the latitude to pursue diverse initiatives that align with their broader ESG objectives.

Issuers benefit from the flexibility to allocate funds broadly, supporting various operational and strategic priorities rather than being confined to specific projects. This allows SLBs to cater to various industries and use cases, enhancing their applicability. To maintain transparency, issuers establish robust monitoring and reporting frameworks, often engaging independent third-party verifiers to assess progress towards SPTs. This verification builds credibility, reassuring investors of the bond's integrity and its alignment with genuine sustainability outcomes.

Investment Lifecycle

The lifecycle of sustainability-linked bonds is meticulously structured to integrate sustainability into each phase, from issuance to repayment. This lifecycle ensures that the financial and impact objectives of the bond are seamlessly aligned and transparently monitored.

Issuance: The lifecycle begins with the issuance of the bond. The issuer identifies and validates sustainability performance targets in collaboration with stakeholders, including potential investors and external certifiers. The bond's financial terms, including the interest rate, maturity, and penalty or reward mechanisms, are established during this phase. To enhance the bond's credibility, many issuers engage third-party certifiers to validate the alignment of SPTs with recognised sustainability frameworks, such as the United Nations Sustainable Development Goals (SDGs). Once the bond structure is finalised, it is marketed and sold to investors through public offerings or private placements, enabling the issuer to raise the desired capital.

Allocation of Proceeds: Unlike traditional green or blue bonds, SLBs do not limit the use of proceeds to specific projects. This flexibility allows issuers to allocate funds towards general corporate purposes while maintaining a commitment to sustainability. The absence of project-specific allocation broadens the applicability of SLBs, enabling businesses to integrate ESG objectives into their core operations and strategies. Issuers must, however, clearly articulate how the proceeds align with their broader sustainability goals to ensure transparency and maintain investor confidence.

Monitoring and Verification: Monitoring is a continuous process that ensures issuers remain accountable for achieving the agreed-upon SPTs. Issuers are required to provide periodic reports detailing their progress, including both quantitative metrics and qualitative assessments. These reports are often subject to independent third-party verification to ensure accuracy and reliability. Monitoring frameworks are designed to track the impact of initiatives, providing stakeholders with a clear understanding of the bond's contribution to sustainability objectives. This phase is critical for maintaining the integrity of the SLB model and for enabling performance-based adjustments to financial terms.

Performance Adjustments: A defining feature of SLBs is the adjustment of financial terms based on the issuer's performance against the SPTs. Issuers that meet or exceed their targets are rewarded with lower coupon rates or other financial benefits, reinforcing their commitment to sustainability. Conversely, failure to achieve the targets may result in higher interest rates or penalties, incentivising issuers to maintain their focus on ESG improvements. This dynamic mechanism ensures that the financial and sustainability aspects of the bond remain intrinsically linked throughout its tenure.

Maturity and Impact Evaluation: The final stage involves the repayment of the bond at maturity, along with an evaluation of its overall impact. This evaluation considers the issuer's success in achieving the SPTs and assesses the broader environmental and social outcomes generated by the initiatives funded through the SLB. Issuers contribute to the growing body of evidence supporting the effectiveness of sustainability-linked finance in driving systemic change by documenting these achievements.

Challenges in Leveraging Bonds in Impact Investing and Their Mitigation

Ensuring Ambitious and Credible Sustainability Performance Targets (SPTs): The credibility and impact of Sustainability-Linked Bonds (SLBs) hinge on the definition of their Sustainability Performance Targets (SPTs). However, issuers often face conflicting pressures: investors demand ambitious targets to ensure meaningful ESG contributions, while issuers may prefer less stringent goals to mitigate the risk of financial penalties. This tension creates a credibility gap that can undermine investor confidence and deter capital flows to impactful projects. Additionally, setting unrealistic or overly ambitious targets may result in penalties that discourage participation in future SLB markets, creating a self-reinforcing cycle that stifles innovation and growth in sustainability-linked finance.

Lack of Standardised Reporting and Verification Frameworks: The absence of universally accepted standards for reporting and verifying progress against SPTs creates significant inconsistencies in the SLB market. Investors rely heavily on issuer disclosures, yet the variability in methodologies and metrics can lead to incomplete or misleading information. This lack of standardisation not only impairs the ability to compare bonds across sectors and geographies but also increases the risk of greenwashing—a practice where issuers exaggerate their sustainability achievements to attract investors. Without robust verification mechanisms, the SLB market risks losing its integrity, undermining trust and reducing its effectiveness as a tool for systemic change.

Limited Applicability in Emerging Markets: SLBs are often inaccessible to issuers in emerging markets due to the high costs of third-party verification, the complexity of ESG reporting, and limited technical expertise. This exclusion is particularly detrimental because emerging markets frequently face the most urgent sustainability challenges, such as climate vulnerability and social inequities. The inability of these regions to access SLB financing perpetuates global inequalities in impact investing, leaving high-need areas underfunded and reducing the overall effectiveness of sustainable finance in addressing systemic issues.

Balancing Flexibility and Accountability: While the flexibility of SLBs to allocate proceeds broadly rather than tying them to specific projects enhances their appeal, it also introduces accountability challenges. Investors may question whether the funds are genuinely contributing to sustainability objectives, especially when issuers fail to provide transparent and detailed disclosures about how proceeds are used. This lack of clarity can erode trust in the SLB model, deterring investors who prioritise accountability and measurable impact, and limiting the model's ability to attract capital at scale.

High Transaction Costs and Complexity: Issuing SLBs involves substantial costs and complexity, including stakeholder engagement, third-party verification, and continuous monitoring. For smaller issuers or those with limited resources, these barriers can be prohibitive, restricting participation and narrowing the market's scope. Furthermore, these high costs reduce the net proceeds available for impactful initiatives, diminishing the overall efficiency of SLBs as a financing tool. This challenge is particularly pronounced in sectors or regions where financial resources are already constrained, further limiting the accessibility and scalability of SLBs.

Opportunities for Mitigation

Redefining SPT Frameworks with Context-Specific Benchmarks: SPTs can be redefined to align more closely with local and sector-specific sustainability challenges. Developing modular frameworks can enable issuers to integrate globally recognised standards, such as the SDGs, while customising metrics to address regional priorities like water scarcity or equitable education. Collaborations between local stakeholders and global institutions can create hybrid benchmarks that are both ambitious and contextually feasible, improving credibility and issuer participation.

Implementing Blockchain for Verification and Reporting: Blockchain technology can revolutionise transparency by enabling immutable, real-time tracking of progress against SPTs. Smart contracts can automate reporting updates, ensuring accuracy while reducing administrative burdens. For instance, emissions data verified through IoT-enabled devices could feed directly into blockchain records, providing investors with consistent and verifiable insights. This technology-driven approach addresses greenwashing concerns and enhances investor confidence in SLBs.

Scaling SLBs in Emerging Markets Through Collaborative Platforms: Development Finance Institutions and multilateral organisations can establish regional SLB facilitation hubs to lower barriers for issuers in emerging markets. These hubs can provide shared resources, including access to standardised templates, third-party verifiers, and capacity-building workshops. Additionally, subsidies for first-time issuers or guarantees against specific risks can de-risk SLBs for both issuers and investors, fostering wider adoption.

Introducing Dynamic Transparency Protocols: Dynamic transparency protocols that combine real-time reporting tools with periodic third-party audits can enhance accountability. Issuers could leverage AI-driven analytics to correlate expenditure data with impact metrics, offering granular insights into fund utilisation. Establishing independent advisory boards to oversee SLB projects can further strengthen trust by ensuring alignment with stated objectives and identifying opportunities for mid-course corrections.

Streamlining Issuance Processes with Digital Platforms: Digital platforms dedicated to SLB issuance can standardise processes, reduce costs, and accelerate timelines. These platforms could integrate AI-driven risk assessments, pre-validated templates, and automated compliance checks to simplify workflows. Crowdfunding-style functionalities could also attract smaller-scale investors, broadening the capital pool and making SLBs accessible to a wider range of issuers.

Innovative Risk-Sharing Mechanisms: Introducing pooled risk-sharing structures, where multiple investors share the financial risks associated with underperformance, can make SLBs more attractive. For example, impact insurance products could mitigate losses for investors if SPTs are unmet due to external factors like economic downturns or climate disruptions. These mechanisms would lower the perceived risk of SLBs, encouraging broader participation and investment.

Sustainability-linked bonds represent a transformative approach to financing global sustainability efforts, connecting financial performance with measurable ESG outcomes. However, their potential is hindered by challenges such as inconsistent reporting, limited applicability in emerging markets, and high transaction costs. Addressing these issues through standardised frameworks, capacity-building initiatives, and innovative technologies can enable SLBs to become a cornerstone of impact investing. The SLB market can achieve greater scale, transparency, and impact through collaborative efforts among issuers, investors, and regulators. These advancements will enhance investor confidence and ensure that sustainable finance reaches the regions and sectors most in need, driving meaningful progress towards a more equitable and sustainable future.

(c) **Microfinance in Impact Investing**

Microfinance is a specialised form of debt financing that provides small loans and financial services to individuals or enterprises typically excluded from traditional banking systems. Within the impact investing landscape, microfinance is a tool to promote financial inclusion, reduce poverty, and empower underserved communities, particularly in developing countries. By granting access to credit, microfinance enables low-income individuals, small businesses, and entrepreneurs to invest in productive activities, enhance livelihoods, and achieve financial independence. It plays a critical role in driving social and economic development, bridging the gap for those left out of conventional financial systems.

Core Structure and Operational Model

Microfinance operates on a framework designed to serve the unique needs of underserved populations. Its structure revolves around providing access to credit and other financial services while fostering economic empowerment and social inclusion. At the heart of microfinance are Microfinance Institutions (MFIs), which include a range of entities such as non-governmental organisations (NGOs), cooperatives, credit unions, and some commercial banks. These institutions act as intermediaries, bridging the gap between financial markets and low-income borrowers. Borrowers typically include individuals, micro-entrepreneurs, and small business owners in low-income communities. Lacking collateral and credit histories, they are assessed based on social factors, local knowledge, and repayment capabilities rather than traditional creditworthiness metrics. Microfinance loans are small, with repayment terms tailored to the borrower's cash flow, allowing for frequent payments, such as weekly or monthly instalments. Interest rates, while higher than those of commercial banks, remain lower than informal moneylenders, balancing affordability with the need to cover operational costs.

Group lending is a notable model within microfinance, where loans are extended to groups of borrowers who collectively ensure each other's repayment. This model relies on social collateral, wherein peer pressure and mutual accountability replace the need for traditional physical collateral, which is often unavailable to low-income borrowers. Group lending has proven effective in fostering financial responsibility and reducing default rates. Alongside group lending, MFIs often offer a range of complementary financial services, including savings accounts, micro-insurance, and remittance options. Many also provide training in financial literacy and business management to help borrowers maximise the benefits of their loans, ensuring long-term financial sustainability and improved livelihoods.

Investment Lifecycle

The lifecycle of microfinance involves several stages, each critical to ensuring its success as a financing instrument:

Loan Origination and Screening: The process begins with identifying eligible borrowers through a thorough screening process. Unlike traditional banks, MFIs rely on local networks, social assessments, and direct engagement to evaluate a borrower's repayment capacity. For group lending, MFIs facilitate the formation of borrowing groups, ensuring mutual trust and accountability among members. The origination phase ensures that loans are extended to individuals or groups with the potential to benefit from microfinance while minimising risk.

Loan Disbursement: Once approved, loans are disbursed swiftly, either in cash or through direct deposit into savings accounts at the MFI. Borrowers typically use these funds for income-generating activities such as purchasing inventory, investing in agricultural inputs, or acquiring equipment for small-scale production. The relatively small loan sizes and quick disbursement timelines enable borrowers to address immediate financial needs and capitalise on economic opportunities.

Monitoring and Support: Active monitoring ensures that borrowers use the funds effectively and remain on track with their repayment schedules. MFIs often deploy field officers to visit borrowers, assess their business progress, and provide support where needed. Group meetings in the group lending model serve as forums for borrowers to share experiences, discuss challenges, and receive mutual encouragement. Many MFIs also provide ongoing training in financial literacy, business skills, and resource management, fostering borrower resilience and enhancing the likelihood of success.

Repayment and Collection: Repayment structures are designed to align with the borrower's cash flow, with frequent, small instalments minimising the burden of larger, infrequent payments. The group lending model's social accountability mechanisms further ensure high repayment rates, as group members collectively bear responsibility for each other's repayments. This creates a strong incentive for borrowers to meet their obligations, reducing default risks.

Impact Evaluation: The final stage involves assessing the social and economic impact of the loans. MFIs use surveys, interviews, and data analysis to evaluate improvements in household income, business growth, education, and overall well-being. These insights not only validate the effectiveness of microfinance but also guide future strategies to refine lending practices and address evolving community needs.

Challenges in Leveraging Microfinance in Impact Investing and Opportunities for Mitigation

Balancing Affordability with Sustainability: The dual mandate of affordability and financial sustainability creates inherent tensions within microfinance. To remain operational, MFIs must cover high costs associated with outreach, monitoring, and providing tailored services to underserved communities. However, charging high interest rates to meet these costs can erode the social mission of microfinance, as borrowers may struggle to repay loans and sustain their livelihoods. This dynamic can lead to a perception of exploitation rather than empowerment, undermining trust in the sector. For impact investing, this challenge reduces the scalability and effectiveness of microfinance as a tool for fostering inclusive economic development.

High Operational Costs and Scalability: The labour-intensive nature of microfinance, including face-to-face interactions, frequent repayments, and ongoing borrower support, significantly increases operational costs. These costs limit the ability of MFIs to expand their services to larger populations or more remote areas. For impact investors, the high overheads dilute financial returns and make it harder to justify investments in microfinance. Moreover, inefficiencies in scaling operations exacerbate the financial exclusion of communities that stand to benefit most from microfinance, limiting its transformative potential within the impact investing ecosystem.

Risk of Over-Indebtedness: Over-indebtedness poses a systemic risk to the microfinance sector, particularly in regions where multiple MFIs operate without coordinated credit checks. Borrowers may take loans from several sources to meet immediate needs, resulting in a debt spiral that jeopardises their financial stability and well-being. This issue harms individual borrowers and damages microfinance's reputation as a responsible financing tool.[11] For impact investing, over-indebtedness undermines the credibility of microfinance, deterring investors and reducing the sector's capacity to attract sustainable capital.

Limited Access to Technology and Financial Literacy: The lack of technological access and financial literacy among low-income borrowers creates significant barriers to the effective utilisation of microfinance. Without digital tools, borrowers rely on manual processes, which are less efficient and more prone to errors. Additionally, low levels of financial literacy hinder borrowers from using loans productively, increasing the risk of default. These challenges reduce the overall impact of microfinance, as funds may not translate into meaningful improvements in livelihoods. For impact investors, this inefficiency diminishes the social and economic returns on investment.

Regulatory and Political Risks: Microfinance operations are often exposed to unstable regulatory environments and political interference. Interest rate caps or sudden loan forgiveness schemes, while intended to protect borrowers, can disrupt the financial viability of MFIs, forcing them to curtail operations or exit markets entirely. For impact investors, such unpredictability increases the perceived risk of investing in microfinance, limiting the flow of capital into the sector. These regulatory uncertainties also hinder long-term planning and innovation, reducing the sector's ability to address systemic financial exclusion.

Opportunities for Mitigation

Innovative Interest Rate Models: MFIs can explore context-sensitive interest rate models incorporating participatory community-driven frameworks. These models could be co-designed with borrowers, ensuring rates align with local economic realities. Additionally, philanthropic funding or government subsidies could offset operational costs, allowing for more affordable interest rates without compromising MFI sustainability. Incorporating scalable risk-pooling mechanisms can also distribute costs equitably, enhancing fairness and operational resilience.

Leveraging Local Expertise for Operational Efficiency: Training local community leaders to act as microfinance facilitators can reduce operational costs significantly. These local agents can handle borrower assessments, repayments, and monitoring, building trust while fostering community ownership of financial processes. Partnerships with cooperatives and local NGOs can enhance outreach, streamline administration, and improve borrower compliance without requiring heavy technological or financial investments.

Building Comprehensive Credit Networks: Creating shared borrower databases across MFIs, regulated by independent oversight bodies, can mitigate risks of over-indebtedness. These networks can integrate manual processes with accessible digital solutions tailored to local contexts, ensuring inclusivity. Collaboration with local financial associations can facilitate the gradual adoption of technology, balancing innovation with usability for communities less familiar with digital tools.

Community-Based Financial Literacy Programmes: Financial literacy programmes can become more engaging and impactful by leveraging peer-to-peer learning groups and role-playing simulations. Borrowers can participate in locally adapted sessions focusing on real-world challenges and opportunities in financial planning. Mobile education units equipped with multilingual and visual learning tools can reach remote areas, ensuring inclusivity. Incorporating feedback loops allows programmes to evolve based on borrower experiences and needs.

Collaborative and Stable Regulatory Frameworks: Policymakers can implement predictable, forward-looking regulations that balance borrower protection with MFI sustainability. Regularly updated regulatory frameworks, co-created with MFIs and impact investors, can establish safeguards that balance the need for borrower protection with the financial sustainability of MFIs. For instance, such frameworks could introduce mechanisms to cap the total debt exposure per borrower, preventing over-indebtedness while ensuring access to necessary credit. Similarly, adaptive repayment models could be developed, allowing borrowers to adjust their schedules based on unforeseen economic hardships or seasonal income fluctuations. These measures, when aligned with regional economic conditions, can foster resilience for borrowers and stability for lenders. Tax incentives for MFIs achieving predefined social impact metrics can encourage sustainable growth. Multi-stakeholder consultations and public reporting can enhance accountability and trust.

Blended Finance Solutions for Capacity-Building: Blended finance initiatives can pool public, private, and philanthropic sector resources to establish regional microfinance innovation hubs. These hubs could fund infrastructure upgrades, capacity-building workshops, and risk mitigation programmes. Dedicated innovation funds could also support research into new microfinance delivery models tailored to underserved regions. By de-risking investments, blended finance can attract more private capital for capacity-building.

Microfinance is a cornerstone of impact investing, offering transformative potential to promote financial inclusion and reduce poverty. However, its effectiveness is hindered by challenges such as high operational costs, over-indebtedness, and regulatory risks. Stakeholders can effectively address these barriers by adopting innovative yet balanced solutions—such as participatory interest rate models, community-based literacy programmes, and collaborative regulatory frameworks. Through thoughtful collaboration among MFIs, investors, policymakers, and communities, microfinance can evolve into a more inclusive, scalable, and impactful instrument. These advancements will ensure that underserved populations gain meaningful access to financial services, empowering them to achieve sustainable livelihoods and contribute to broader economic development.

Hybrid Instruments

Hybrid instruments blend debt and equity financing characteristics, offering flexible solutions that align financial returns with social and environmental impact. These instruments are particularly suited for projects or enterprises that require adaptable financing terms to achieve impact objectives while managing financial risks. Hybrid instruments are used to fill the gaps where traditional equity or debt financing may not fully meet the needs of impact-focused enterprises. They allow investors to participate in the upside potential of a project (similar to equity) while also securing downside protection (as with debt).

Hybrid instruments have gained traction in recent years as impact investors seek innovative financial tools to enhance flexibility and scalability. In particular, they have played a significant role in the development of the impact investing industry by addressing the diverse and unique needs of impact-driven enterprises that may not fit neatly into conventional financing structures. Allocations through hybrid instruments have grown significantly, with increasing interest from both institutional and individual impact investors. Investors using hybrid instruments often include those focused on below-market-rate returns, as well as those seeking a balance between financial and impact objectives. The instruments have grown substantially, with a Compound Annual Growth Rate (CAGR) of 104 per cent from 2019 to 2024. This growth indicates that investors increasingly seek flexible and adaptive financing structures that align financial returns with social and environmental outcomes. Hybrid models are particularly attractive for projects in emerging markets, where traditional financing may be challenging due to higher risks and market instability.

Structuring hybrid instruments can be complex, requiring financial engineering and impact measurement expertise. In some regions, regulatory frameworks may not fully accommodate hybrid instruments, adding legal and administrative hurdles. Stakeholders need greater awareness of how hybrid instruments function and the value they add to impact investing. Governments and regulatory bodies can also play a role in facilitating the use of hybrid instruments by addressing legal and regulatory barriers. Nonetheless, hybrid instruments represent a promising avenue for scaling impact investing by combining the strengths of debt and equity financing. Their versatility and alignment with impact objectives make them a valuable tool for addressing diverse needs across markets and sectors.

(i) **Convertible Debt in Impact Investing**

Convertible debt is a hybrid financial instrument that starts as a loan and includes an option for the lender to convert the outstanding principal into equity under predefined conditions. This financing model is particularly attractive in impact investing as it combines downside protection for investors with the potential for upside if the enterprise achieves its financial and social or environmental goals. It provides mission-driven organisations with the capital they need while delaying valuation discussions until their operations are more stable. This instrument fosters alignment between financial returns and impact outcomes, offering flexibility for both investors and borrowers.

Core Structure and Operational Model

Convertible debt operates as a bridge between debt and equity, providing flexibility that is especially valuable for early-stage impact enterprises. The loan aspect includes a principal amount, interest rate, and maturity date, offering downside protection to the investor as a creditor. However, the distinguishing feature is the conversion option, allowing the debt to be converted into equity under specific conditions such as a new funding round, achievement of milestones, or a predetermined date. This model is particularly useful for enterprises operating in emerging sectors or tackling societal challenges such as renewable energy access or healthcare provision. Investors often structure agreements to include a valuation cap, ensuring that their equity conversion occurs at a favourable valuation. Discount rates on share prices at conversion are another common feature, compensating investors for early-stage risk. Interest rates are typically set lower than traditional loans, reflecting the hybrid nature of the instrument and the additional equity potential.

Impact-linked clauses can be embedded in the terms, adjusting the conversion or repayment conditions based on the achievement of social or environmental milestones. For instance, a borrower improving access to education for underserved communities might trigger more favourable conversion terms. Risk mitigation mechanisms, such as guarantees or blended finance models, are frequently used to make convertible debt more appealing to private investors, particularly in nascent markets.

Investment Lifecycle of Convertible Debt

Origination and Structuring: The lifecycle of convertible debt begins with origination, where potential investors and borrowers establish terms that align with both financial and impact objectives. This stage involves detailed negotiations on key terms such as the loan principal, interest rate, maturity date, conversion triggers, and valuation caps. Investors often include impact-linked clauses to ensure that social or environmental milestones are explicitly tied to financial terms. Agreements may also incorporate discount rates on equity conversion, compensating investors for the risks associated with early-stage investments. Blended finance models or guarantees are sometimes embedded to reduce perceived risks for investors.

Disbursement and Deployment: Once terms are agreed upon, funds are disbursed to the borrower. Enterprises typically use this capital to scale operations, develop products, or address key societal challenges. During this phase, the focus is on the effective deployment of resources to achieve both financial and impact objectives. Borrowers often allocate funds to areas like research and development, infrastructure expansion, or market entry, ensuring alignment with agreed-upon milestones.

Monitoring and Impact Tracking: Monitoring is a critical component of the convertible debt lifecycle. Investors track financial performance, operational progress, and impact metrics to ensure alignment with the predefined terms. Regular reporting from the borrower—including financial statements and impact assessments—is essential. Third-party verification may be used to validate social or environmental outcomes, bolstering transparency and accountability. This phase ensures that both parties remain committed to the dual objectives of financial returns and societal impact.

Conversion or Repayment Decision: As the debt approaches maturity or other conversion triggers are met, investors decide whether to convert the outstanding principal into equity or to demand repayment. Conversion is often triggered by events such as a new funding round, the achievement of impact milestones, or the end of the loan term. If conversion is chosen, the terms—including valuation caps and discounts—determine the equity stake acquired by the investor. This stage represents a transition from debt obligations to equity-based partnerships for borrowers, which may influence governance and decision-making structures.

Exit and Value Realisation: For investors, the final stage involves realising value through either an equity exit (via IPOs, acquisitions, or secondary sales) or the repayment of the loan. The outcomes depend on the enterprise's financial performance and ability to scale its impact. Successful exits provide financial returns and validate the use of convertible debt as a mechanism for advancing impact investing goals. For borrowers, a successful exit signifies enhanced financial sustainability and the achievement of meaningful societal contributions.

Challenges in Leveraging Convertible Debt in Impact Investing

Complexity in Structuring and Negotiation: Convertible debt agreements are inherently complex, requiring detailed negotiations to balance financial and impact objectives. The inclusion of impact-linked terms, valuation caps, and conversion discounts further complicates the structuring process. This complexity can deter both investors and enterprises, particularly those with limited legal or financial expertise. For impact investing, these complications can lead to longer deal cycles and higher transaction costs, reducing the attractiveness of the instrument. Moreover, poorly structured agreements risk misalignment of incentives, undermining the potential for achieving both financial and social outcomes.

Uncertainty in Valuation and Conversion: Delayed valuation discussions—a defining feature of convertible debt—introduce significant uncertainty for both investors and borrowers. For investors, undervaluation at the time of conversion can dilute potential returns, while borrowers may feel pressured to accept unfavourable terms to secure financing. This tension risks creating adversarial relationships, hindering long-term collaboration. Furthermore, reliance on external market conditions to determine valuations exposes both parties to volatility, making convertible debt less predictable and reliable as a financing tool for impact enterprises.

Balancing Impact with Financial Returns: Convertible debt's dual mandate to deliver both financial and social returns often creates conflicting priorities. Investors seeking competitive financial returns may push for aggressive growth strategies that conflict with the enterprise's mission-driven objectives. Conversely, enterprises prioritising impact over profitability risk alienating investors, reducing their ability to secure future funding. This challenge is particularly detrimental in sectors where achieving meaningful impact requires long-term investments with uncertain financial payoffs, such as renewable energy or public health initiatives.

Risk of Default or Conversion Avoidance: In cases where enterprises fail to meet financial or impact milestones, the risk of default increases, particularly if the enterprise operates in high-risk or nascent markets. Investors may face losses if borrowers are unable to repay loans, while unmet milestones may also deter conversion. Conversely, investors opting for conversion despite missed impact targets may undermine the credibility of convertible debt as an impact-focused instrument. This dynamic creates reputational risks for both investors and enterprises, reducing the overall effectiveness of the model.

Limited Adoption in Emerging Markets: Convertible debt remains underutilised in emerging markets due to a lack of financial expertise, inadequate legal frameworks, and high perceived risks. Enterprises in these regions often lack the capacity to negotiate and manage complex financing instruments, while investors face challenges in assessing risks and ensuring accountability. This limited adoption restricts the scalability of convertible debt, leaving many high-impact sectors and geographies underserved.

Opportunities for Mitigating Challenges

Standardising Convertible Debt Agreements: Standardised templates for convertible debt agreements could evolve into smart contract-enabled frameworks. These frameworks could integrate automated clauses using blockchain technology, ensuring transparency and efficiency in execution. For instance, impact milestones and valuation caps could trigger automated adjustments or conversions, reducing negotiation timelines and potential disputes. This innovation could lower transaction costs, attract more participants, and ensure robust compliance with agreed terms.

Enhancing Valuation Transparency: Utilising AI-driven valuation models can improve the transparency and accuracy of valuations at conversion points. These models could analyse market trends, enterprise performance, and sector-specific metrics in real time, offering objective and data-driven insights. AI-based tools could also be used for predictive analytics, helping investors and enterprises anticipate valuation shifts and align expectations more effectively. This approach ensures fairness and builds trust between stakeholders.

Incorporating Flexible Impact Metrics: Dynamic impact metrics tied to real-time data collection can offer more adaptable frameworks for tracking progress. For example, IoT devices and satellite imagery could monitor environmental impacts like deforestation rates or energy efficiency, feeding into an enterprise's performance metrics. Flexible metrics could also integrate feedback loops from beneficiaries, ensuring that social outcomes remain aligned with the needs of target communities. These innovations enable a more responsive approach to measuring and achieving impact.

Leveraging Blended Finance Models: Blended finance structures could incorporate novel insurance products that hedge risks associated with high-impact, high-risk sectors. For example, parametric insurance could provide payouts based on predefined triggers, such as climate-related disruptions affecting agricultural enterprises. This approach reduces risks for private investors while enabling enterprises to recover quickly from external shocks. Blended finance could also explore securitisation of convertible debt portfolios, pooling multiple investments to attract institutional capital and enhance scalability.

Building Capacity in Emerging Markets: Forward-thinking capacity-building initiatives could focus on creating digital platforms for financial literacy and contract negotiation training. These platforms could use gamification to engage entrepreneurs and simulate real-world scenarios for managing convertible debt. Partnerships with local universities and global institutions could foster the development of bespoke curricula, combining traditional financial training with insights into impact investing trends. Regional hubs equipped with these resources could serve as innovation centres, accelerating the adoption of convertible debt in emerging markets.

Convertible debt offers a unique blend of flexibility, downside protection, and alignment with impact objectives, making it a valuable tool in the impact investing ecosystem. However, its complexity, valuation uncertainties, and limited adoption present significant challenges. Integrating innovative solutions, such as smart contract-enabled agreements, AI-driven valuation tools, dynamic impact metrics, and novel blended finance mechanisms, can enable stakeholders to enhance the efficiency and scalability of convertible debt. Through collaborative efforts and strategies, convertible debt can evolve into a transformative instrument that bridges the gap between financial sustainability and impactful change. These advancements will empower mission-driven enterprises to thrive while achieving measurable and lasting societal benefits.

(ii) **Mezzanine Finance**

Mezzanine finance is a hybrid debt instrument that provides capital to established social enterprises or impact-focused organisations, typically to finance expansion, acquisitions, or other growth initiatives. As a hybrid debt instrument, it occupies a middle ground between senior debt and equity in the capital structure. This positioning reflects its dual nature—offering the stability of debt and the growth potential of equity. It offers an attractive solution for enterprises with predictable cash flows but a need for additional capital. In impact investing, the terms of mezzanine finance can be structured to align with the impact goals of the investment.

Core Structure and Operational Model

It provides a nuanced solution for organisations looking to expand, acquire, or undertake transformative growth without relinquishing substantial ownership or securing hard collateral. They are typically used by more mature companies who need more capital than traditional bank loans can provide but does not want to dilute ownership too much with equity financing. Mezzanine debt is subordinated to senior debt (like bank loans) but senior to equity. This means that in case of liquidation, senior debt holders are paid first, then mezzanine debt holders, and finally equity holders. This form of financing carries higher interest rates than senior debt due to its higher risk. It functions like debt by offering lenders fixed interest payments, while also incorporating equity-like features, such as warrants or options, that allow lenders to share in the borrower's upside potential. This dual nature creates a flexible financial instrument, appealing to organisations with stable cash flows but limited collateral or a desire to retain control over their operations. An interesting facet of mezzanine finance is its subordinated position in the capital stack. In case of liquidation or bankruptcy, senior lenders are paid first, leaving mezzanine lenders to recover their investments only after these obligations are met. This increased risk is offset by a blend of predictable returns from interest and the possibility of higher rewards from equity-linked components.

Mezzanine finance is tailored to suit growth-oriented organisations. Typically, these are established enterprises, such as social businesses or impact-focused organisations, with a proven track record of cash flow stability. Borrowers often use mezzanine finance to fund critical initiatives like expansion, acquisitions, or innovative projects, making it an ideal tool for bridging funding gaps without incurring the full dilution effects of issuing new equity. The due diligence process in mezzanine finance is rigorous. Lenders carefully evaluate the borrower's financial health, focusing on cash flow predictability and growth potential. In impact-focused scenarios, additional scrutiny is placed on how well the borrower's objectives align with measurable social or environmental outcomes. This alignment often extends into the loan terms, where innovative provisions might link interest rates or repayment flexibility to the achievement of specific impact milestones.

Key components of mezzanine finance include equity kickers, such as warrants or options to purchase equity at a predetermined price, and profit-sharing arrangements. These features allow investors to participate in the borrower's upside potential. Repayment terms are flexible and may include cash interest, payment-in-kind (PIK) interest, or a blend of both. This flexibility ensures the financing adapts to the borrower's cash flow, making it particularly suitable for enterprises with seasonal or variable revenue streams. The terms of mezzanine finance are often negotiated to reflect both financial and impact objectives. In impact investing, agreements can incorporate impact-linked features, such as adjustments to interest rates or equity conversion terms based on the achievement of predefined social or environmental milestones. For example, a borrower achieving significant reductions in greenhouse gas emissions might benefit from reduced interest rates or extended repayment periods. Mezzanine finance also includes covenants to protect investors, such as financial performance thresholds or additional borrowing restrictions. These safeguards provide risk mitigation while maintaining sufficient flexibility for the borrower to pursue growth initiatives.

Mezzanine finance providers include specialised funds, private equity firms, Development Finance Institutions (DFIs), and impact investors. Borrowers typically consist of mid-stage enterprises, social businesses, or infrastructure projects in sectors like renewable energy, healthcare, and education. These organisations often possess predictable cash flows but lack access to traditional funding sources, making mezzanine finance an essential tool for their expansion. This hybrid finance offers a compelling combination of flexibility, risk-adjusted returns, and impact alignment. By straddling the line between debt and equity, it provides a dynamic financial tool that supports growth while fostering innovation and measurable impact. Lenders may recover their investments through scheduled repayments, the borrower's cash flow, or by converting debt into equity under certain conditions, allowing them to participate in the long-term success of the business. This adaptability makes it a powerful instrument for organisations navigating complex financial landscapes and seeking to create transformative value.

Investment Lifecycle of Mezzanine Finance

The investment lifecycle of mezzanine finance typically follows a structured sequence, from origination to exit, and involves distinct stages that cater to both the borrower's needs and the lender's risk-return objectives. Below is a detailed examination of each stage:

Identification of Opportunities: The lifecycle begins with identifying businesses or projects with robust cash flow potential and growth prospects. Ideal candidates for mezzanine financing include established companies planning expansions, acquisitions, or transformative initiatives that traditional debt or equity financing cannot fully support. In the impact investing space, potential borrowers are often organisations aiming to generate measurable social or environmental impact alongside financial returns.

Due Diligence: At this stage, lenders undertake comprehensive due diligence to evaluate the borrower's creditworthiness, cash flow stability, and operational risks. Key metrics include historical financial performance, projected cash flows, industry dynamics, and alignment with lenders' impact and financial objectives. For impact-focused organisations, additional assessments of their impact potential and the feasibility of achieving stated goals are critical.

Structuring the Deal: Based on due diligence findings, lenders and borrowers negotiate the terms of the mezzanine financing. These terms include interest rates, repayment schedules, and equity-linked features such as warrants, options, or convertible debt. In impact investing, agreements may include impact performance targets, with incentives like reduced interest rates or extended repayment terms for achieving these targets.

Funding: Capital is disbursed to the borrower upon finalising the deal structure. Mezzanine financing typically serves as a complementary funding source, bridging the gap between senior debt and equity. Borrowers often use these funds for specific objectives, such as expanding operations, acquiring new assets, or scaling impact-driven initiatives.

Monitoring and Governance: Throughout the deployment stage, lenders actively monitor the borrower's financial health and adherence to agreed-upon terms. For impact investments, lenders may require periodic reporting on financial and impact metrics. Regular communication ensures early detection of potential issues and fosters accountability. Lenders may also participate in governance through board representation or advisory roles, particularly when equity-linked features are involved.

Repayment: Borrowers repay mezzanine finance through a mix of scheduled interest payments, profit-sharing (if applicable), and a final balloon payment at the end of the loan term. Flexible repayment structures—such as deferring interest payments until cash flows stabilise—help borrowers manage financial commitments without hampering growth.

Exit: The exit strategy is a critical phase for lenders. Common exit mechanisms include repayment of principal and interest, conversion of debt into equity, or the sale of equity stakes if conversion rights are exercised. For impact-focused investments, the exit also involves assessing the borrower's success in achieving predefined impact objectives. Exits are often coordinated with other stakeholders to ensure a smooth transition and continued operational stability for the borrower.

Assessing Outcomes: After exiting the investment, lenders evaluate both financial returns and the achievement of impact goals. This stage provides insights into the effectiveness of the investment strategy and highlights lessons for future mezzanine financing deals. Lenders may disseminate their findings through reports or case studies, contributing to industry-wide knowledge on best practices and challenges in leveraging mezzanine finance for impact.

Challenges in Leveraging Mezzanine Finance in Impact Investing

Complex Structuring and Negotiation: Mezzanine finance deals are inherently complex, requiring meticulous structuring to balance the needs of both lenders and borrowers. For impact investing, the additional layer of aligning financial returns with measurable impact goals further complicates negotiations. Borrowers often lack expertise in negotiating equity-linked terms, such as warrants or convertible options, which can lead to misalignment of incentives or unfavourable terms for either party. These complexities can increase transaction costs and lengthen deal cycles, reducing the attractiveness of mezzanine finance as a viable option for impact-focused enterprises.

High Risk for Lenders: The subordinated position of mezzanine finance in the capital structure places lenders at significant risk. In the event of default or liquidation, mezzanine lenders are repaid only after senior debt obligations are fulfilled. This high-risk profile is exacerbated in impact investing, where borrowers often operate in emerging or underserved markets with volatile cash flows and heightened operational risks. Without adequate safeguards, the potential for financial loss can deter investors, limiting the scalability of mezzanine finance in the impact space.

Challenges in Impact Measurement: Integrating impact metrics into mezzanine finance agreements presents unique challenges. Many enterprises lack the capacity to track and report on social or environmental outcomes effectively, leading to gaps in accountability. For lenders, this creates difficulties in evaluating the success of their investments beyond financial returns. The absence of standardised frameworks for measuring impact further complicates comparisons across sectors and geographies, reducing transparency and trust in the instrument's ability to drive meaningful change.

Limited Access for Mid-Sized Enterprises: While mezzanine finance is ideally suited for mid-sized enterprises, many such businesses in developing markets lack access due to insufficient credit histories, weak financial infrastructure, or perceived risk profiles. Impact-focused enterprises operating in high-need sectors, such as renewable energy or healthcare, often face additional hurdles, including high transaction costs and limited knowledge of hybrid financing instruments. This exclusion perpetuates funding gaps for enterprises that could drive significant societal or environmental benefits.

Balancing Flexibility with Accountability: The flexibility of mezzanine finance—a key feature that attracts borrowers—can sometimes lead to challenges in ensuring accountability. Features like payment-in-kind (PIK) interest or deferred repayments may reduce short-term financial pressure on borrowers but can also result in lax financial discipline. For impact investors, this flexibility must be balanced with mechanisms that ensure borrowers remain committed to achieving both financial and impact goals, which can be difficult to enforce in practice.

Opportunities for Mitigation

Standardised Hybrid Financing Templates: Developing modular and digital templates for mezzanine finance agreements can reduce transaction costs and streamline negotiations. These templates could integrate AI-driven customisation options to adjust terms like interest rates, repayment schedules, and impact-linked clauses in real-time based on sector-specific or geographic needs. Stakeholders can also foster consistency and innovation across the industry by creating open-source platforms for sharing and refining these templates.

Risk Mitigation through Collaborative Blended Finance Models: Blended finance models can incorporate advanced insurance products like parametric insurance to de-risk investments. For example, triggers based on predefined milestones (e.g., achieving revenue thresholds or impact metrics) could activate payouts to lenders, reducing potential losses. Multi-stakeholder funds that pool capital from private investors, DFIs, and philanthropic sources can create diversified risk-sharing structures that attract broader participation in mezzanine finance.

Leveraging Digital Ecosystems for Impact Tracking: Adopting blockchain-based ecosystems can revolutionise impact measurement by ensuring transparency and immutability in reporting. Smart contracts could automatically adjust financial terms (e.g., interest rate reductions) when impact milestones are verified, incentivising performance alignment. Digital dashboards powered by IoT and AI analytics can provide lenders and investors with real-time insights into impact achievements, reducing administrative burdens and enhancing accountability.

Capacity-Building Through Virtual Learning Platforms: Building the capacity of borrowers can be accelerated through virtual learning platforms that offer tailored modules on financial literacy, hybrid financing, and impact measurement. These platforms could include gamified scenarios to simulate real-world mezzanine finance negotiations and decision-making. Collaborations with local institutions and tech firms can enhance accessibility and relevance, ensuring that mid-sized enterprises in underserved markets are better prepared to engage with mezzanine finance.

Sector-Focused Innovation Labs: Creating sector-specific innovation labs for mezzanine finance can catalyse tailored solutions for industries like renewable energy, healthcare, and education. These labs could explore innovative instruments such as revenue-linked mezzanine loans or equity-sharing mechanisms tied to community outcomes. By involving industry experts, impact investors, and policymakers, these labs can test and scale best practices, driving efficiency and effectiveness in high-impact sectors.

Dynamic Incentive Structures Linked to AI Insights: Incentive structures can evolve to incorporate AI-driven predictions of borrower performance and market conditions. For example, interest rates or equity conversion terms could adjust dynamically based on AI-generated forecasts of cash flow stability or impact milestones. These predictive analytics can help align borrower incentives with lender expectations while reducing risks and improving overall investment outcomes.

Mezzanine Finance is uniquely positioned in the impact investing ecosystem, offering a flexible and dynamic tool for funding growth-oriented enterprises. However, challenges such as high risks, complex structuring, and difficulty measuring impact often hinder its full potential. Strategies such as modular financing templates, blockchain-based tracking, and sector-focused innovation labs stakeholders can mitigate these challenges effectively. Mezzanine finance can become a cornerstone of sustainable development through collaboration and innovation, bridging funding gaps for mid-sized enterprises and driving measurable social and environmental impact. These efforts will ensure that mezzanine finance continues to evolve as a versatile and impactful instrument in the global effort to achieve equitable and sustainable growth.

(iii) **Revenue-Based Financing**

Revenue-Based Financing (RBF), or royalty-based financing, is a flexible funding model where investors provide capital upfront to organisations for a fixed percentage of future revenues. This arrangement continues until a predefined repayment multiple of the initial investment is achieved. In the context of impact investing, RBF is particularly suited for mission-driven enterprises with predictable revenue streams but limited access to traditional debt or equity financing. By linking repayments to revenue generation, RBF aligns investor returns with the financial and operational realities of the borrowing enterprise, offering a sustainable pathway to scale while preserving ownership and control.

Core Structure and Operational Model

The RBF structure is designed to balance adaptability and predictability for both investors and borrowers. Investors typically agree to finance enterprises with established or predictable revenue streams, providing capital that can be used for scaling operations, expanding market reach, or investing in social or environmental projects. Borrowers agree to remit a percentage of their revenues—usually gross revenues—at regular intervals (monthly or quarterly) until the repayment cap is reached. A central element of RBF is the predefined repayment cap, often set at a multiple of the initial investment (e.g., 1.5x or 2x). This cap clarifies and ensures the revenue-sharing agreement has a fixed endpoint. Unlike traditional loans, RBF does not impose fixed payments, enabling repayments to fluctuate with the borrower's revenue performance. This flexibility alleviates financial strain during periods of low revenue and creates a repayment structure that supports operational sustainability. RBF is inherently non-dilutive, meaning borrowers do not have to relinquish equity stakes or ownership control. This characteristic makes it an attractive option for mission-driven enterprises and startups prioritising decision-making autonomy. Additionally, the model allows for customisation to align financial terms with social or environmental objectives. For example, clauses may adjust the repayment cap or revenue share percentage based on the achievement of impact milestones such as reducing carbon emissions or improving access to essential services.

Investment Lifecycle

The lifecycle of an RBF investment begins with the origination and structuring phase. Investors identify enterprises with predictable revenue streams and strong potential for impact. Due diligence assesses the enterprise's financial health, operational efficiency, and alignment with impact objectives. Negotiations then establish terms such as the revenue share percentage, repayment cap, and duration of the agreement, ensuring they reflect the enterprise's growth trajectory and capacity. After finalising the agreement, capital is disbursed to the borrower, either as a lump sum or in tranches tied to milestones. This phase often supports initiatives such as market entry, infrastructure development, or scaling mission-driven operations. Investors closely monitor the utilisation of funds to ensure alignment with agreed-upon objectives. During the repayment phase, borrowers remit a percentage of their revenues to investors. Payments are dynamic, scaling with revenue performance and accommodating seasonal or cyclical fluctuations. Transparency is maintained through regular financial and impact reporting, allowing investors to track progress and provide strategic support where necessary. The flexibility inherent in this phase ensures that repayment obligations do not compromise the enterprise's ability to achieve its mission. The agreement concludes once the repayment cap is achieved, marking a successful exit for investors. At this stage, enterprises retain full ownership and operational control, while investors realise returns on their capital. A final evaluation assesses both financial outcomes and impact achievements, offering insights to refine future RBF agreements and strategies.

Challenges in Leveraging Revenue-Based Financing (RBF) in Impact Investing

Revenue Volatility: RBF's reliance on revenue streams introduces inherent uncertainty for both borrowers and investors. Seasonal fluctuations, market volatility, or external shocks, such as economic downturns or supply chain disruptions, can significantly affect revenue levels. For impact-driven enterprises operating in high-risk sectors or underserved markets, these challenges are amplified. For example, an agricultural enterprise may experience drastic revenue drops due to climate-related events, directly affecting its ability to meet repayment obligations. This uncertainty not only delays repayment timelines but also creates significant risks for investors, who may be deterred from supporting enterprises in unpredictable markets. This systemic aversion to volatility restricts the flow of capital into high-impact but volatile sectors, stifling innovation and limiting transformative change in vulnerable communities.

Misalignment of Revenue Metrics with Impact Objectives: While RBF ties repayments to revenues, there is often a disconnect between financial performance and the achievement of impact goals. Enterprises prioritising mission-driven objectives, such as community development or environmental restoration, may face challenges balancing revenue growth with long-term impact outcomes. For instance, a healthcare enterprise focused on providing affordable treatments may struggle to maintain low service costs while achieving the revenue levels required for RBF repayments. This trade-off risks forcing enterprises to compromise their social missions to meet financial obligations, undermining the credibility of impact investing as a tool for equitable development. The persistent tension between revenue generation and impact delivery can also erode trust among stakeholders, limiting the willingness of both investors and enterprises to engage with RBF models.

Limited Applicability for Early-Stage Enterprises: RBF is best suited for enterprises with predictable and recurring revenue streams, excluding early-stage organisations that have yet to establish consistent income. This limitation disproportionately affects high-potential sectors like renewable energy and education, where significant upfront investments are required to build infrastructure or develop services before revenue stability can be achieved. For example, a clean energy startup may need years of operational build-up before generating reliable cash flows, rendering it ineligible for RBF. This exclusion narrows the pool of eligible borrowers and deprives the impact investing ecosystem of the innovation and long-term value creation that early-stage enterprises can offer. RBF fails to fully address the systemic financing gaps in high-need areas by sidelining these nascent ventures.

High Transaction Costs for Small Enterprises: For smaller enterprises, the administrative burden of structuring, monitoring, and reporting RBF agreements can be disproportionately high. This issue is exacerbated in underserved markets where access to financial expertise and technological infrastructure is limited. For instance, a rural cooperative may lack the capacity to generate accurate financial reports or maintain consistent communication with investors, leading to inefficiencies and delays. High transaction costs not only reduce the net benefit of RBF for borrowers but also diminish investor returns, making the model less attractive for both parties. This creates a structural barrier that excludes grassroots initiatives and micro-enterprises, undermining the inclusive ethos of impact investing.

Challenges in Setting Appropriate Terms: Determining the revenue share percentage, repayment cap, and duration of the agreement requires careful calibration to ensure fairness and feasibility. Overly aggressive terms can strain borrowers, while conservative terms may limit investor returns. For example, an enterprise with cyclical revenues may struggle to meet consistent repayment expectations during off-peak seasons, leading to financial distress. Conversely, overly lenient terms could disincentivise investors from participating, reducing the availability of capital. This delicate balancing act becomes even more challenging in volatile markets, where revenue projections are inherently uncertain. Misaligned terms can erode trust between stakeholders and hinder the adoption of RBF as a scalable and sustainable financing tool in impact investing.

Opportunities for Mitigation

Dynamic Revenue Sharing Models: Innovative RBF agreements can incorporate dynamic revenue-sharing models that adjust repayment terms based on real-time performance data. For instance, leveraging AI-driven predictive analytics can enable investors to anticipate revenue fluctuations and adjust the revenue share percentage accordingly. This flexibility reduces pressure on borrowers during low-revenue periods while ensuring that investors achieve their returns over time. Such models align repayment structures with enterprise realities, fostering resilience and sustainability.

Blended Finance Solutions: Blended finance can enhance the applicability of RBF for early-stage and high-impact enterprises. Combining concessional funding or grants from philanthropic sources with commercial RBF capital can enable enterprises to access initial funding to stabilise operations and build revenue streams. Concessional capital can also absorb initial losses, de-risking investments for commercial lenders and expanding the reach of RBF to underserved sectors and geographies.

Integrated Impact and Revenue Metrics: RBF agreements can include integrated metrics that link financial performance with impact milestones to address the disconnect between revenue generation and impact outcomes. For example, repayment caps could be adjusted based on the achievement of predefined impact goals, such as increasing access to education or reducing greenhouse gas emissions. This alignment incentivises enterprises to prioritise mission-driven objectives without compromising financial sustainability.

Digital Platforms for Streamlined Processes: Developing digital platforms tailored for RBF can reduce transaction costs and enhance accessibility for small enterprises. These platforms could automate tasks such as revenue tracking, impact reporting, and compliance monitoring. Blockchain technology could further enhance transparency and accountability by creating immutable records of revenue flows and impact data.

Such innovations simplify administrative processes, making RBF more efficient and scalable.

Capacity-Building for Enterprises: Targeted capacity-building programmes can empower enterprises to engage effectively with RBF. These programmes could focus on improving financial literacy, revenue forecasting, and impact measurement capabilities. Collaborations with local organisations and financial intermediaries can ensure that these programmes are contextually relevant and accessible, particularly in underserved markets. Building enterprise capacity not only enhances the effectiveness of RBF agreements but also fosters long-term financial and operational resilience.

Sector-Specific RBF Funds: Creating sector-specific RBF funds can address the unique challenges and opportunities within industries such as renewable energy, healthcare, and education. These funds can pool resources and expertise to develop tailored financing solutions that account for industry-specific revenue patterns and impact metrics. For example, an RBF fund for renewable energy projects could incorporate metrics tied to energy output and carbon reduction, ensuring alignment with both financial and environmental goals.

Revenue-Based Financing offers a flexible and innovative solution for impact investing, bridging the gap between financial sustainability and mission-driven objectives. However, its effectiveness is limited by challenges such as revenue volatility, high transaction costs, and misaligned metrics. Dynamic revenue-sharing models, blended finance solutions, and digital platforms can enable stakeholders to enhance the scalability and impact of RBF. Through collaboration and innovation, RBF can evolve into a cornerstone of impact investing, empowering mission-driven enterprises to achieve sustainable growth while delivering measurable social and environmental benefits. These advancements will ensure that RBF continues to align financial returns with the broader goals of equitable and sustainable development.

(iv) **Participating Loans**

Participating loans, sometimes known as profit-sharing loans, offer a hybrid approach to financing by combining fixed repayment obligations with a variable component linked to the borrower's financial performance.

In impact investing, these loans are uniquely structured to support organisations that seek to balance financial returns with measurable social or environmental outcomes. Tying investor returns to the borrowing entity's performance allows participating loans to provide essential capital and foster alignment between the goals of lenders and borrowers, ensuring that financial success is coupled with the achievement of mission-driven objectives. Participating loans operate at the intersection of flexibility and accountability. Borrowers benefit from a financial structure that adjusts to their performance, providing a safety net during economic uncertainty. This is achieved through a combination of fixed interest payments and a variable profit- or revenue-sharing component. The flexibility inherent in this model allows organisations to maintain cash flow stability without the rigidity of conventional debt instruments. A defining advantage of participating loans is their non-dilutive nature. Unlike equity financing, these loans do not require organisations to cede ownership or control. The preservation of autonomy is invaluable for mission-driven entities and social enterprises, enabling them to pursue strategic goals without external interference. Another feature of participating loans is their potential to incentivise collaboration. Lenders can become stakeholders in the borrower's success by linking returns to profitability or revenue. This alignment encourages lenders to offer advisory support and strategic guidance, fostering a partnership dynamic that prioritises financial and impact objectives. Participating loans in impact investing can incorporate innovative terms that directly link financial metrics with impact milestones. Such provisions underscore the adaptability of participating loans in promoting dual returns.

Core Structure and Operational Model

At the heart of participating loans lies their dual structure. The fixed component provides lenders a baseline return, typically through regular interest payments. This stability is complemented by a variable element, often linked to the borrower's revenue, profit, or other financial metrics. This performance-based mechanism introduces flexibility, allowing repayments to adjust according to the borrower's financial health. During periods of strong financial performance, the variable component ensures lenders benefit proportionally, while in downturns, it prevents undue strain on the borrower's cash flow. This adaptability is a defining feature, making participating loans particularly attractive to mission-driven organisations navigating uncertain economic landscapes. Participating loans foster a sense of partnership and alignment that benefits lenders and borrowers. The operational model of participating loans reflects this partnership-driven ethos.

The lifecycle begins with a rigorous origination process, where lenders identify organisations with strong growth potential and a clear commitment to achieving measurable impact. Due diligence is a critical step involving an in-depth analysis of the borrower's financial health, business model, and impact objectives. For impact investments, this evaluation also includes assessing the feasibility and credibility of the borrower's social or environmental goals. Capital is deployed once the loan terms are negotiated, including fixed repayment rates and the thresholds for the variable profit- or revenue-sharing component. Participating loans often serve as part of a blended finance strategy, complementing grants, equity, or senior debt. Borrowers typically use the funds to scale operations, invest in infrastructure, or implement initiatives that align with their impact mission. Throughout the loan term, lenders actively monitor the borrower's financial and impact performance. Regular reporting ensures transparency and allows for course corrections if necessary. The profit-sharing element further strengthens the lender-borrower relationship, as lenders are vested in the borrower's success. This alignment can lead to enhanced collaboration, with lenders offering support beyond capital, such as advisory services or market insights. Repayment in participating loans is structured to balance stability with opportunity. Borrowers meet their fixed obligations while fulfilling the variable component based on their financial performance. This model provides lenders a diversified return profile—a steady income stream from the fixed component and upside potential through the variable element. The flexible repayment mechanism also reduces the risk of default, as payments adjust to the borrower's actual performance rather than rigid schedules. At the end of the loan term, lenders evaluate the overall outcomes, including both financial returns and the achievement of impact goals. This phase often involves assessing how the borrower's initiatives have contributed to social or environmental progress. Insights from these evaluations inform future investments and highlight best practices for leveraging participating loans to achieve dual returns.

Investment Lifecycle of Participating Loans

Origination and Structuring: The investment lifecycle begins with lenders identifying potential borrowers who exhibit robust financial health, growth potential, and alignment with the lender's financial and impact objectives. During this phase, lenders comprehensively evaluate the borrower's financial performance, business model, and operational risks. For impact investments, additional assessments focus on the borrower's ability to achieve measurable social or environmental outcomes. The terms of the participating loan are tailored to the borrower's circumstances, including fixed repayment terms and variable performance-based components. Metrics for profit-sharing or revenue-sharing thresholds are clearly defined, along with key impact milestones where applicable.

Capital Deployment: Once the loan is structured, capital is disbursed to the borrower. Participating loans often serve as part of a broader blended finance strategy, complementing other funding sources such as grants, senior debt, or equity. Borrowers use the funds for initiatives such as scaling operations, implementing impact-driven programmes, or investing in infrastructure. The planned use of funds aligns with financial and impact objectives to ensure optimal outcomes.

Monitoring and Collaboration: Lenders actively monitor the borrower's financial and operational metrics to track performance against loan terms. Impact-focused borrowers must also report progress towards predefined social or environmental goals. The participatory nature of these loans encourages lenders to provide ongoing strategic guidance, technical assistance, and access to networks to support the borrower's success. This collaboration is particularly valuable in navigating challenges and capitalising on growth opportunities.

Repayment and Returns: Borrowers fulfil repayment obligations through fixed interest payments and variable profit- or revenue-sharing contributions. The adaptive structure of repayments, linked to financial performance, ensures that borrowers can maintain financial stability even during periods of volatility. For lenders, participating loans offer a diversified return structure, combining predictable fixed payments with the potential for higher returns through the variable component. This balance enhances the overall attractiveness of the investment.

Exit and Evaluation: After the loan period, lenders exit the investment. This may involve a final assessment of the borrower's financial health, repayment completion, and achievement of impact milestones. Lenders evaluate the dual outcomes of the loan—financial returns and the borrower's impact achievements. These evaluations often involve detailed analyses that inform future lending strategies and highlight the effectiveness of the participating loan model. Insights gained during the lifecycle are shared with the broader impact investing community, contributing to industry-wide learning and best practices for leveraging participating loans to drive social and environmental impact.

Challenges of Leveraging Participating Loans in Impact Investing

Participating loans represent a promising avenue for financing mission-driven organisations. However, inherent challenges must be addressed to maximise their effectiveness within the impact investing ecosystem. These challenges are primarily financial, operational, and strategic.

Complex Structuring and Negotiation: Participating loans require a detailed and transparent agreement that balances the interests of lenders and borrowers. Negotiating the fixed and variable components, along with the performance metrics, is complex and time-intensive. This complexity can deter both parties from engaging in such arrangements. For borrowers, particularly those in the early stages of growth or operating in underserved markets, the intricate terms can appear overly burdensome, leading them to opt for simpler financing options. From the lender's perspective, the extensive due diligence required increases transaction costs, reducing the attractiveness of this financing model. When these hurdles are not adequately addressed, the potential for participating loans to facilitate impactful projects is diminished, leaving many promising organisations underfunded.

Monitoring and Reporting Requirements: The dual repayment structure necessitates rigorous and ongoing monitoring of the borrower's financial and impact performance. Borrowers often lack the technical expertise or resources to implement robust systems for tracking and reporting complex financial and impact data. This can result in inconsistent or incomplete reporting, eroding trust between lenders and borrowers. For lenders, the need for continuous oversight adds administrative burdens and costs, which can limit their willingness to engage with smaller or less-established borrowers. These challenges create a systemic barrier to scaling participating loans, particularly in regions where impact investing is most needed, such as low-income or underserved markets.

Economic Volatility and Uncertainty: Borrowers in impact investing often operate in sectors or geographies exposed to significant economic volatility. While the flexible repayment mechanism of participating loans helps mitigate some risk, it cannot fully shield borrowers during sustained downturns. For example, organisations reliant on seasonal revenues or operating in politically unstable regions may face prolonged periods of reduced income, rendering even the fixed obligations untenable. This creates a ripple effect, where lenders experience reduced returns or defaults, leading to a reluctance to extend participating loans in high-risk contexts. Such hesitation undermines the ability of impact investing to channel funds into transformative sectors that desperately need capital.

Balancing Financial Returns and Impact: Achieving dual returns is a core objective of participating loans, but the inherent tension between financial and impact goals can lead to suboptimal outcomes. Lenders driven by profit motives may prioritise borrowers with higher financial returns over those with greater social or environmental impact potential. Conversely, borrowers may feel pressured to prioritise financial performance to meet repayment terms, potentially compromising their mission-driven objectives. This misalignment risks perpetuating a cycle where participating loans fail to achieve their intended purpose of fostering both financial sustainability and measurable impact, thus limiting their transformative potential.

Limited Market Adoption and Understanding: Participating loans remain underutilised in the impact investing market due to a lack of familiarity and understanding among stakeholders. Borrowers often view these instruments with scepticism, fearing hidden risks or excessive lender control. Lenders, on the other hand, may lack the expertise to structure and manage these loans effectively. The limited adoption stifles innovation and experimentation, which are critical for refining the model and building trust in its efficacy. This hesitancy also limits the pool of available capital, restricting the ability of participating loans to address large-scale social or environmental challenges.

Legal and Regulatory Barriers: In some jurisdictions, regulatory frameworks may not accommodate the hybrid structure of participating loans. Legal uncertainties around profit-sharing mechanisms, tax implications, and compliance requirements create significant hurdles. Borrowers may face additional legal costs or delays in securing approvals, while lenders may perceive these risks as insurmountable. This regulatory ambiguity disproportionately affects organisations in emerging markets, where impact investing could have the greatest effect. Without clear legal frameworks, participating loans risk remaining a niche instrument, unable to achieve widespread adoption or scalability.

Opportunities for Mitigating Challenges

Despite these challenges, participating loans offer significant opportunities for fostering innovation and collaboration in impact investing. This section discusses some strategies to overcome the barriers and unlock the transformative potential of participating loans.

Streamlined Structuring and Negotiation Processes: Creating a digital platform dedicated to standardising participating loan agreements can dramatically simplify the structuring process. Using AI-driven tools to assess borrower profiles and recommend tailored terms based on historical data and predictive analytics can reduce negotiation timelines. Additionally, introducing blockchain technology for smart contracts can automate compliance, ensuring transparency and fairness while minimising disputes. These innovations can serve as catalysts for wider adoption and replication across sectors.

Capacity-Building for Monitoring and Reporting: Adopting advanced monitoring technologies, such as IoT devices and blockchain, can revolutionise how borrowers track and report metrics. Real-time dashboards accessible to both borrowers and lenders can enhance transparency and trust. Establishing specialised impact reporting hubs, where borrowers can outsource complex tracking tasks to expert intermediaries, can alleviate the operational burden. Moreover, gamified training modules can help borrowers build internal capacity while making the learning process engaging and sustainable.

Risk Mitigation Mechanisms: Innovative insurance products tailored to participating loans, such as impact-linked guarantees, can protect lenders and borrowers against economic volatility. Collaborating with fintech companies to offer dynamic financial modelling and predictive analytics can enable lenders to anticipate risks and adapt terms proactively. Additionally, creating multi-stakeholder risk-sharing pools—drawing contributions from development banks, private investors, and philanthropic organisations—can distribute the financial burden, ensuring sustainability.

Alignment of Incentives: Developing a dual-return scoring system that integrates financial performance and impact metrics can guide borrowers and lenders in achieving balanced objectives. This system could be supported by AI-driven simulations that project the long-term outcomes of various financial-impact trade-offs. Encouraging equity-like rewards for lenders, such as preferential access to the borrower's future projects or markets, can also deepen their commitment to impact alignment.

Awareness and Education: Establishing virtual reality (VR) experiences that simulate the lifecycle of participating loans can demystify the instrument for stakeholders, fostering greater understanding and trust. A global knowledge-sharing platform that connects borrowers, lenders, and intermediaries to exchange case studies and best practices can accelerate market maturity. Additionally, launching certification programmes that validate lender and borrower proficiency in managing participating loans can build credibility and confidence.

Policy Advocacy and Regulatory Support: Engaging in forward-thinking policy dialogues with governments to introduce sandboxes for testing participating loan models can accelerate regulatory acceptance. Incentivising green or impact-linked bonds as complementary tools to participating loans can broaden their appeal. Moreover, advocating for tax credits or deductions tied to the achievement of impact milestones can make these instruments more attractive to a diverse range of investors.

Participating loans offer a unique and flexible financing option that aligns financial success with measurable social and environmental impact. Embracing innovative technologies, dynamic partnerships, and progressive policies, can enable the impact investing community to transform the challenges of structuring, monitoring, and adoption of participating loans into opportunities.

(v) **Venture Debt with Warrants**

Venture debt with warrants is a hybrid financing instrument commonly used to support high-growth enterprises, including those in the impact investing space. Like most hybrid instruments, it combines traditional loan structures with equity-linked features, offering lenders an opportunity to participate in the enterprise's future success. The debt component provides the borrower with immediate capital to fund operations or growth initiatives, while the attached warrants give lenders the right to purchase equity in the enterprise at a predetermined price within a specified timeframe. The dual structure allows enterprises to access growth capital while retaining ownership and control and provides lenders with both downside protection and upside potential if the enterprise thrives. Venture debt with warrants is similar to mezzanine financing, except whereas Mezzanine is primarily used by matured companies for the purpose of expansion, acquisitions, or leveraged buyouts (LBOs), venture debt with warrants focuses on early-stage, high-growth companies (startups) soon after venture capital funding. They are considered riskier than mezzanine financing. Nonetheless, it provides additional capital without significant equity dilution. In impact investing, venture debt with warrants is tailored to align with both financial and social or environmental objectives. The instrument's flexibility makes it particularly suitable for enterprises with predictable cash flows and measurable impact outcomes, such as those in renewable energy, healthcare, or education.

Core Structure and Operational Model

Venture debt with warrants is designed to balance risk and reward for both borrowers and investors. The loan component typically features a principal amount, fixed or floating interest rates, and a defined repayment schedule. Unlike traditional loans, it does not require immediate equity dilution, which is especially attractive for founders of mission-driven enterprises. The debt provides borrowers with non-dilutive growth capital, ensuring that they can fund operations or expansion while maintaining strategic autonomy. The warrants element represents the equity-linked feature of the instrument. Warrants give lenders the right, but not the obligation, to purchase a predefined number of shares in the borrower's enterprise at a fixed price (the exercise price) within a set period. This feature provides lenders with potential equity upside if the borrower's valuation increases. For impact enterprises, this equity participation can incentivise lenders to support long-term growth and sustainability initiatives.

The structure of venture debt with warrants is often customised to align with the borrower's financial and impact goals. Interest rates may be lower than traditional venture loans due to the added value of the warrants. Repayment schedules can be tailored to accommodate the enterprise's cash flow dynamics, ensuring financial sustainability. Warrants can also be structured to align with impact milestones, rewarding lenders with enhanced equity participation for achieving predefined social or environmental outcomes. Risk mitigation mechanisms are frequently employed to make the instrument more attractive to private investors. These may include first-loss guarantees, co-investment models, or subordinated debt tranches. Such mechanisms help de-risk the investment while broadening access to capital for impact-focused enterprises.

Investment Lifecycle of Venture Debt with Warrants

Origination and Structuring: The origination phase focuses on identifying suitable borrowers, often early-stage companies that have recently secured venture capital funding. These companies typically exhibit high growth potential and operate in sectors where scalability and innovation drive success. For impact investing, borrowers are often mission-driven enterprises aiming to deliver both financial returns and measurable social or environmental benefits. They are evaluated based on their market position, competitive edge, and the scalability of their business model. Critical considerations include the enterprise's alignment with defined impact objectives and track record of delivering meaningful outcomes. A rigorous due diligence process is conducted to assess the borrower's financial health, operational capacity, and potential for future growth. Key aspects include historical revenue trends, profitability, cash flow predictability, and the competence of the management team as well as the credibility of their mission and their ability to execute on both financial and impact goals. This phase helps lenders assess risk and align the loan structure with the borrower's unique needs. The debt component is designed to offer non-dilutive capital, featuring a principal amount, fixed or floating interest rates, and a repayment schedule tailored to the borrower's cash flow dynamics. Warrants are structured to provide lenders with the right to purchase equity at a predetermined exercise price within a specific timeframe. Warrants can be aligned with specific milestones, such as achieving sustainability goals or expanding services to underserved communities, creating incentives for long-term collaboration.

Capital Deployment: Once the loan structure is finalised, capital is disbursed to the borrower. This funding is typically used for strategic initiatives such as scaling operations, launching new products, expanding into new markets, or building critical infrastructure, inherent with measurable social or environmental outcome. Risk mitigation strategies are often incorporated whether first-loss guarantees (where a third party absorbs initial losses), co-investment models (where multiple investors share risk), or subordinated debt tranches (where lenders accept lower repayment priority). These mechanisms reduce the perceived risk for investors while ensuring that borrowers can access the capital they need.

Monitoring and Collaboration: During the loan term, lenders maintain regular oversight of the borrower's financial and operational performance. This involves tracking key metrics such as revenue growth, profit margins, and adherence to repayment schedules, while assessing progress towards predefined social or environmental milestones. Transparent reporting and consistent communication between lenders and borrowers help identify potential challenges early, enabling timely interventions to mitigate risks. The inclusion of warrants fosters a partnership dynamic, as lenders have a vested interest in the borrower's long-term success. Beyond providing capital, lenders often contribute strategic guidance, market insights, and access to networks that can enhance the borrower's growth trajectory, refine impact strategies, identify new opportunities for creating value. The supportive relationship enhances the likelihood of achieving both financial and impact objectives.

Repayment and Returns: Borrowers meet their repayment obligations through a combination of scheduled interest payments and principal repayments. The repayment structure is carefully calibrated to align with the borrower's cash flow, ensuring financial stability without compromising growth.

Exit and Evaluation: At the end of the loan period, lenders assess their financial and equity returns. For warrants that have appreciated in value, lenders may choose to exercise them or sell them for a profit. Borrowers benefit from completing the loan obligations without significant equity dilution. The exit phase also involves evaluating the borrower's success in achieving their impact objectives. This may include measuring key sustainability indicators outcomes. The evaluation provides insights into the effectiveness of venture debt with warrants in balancing financial returns with measurable impact. Insights gained during the lifecycle are documented and shared with the broader investment community. These lessons help refine future investment strategies and contribute to industry-wide innovation in structuring venture debt with warrants.

Challenges of Leveraging Venture Debt with Warrants in Impact Investing

Venture debt with warrants represents a powerful hybrid financing instrument for supporting high-growth enterprises, particularly in the impact investing space. Despite its potential, the model faces significant challenges that hinder its broader adoption and optimal use in driving financial and social/environmental returns. These challenges span financial, operational, and strategic dimensions.

Complex Structuring and Negotiation: Structuring venture debt with warrants is inherently complex, requiring a balance between the immediate financial needs of borrowers and the risk-reward expectations of lenders. Negotiating the terms of both the debt component and the equity-linked warrants often involves intricate legal, financial, and operational considerations. The challenges include setting appropriate interest rates, repayment schedules, and warrant exercise prices while aligning these terms with the borrower's growth trajectory and impact objectives. This complexity increases transaction costs and deters smaller enterprises from pursuing this instrument. The lack of standardised frameworks further complicates the process, leading to inefficiencies and prolonging deal closures.

Equity Dilution Concerns: While venture debt is designed to minimise immediate equity dilution, the warrants component introduces potential long-term dilution. For mission-driven enterprises, particularly those led by founders deeply invested in their social or environmental objectives, this can be a significant concern. The prospect of equity dilution may discourage such enterprises from engaging with the instrument, limiting its applicability in the impact investing space. Additionally, the dilution risk can create tension between borrowers and lenders, especially if the enterprise's valuation increases significantly, amplifying the equity cost.

Alignment of Financial and Impact Objectives: Balancing financial returns with measurable social or environmental impact remains a critical challenge. Lenders may prioritise financial metrics, such as revenue growth or profitability, over impact outcomes, potentially leading borrowers to compromise on their mission-driven goals. This misalignment can undermine the core purpose of impact investing, creating a dynamic where financial performance is achieved at the expense of sustainability or social equity.

High Risk for Early-Stage Enterprises: Venture debt with warrants is often deployed for early-stage, high-growth enterprises that exhibit significant potential but also face high levels of risk. These enterprises frequently operate in volatile sectors or geographies, exposing them to market, operational, and regulatory uncertainties. The high-risk profile can deter lenders from engaging or result in onerous terms that exacerbate the borrower's financial pressures. For borrowers, managing the dual obligations of debt repayment and achieving equity valuation targets can strain resources and divert attention from impact-driven initiatives.

Monitoring and Reporting Challenges: The hybrid nature of venture debt with warrants demands rigorous monitoring of financial performance and impact metrics. Borrowers often lack the infrastructure or expertise to produce consistent, high-quality reports, particularly when operating in resource-constrained environments. This can result in delays, inaccuracies, or a lack of transparency, eroding trust between lenders and borrowers. For lenders, the increased oversight requirements translate into higher administrative costs and potential inefficiencies, reducing the overall attractiveness of the instrument. Limited Market Awareness and Understanding: Venture debt with warrants remains a relatively niche instrument, particularly in the impact investing landscape. A lack of familiarity among borrowers, investors, and intermediaries limits its adoption. Many enterprises are unaware of how to structure and leverage this tool effectively, while lenders may lack the expertise to evaluate and execute these deals. This knowledge gap stifles innovation and reduces the flow of capital to high-potential, impact-driven enterprises.

Regulatory and Legal Barriers: The legal and regulatory environment for venture debt with warrants varies widely across jurisdictions. In many regions, the hybrid structure of this instrument does not fit neatly within existing financial regulations, creating compliance challenges. Borrowers and lenders may face legal ambiguities around the tax treatment of warrants, the enforceability of agreements, or restrictions on equity ownership. These regulatory hurdles disproportionately affect enterprises in emerging markets, where the need for impact-driven capital is most acute.

Despite these challenges, venture debt with warrants presents significant opportunities for fostering innovation and scalability in impact investing. Proactive and forward-thinking strategies can help overcome these barriers and unlock the full potential of this instrument.

Opportunities for Mitigating the Challenges

Modular Frameworks for Structuring and Negotiation: Rather than rigid standardisation, adopting modular frameworks offers a flexible yet efficient approach. These frameworks provide core templates for structuring debt and warrants while including optional customisation modules tailored to specific sectors, geographies, or borrower profiles. This modularity allows borrowers and lenders to adapt the terms to their unique circumstances without starting from scratch. For example, templates could include baseline financial terms, while add-on modules could address region-specific legal considerations or sector-specific impact metrics. Digital platforms powered by AI could further streamline the process by offering tailored recommendations based on the borrower's data and the lender's objectives, ensuring efficiency without sacrificing nuance.

Innovative Equity Management Solutions: To address equity dilution concerns, lenders can offer flexible warrant structures, such as capped or performance-linked warrants, that limit potential dilution while aligning with impact milestones. For example, capped warrants could ensure borrowers retain more ownership even if their valuation increases dramatically. Introducing equity buyback options allows borrowers to regain equity ownership under favourable conditions, preserving their long-term autonomy and mission alignment. These solutions reduce friction and build trust between borrowers and lenders.

Integrated Impact and Financial Metrics: Developing customised, integrated dashboards that combine financial performance with real-time impact tracking can significantly improve alignment between borrowers and lenders. These dashboards allow both parties to monitor progress effectively and anticipate potential misalignments through visual, data-driven insights. Predictive analytics powered by AI can further enhance this by forecasting the long-term outcomes of decisions, ensuring a balanced focus on dual returns. Lenders can also tie bonuses or improved terms to measurable impact achievements, reinforcing a commitment to shared goals.

Risk Mitigation Mechanisms and Collaborative Models: Innovative risk-sharing mechanisms such as blended finance structures, which include first-loss guarantees or subordinated tranches, can lower the perceived risk for private investors. Collaborative models, where multiple stakeholders co-invest, can also enhance the scalability of venture debt with warrants. For instance, partnerships with Development Finance Institutions (DFIs) or philanthropic organisations can de-risk investments while improving access to capital for impact enterprises. These models also foster shared expertise, enabling more robust financial and impact outcomes.

Capacity-Building and Knowledge-Sharing: Investing in capacity-building programmes for borrowers ensures they are better equipped to manage complex hybrid instruments. Offering hands-on training, case studies, and financial literacy workshops can empower enterprises to understand better and navigate venture debt with warrants. Establishing an open-access knowledge hub featuring real-world examples, templates, and technical guidance can demystify the instrument for both borrowers and lenders. Additionally, organising global conferences and industry forums fosters cross-sector collaboration and accelerates adoption.

Policy Advocacy and Regulatory Innovation: Proactive engagement with policymakers can address regulatory barriers by advocating for the integration of venture debt with warrants into broader national and regional development strategies. For instance, tax incentives could be offered to lenders for achieving impact milestones, while streamlined approval processes could simplify compliance for borrowers. Establishing regulatory sandboxes can allow lenders and borrowers to test innovative financial structures under monitored conditions, enabling gradual acceptance of these hybrid instruments.

Dynamic Partnership Models: Encouraging long-term partnerships between borrowers, lenders, and intermediaries can create a collaborative ecosystem that maximises the potential of venture debt with warrants. These partnerships could involve co-designing financing structures, sharing market insights, and aligning long-term strategies to prioritise both financial and impact outcomes. Such models foster a spirit of trust and innovation, enabling enterprises to scale effectively while meeting their dual objectives.

Venture debt with warrants offers a dynamic and versatile financing solution tailored to the needs of high-growth, impact-driven enterprises. Modular frameworks, innovative risk-sharing mechanisms, and integrated metrics can significantly enhance the ability of impact investing to drive measurable change. With a strong focus on collaboration, capacity-building, and policy support, venture debt with warrants can enable enterprises to achieve transformative outcomes in both financial and social dimensions.

(vi) **Profit-Participating Notes (PPNs)**

Profit-Participating Notes (PPNs) are another hybrid financial instruments that combine features of debt and equity to provide capital to enterprises while aligning investor returns with the borrower's financial performance. In impact investing, PPNs are particularly appealing as they enable investors to participate in an enterprise's success without taking an ownership stake. The instrument provides a loan with a fixed repayment obligation, often supplemented by a variable component tied to the enterprise's profitability or revenues. PPNs balance risk and reward effectively, supporting mission-driven enterprises in scaling their operations while delivering measurable social or environmental outcomes.

Core Structure and Operational Model

PPNs are debt instruments with a fixed interest rate, ensuring baseline returns for investors. However, they also include a profit-sharing mechanism, whereby borrowers agree to pay an additional return based on their financial performance. This variable component is often calculated as a percentage of the borrower's profits or revenues, aligning investor incentives with enterprise success. The instruments are well-suited for enterprises operating in sectors with predictable cash flows and clear profitability metrics. Borrowers benefit from the non-dilutive nature of PPNs, as the instrument does not require them to cede ownership or control. This is particularly valuable for mission-driven organisations prioritising autonomy while accessing growth capital. In impact investing, PPNs may include terms linked to the achievement of predefined social or environmental milestones. For example, an enterprise might agree to reduce the profit-sharing percentage if it achieves specific impact outcomes. These impact-linked adjustments ensure that the instrument remains aligned with the borrower's mission. Risk mitigation mechanisms, such as subordinated tranches or blended finance models, are often employed to make PPNs more attractive to private investors. These safeguards provide additional layers of security, encouraging investment in high-impact but potentially risky ventures.

Investment Lifecycle

Origination and Structuring: Lenders seek enterprises with strong growth potential, stable cash flow patterns, and a clear alignment with financial and impact objectives. Typical borrowers include mission-driven organisations in sectors like renewable energy, sustainable agriculture, healthcare, and education. Such borrowers often focused on delivering measurable sustainable outcomes, in the case of impact investing. A comprehensive evaluation of their financial health, operational model, and impact potential is conducted as well as the assessment of the borrower's impact measurement frameworks and track record of achieving social or environmental goals. The structuring phase is pivotal, as it determines the balance between risk and reward for both parties. During this phase, the fixed component of the PPN, such as setting the interest rate, repayment schedule, and loan amount is finalised to ensure a baseline return for investors. The profit-sharing mechanism is designed tying additional returns to the borrower's financial performance. For instance, a percentage of profits or revenues is agreed upon, creating a variable return that incentivises lenders to support the borrower's success. Debt warrants are incorporated, giving lenders the right to purchase equity at a predefined price within a specific timeframe. The terms of the warrants, such as the exercise price and duration, are calibrated to reflect the borrower's growth trajectory and risk profile. Impact-linked adjustments are negotiated, allowing for reduced profit-sharing obligations or other benefits if the borrower achieves specific social or environmental milestones. These features reinforce the alignment of financial goals with the borrower's mission. Risk mitigation measures, such as subordinated tranches or co-investment frameworks, are embedded to protect investors while ensuring accessible terms for the borrower.

Capital Deployment: Capital is deployed to the borrower for specific purposes, such as scaling operations, developing new products, or expanding market presence. For impact enterprises, funds may also support initiatives like building renewable energy infrastructure or improving healthcare access in underserved regions. Risk mitigation strategies such as first-loss guarantees, subordinated tranches, or blended finance models are employed to de-risk the investment, making it more appealing to a broad range of investors while ensuring capital availability for high-impact ventures.

Monitoring and Collaboration: Lenders monitor the borrower's financial and operational metrics assessing progress towards predefined social or environmental goals, ensuring alignment with mission-driven objectives. Regular reporting ensures transparency and allows lenders to track adherence to loan agreements. The profit-sharing component fosters a collaborative relationship between lenders and borrowers, with lenders often providing strategic guidance, technical assistance, or market insights to enhance the borrower's performance and impact delivery.

Repayment and Returns: Borrowers meet their repayment obligations through a combination of fixed interest payments and profit-sharing contributions based on financial performance. During this phase, lenders receive predictable returns through the fixed component while benefiting from the borrower's financial growth via the profit-sharing arrangement. The repayment structure is designed to align with the borrower's cash flow, ensuring that repayments are proportional to their financial performance. This reduces financial strain during periods of lower profitability while allowing borrowers to prioritise long-term growth.

Exit and Evaluation: At the end of the loan period, lenders assess their financial returns and decide whether to exercise their warrants or exit the investment through other means, such as selling their equity stake. Lenders evaluate the borrower's financial performance and the achievement of a predefined impact objective. This phase provides insights into the effectiveness of the PPN in delivering dual returns. Lessons learned during the lifecycle are documented and shared within the investment community. These insights help refine future PPN structures and contribute to adopting innovative financing instruments in impact investing.

Challenges of Leveraging Profit-Participating Notes (PPNs) in Impact Investing

The adoption of PPN in impact investing is hindered by several challenges spanning financial, operational, and strategic dimensions.

Complex Structuring and Negotiation: Structuring PPNs demands a high level of expertise to balance the fixed and variable components effectively. Determining the profit-sharing percentage, repayment schedules, and the mechanisms for tying returns to financial performance often requires extensive legal and financial consultation. For impact enterprises, additional complexity arises when integrating impact-linked milestones, which must be measurable, verifiable, and agreed upon by both parties. The absence of standardised frameworks for structuring PPNs exacerbates these challenges, leading to inconsistencies and delays. Smaller organisations often lack the resources to navigate these complexities, making PPNs an inaccessible financing option for many mission-driven enterprises. This issue creates a systemic bias towards larger, well-established borrowers, leaving smaller, high-potential ventures underfunded.

Alignment of Financial and Impact Objectives: Balancing the dual objectives of financial returns and impact outcomes remains a significant challenge. Investors may prioritise financial profitability, particularly when their returns are tied to revenue or profit growth. This focus can inadvertently pressure borrowers to prioritise financial performance over achieving their social or environmental goals. Conversely, during periods of financial strain, borrowers may deprioritise impact objectives to meet their financial obligations, diluting the transformative potential of the investment. Such misalignments risk creating a transactional relationship rather than a partnership, undermining the core ethos of impact investing.

High Dependence on Predictable Cash Flows: PPNs are most effective for enterprises with stable, predictable cash flows, which are often found in sectors like renewable energy or healthcare. However, many impact-driven enterprises operate in emerging markets or volatile sectors where revenue predictability is low. This dependency on cash flow stability limits the applicability of PPNs, as lenders may perceive these ventures as too risky. Early-stage enterprises, which often have uncertain revenue streams but high potential for social impact, struggle to qualify for PPN financing. This creates a financing gap that disproportionately affects smaller and newer impact ventures operating in challenging markets.

Monitoring and Reporting Burden: The dual repayment structure of PPNs necessitates robust monitoring systems to track both financial performance and impact milestones. Borrowers often lack the infrastructure, expertise, or resources to produce consistent, high-quality reports, particularly in resource-constrained settings. This results in delays, inaccuracies, or incomplete reporting, eroding trust between borrowers and lenders. For lenders, the heightened oversight requirements translate into increased administrative costs, reducing the overall appeal of PPNs. Additionally, in regions with weak institutional frameworks, the lack of reliable data can further complicate the monitoring process, impeding the scalability of PPNs.

Limited Market Awareness and Understanding: Despite their potential, PPNs remain a niche instrument in the impact investing landscape. Borrowers and investors often lack a clear understanding of how PPNs function, their benefits, and how to structure them effectively. This knowledge gap discourages enterprises from exploring PPNs as a viable financing option and deters investors unfamiliar with the instrument's mechanics. Intermediaries, who play a crucial role in bridging this gap, may also lack the technical expertise to guide stakeholders effectively. This limited awareness stifles innovation and prevents the broader adoption of PPNs, reducing their impact potential.

Regulatory and Legal Constraints: The hybrid nature of PPNs presents significant regulatory challenges. Legal ambiguities surrounding profit-sharing mechanisms, tax treatments, and compliance requirements vary widely across jurisdictions. These ambiguities increase the cost and complexity of implementing PPNs, particularly in emerging markets where financial regulations may be underdeveloped or inconsistent. Borrowers face additional hurdles, such as navigating unclear legal frameworks or incurring high compliance costs, while lenders may perceive these challenges as insurmountable risks. This regulatory uncertainty further limits the scalability and adoption of PPNs in markets where they could have the greatest impact.

Opportunities for Mitigating Challenges

Despite these challenges, PPNs hold significant promise for impact investing. This section discusses some strategies for overcoming these barriers and unlocking the full potential of PPNs to drive financial and impact returns.

Flexible and Adaptive Structuring Models: Adopting flexible structuring models tailored to the borrower's circumstances can simplify the negotiation process. Modular frameworks similar to those used in other hybrid instruments can include baseline templates for fixed and profit-sharing components while allowing for adjustments based on sector-specific or regional needs. Digital platforms leveraging AI can assist in customising terms by analysing borrower profiles, projected cash flows, and impact metrics, ensuring efficiency without sacrificing nuance.

Integrated Impact and Financial Metrics Dashboards: Developing integrated dashboards that combine financial performance with impact metrics can help align stakeholder objectives. Real-time monitoring tools can improve transparency and trust by enabling both lenders and borrowers to track progress against predefined goals. AI-driven analytics can predict potential misalignments and provide early warnings, allowing stakeholders to make proactive adjustments. Such tools also reduce the administrative burden of manual reporting.

Risk Mitigation Mechanisms and Collaborative Financing Models: Blended finance approaches, including first-loss guarantees and co-investment models, can de-risk PPNs for private investors. Establishing sector-specific risk-sharing pools supported by development banks or philanthropic organisations can provide additional layers of security. Collaborative financing structures that involve multiple stakeholders can spread risks, attract diverse capital sources, and enhance the scalability of PPNs.

Capacity-Building and Education: Investing in borrower education programmes can help enterprises better understand and leverage PPNs. Workshops, case studies, and mentorship programmes can equip mission-driven organisations with the skills needed to navigate complex financing structures. Creating an open-access repository of resources, including templates, guides, and best practices, can demystify PPNs for a broader audience.

Regulatory Engagement and Advocacy: Engaging policymakers to create clear regulatory frameworks for PPNs can address legal ambiguities and reduce compliance costs. Advocacy efforts should focus on highlighting the role of PPNs in achieving Sustainable Development Goals, encouraging governments to provide tax incentives or other supportive measures. Introducing regulatory sandboxes can enable experimentation and innovation within a controlled environment.

Dynamic Partnership Models: Encouraging partnerships between borrowers, lenders, and intermediaries can foster a collaborative ecosystem that supports the effective implementation of PPNs. Such partnerships can facilitate knowledge-sharing, streamline structuring processes, and ensure alignment of financial and impact objectives. By leveraging the expertise of multiple stakeholders, dynamic partnerships can enhance the success rate and scalability of PPNs.

Profit-Participating Notes (PPNs) offer a unique and flexible financing option for impact investing, aligning financial returns with enterprise success while preserving borrower autonomy. Addressing the challenges of structuring, monitoring, and adoption requires innovative solutions such as modular frameworks, risk-sharing mechanisms, and integrated monitoring tools. By fostering education, collaboration, and regulatory innovation, PPNs can unlock new possibilities for scaling impact-driven enterprises, driving transformative outcomes in both financial and social dimensions.

Grant and Donation-Based Models

Grant and donation-based models are foundational components of the impact investing landscape, serving as a critical source of non-repayable funding for ventures that address pressing social and environmental challenges. These models are particularly significant for early-stage projects or initiatives deemed too risky or unprofitable for traditional capital markets. Providing seed funding and operational support enables grant and donation-based models to foster innovative solutions, often laying the groundwork for future investments and scaling opportunities within the broader impact investing ecosystem. Grants are typically issued by governments, foundations, multilateral organisations, or international development agencies. These funds are allocated to achieve specific objectives, such as advancing educational initiatives, mitigating climate change, or addressing public health crises. Grants often come with predefined conditions that ensure the funding is directed towards achieving measurable social or environmental outcomes. In most cases, government grants tend to focus on fostering renewable energy projects, while foundation grants support research and development in underserved communities. Conversely, donations are usually voluntary contributions made by individuals, philanthropic organisations, or corporate entities. Unlike grants, donations are often less restrictive, allowing recipients to allocate resources according to their immediate needs. The flexibility is vital for grassroots organisations or small-scale projects that may not fit into rigid grant frameworks but still require financial support to sustain their operations and mission.

The importance of these models extends beyond their immediate financial contributions. They play a pivotal role in blended finance, to de-risk projects and attract commercial investments. Grant- and donation-based funding creates a more favourable environment for private-sector participation by covering initial costs and mitigating risks. An emerging trend in this domain is the increasing emphasis on accountability and measurable outcomes. Donors and grant-making organisations are increasingly adopting impact measurement frameworks to assess the effectiveness of their funding. This shift aligns with broader trends in the impact investing space, where data-driven approaches and performance metrics are used to evaluate the social and environmental returns on investment. As the field continues to evolve, the integration of impact measurement practices and innovative funding mechanisms is likely to enhance the efficiency and effectiveness of these foundational models. Program-Related Investments (PRIs) and traditional grants are notable examples of mechanisms that blend philanthropic objectives with impact-oriented outcomes.

(i) **Traditional Grants in Impact Investing**

Traditional grants are non-repayable financial contributions awarded by entities such as governments, philanthropic organisations, foundations, and international development agencies to support initiatives addressing social, environmental, or cultural challenges. These grants have a long-standing history as a key mechanism for advancing public welfare, enabling stakeholders to mobilise resources to tackle issues ranging from poverty alleviation and education to healthcare and environmental conservation. The concept of grants has its roots in philanthropy and public service, where early models were designed to support charitable causes and provide relief to underserved communities. In the contemporary context, traditional grants serve as a structured approach to funding that emphasises accountability, targeted outcomes, and alignment with donor priorities. Unlike loans or investments, grants do not require repayment, which makes them particularly suitable for projects that may lack immediate financial returns but promise significant societal benefits. These funds are typically allocated to non-profit organisations, academic institutions, or social enterprises to enable the execution of specific projects or programmes that align with the donor's mission and strategic goals. A defining characteristic of traditional grants is their targeted nature. Grantmakers specify the intended use of the funds, often with detailed requirements for how the money should be spent and what outcomes should be achieved. This ensures that resources are utilised effectively to create measurable impact. For example, a foundation focused on improving educational outcomes might provide grants to schools or NGOs to develop literacy programmes in underserved communities. Traditional grants catalyse innovation and capacity-building within their respective domains by addressing funding gaps in areas where private investment is scarce. The evolution of traditional grants has also seen the integration of performance metrics and evaluation frameworks, allowing grantmakers to assess the effectiveness of funded initiatives. This approach fosters transparency and enables the continuous improvement of grant programmes to maximise their social or environmental impact. In essence, traditional grants remain a cornerstone of philanthropic and public-sector funding, bridging the gap between societal needs and sustainable solutions.

Core Structure and Operational Model

Traditional grants are characterised by their systematic structure, designed to ensure alignment between grantmakers and the recipients of funding. At the heart of this structure is the role of grant-making organisations, which include governments, foundations, and international agencies. Each of these entities brings its unique priorities and resources to the table. For example, governments often focus on public welfare programmes such as healthcare and infrastructure, while foundations might prioritise specific causes like climate action or social equity. International agencies, on the other hand, address cross-border challenges such as poverty and environmental conservation, leveraging their global reach and influence. Eligibility criteria are fundamental to the grant-making process. These criteria determine which organisations or projects qualify for funding. They are carefully crafted to align with the strategic objectives of the grantmaker, ensuring that the resources are directed towards initiatives with the potential for meaningful impact. For instance, a grant focused on biodiversity might prioritise applications from organisations working in ecologically sensitive areas, whereas a grant targeting educational disparities might focus on regions with low literacy rates. Another critical component of traditional grants is the framework for accountability and transparency. Grantmakers establish clear guidelines for how the funds should be used, including requirements for progress reports and financial audits. This ensures that the grant money is spent as intended and provides a mechanism for evaluating the effectiveness of funded projects. Additionally, grantmakers often engage with grantees throughout the project's lifecycle, offering guidance and support to address challenges and maximise outcomes.

The operational model of traditional grants also emphasises the importance of fostering partnerships. Many grant programmes encourage stakeholder collaboration, including local communities, government agencies, and private-sector actors. These partnerships can enhance the sustainability of funded projects by pooling resources, sharing expertise, and building local capacity. For example, a renewable energy grant might involve partnerships between local governments, non-profits, and private companies to ensure the successful implementation and long-term maintenance of solar power systems. Maintaining a robust structure that prioritises accountability, transparency, and collaboration enables traditional grants to create an enabling environment for addressing complex social and environmental challenges. This structured approach allows grantmakers to allocate resources effectively while empowering grantees to deliver impactful and sustainable solutions.

Investment Lifecycle of Traditional Grants

The lifecycle of traditional grants encompasses several phases, each critical to ensuring their effectiveness and alignment with intended goals:

Initiation and Planning: The grant-making process begins with the identification of a specific issue or area of focus. This phase involves extensive research and consultations with stakeholders to determine pressing needs and define strategic priorities. For instance, a government agency targeting food insecurity may conduct studies to identify regions most affected by malnutrition and assess potential intervention strategies. Planning during this phase includes setting clear objectives, determining the size of the grant pool, and outlining the expected outcomes.

Proposal Solicitation: Once the focus areas are identified, grantmakers issue a formal call for proposals. This document serves as a comprehensive guide for potential applicants, detailing eligibility criteria, application procedures, timelines, and evaluation metrics. For example, a foundation supporting renewable energy initiatives might specify requirements such as geographic focus, project scalability, and alignment with sustainability goals. The proposal solicitation phase ensures that applicants have a clear understanding of what is expected and fosters transparency in the process.

Proposal Submission and Evaluation: Eligible organisations submit detailed proposals that outline their project goals, methodologies, anticipated impacts, and budgets. These submissions are rigorously evaluated by expert panels or review committees against predefined criteria. The evaluation process may include scoring mechanisms to assess feasibility, innovation, alignment with grant objectives, and potential for measurable outcomes. For instance, a proposal for a public health grant might be evaluated based on its ability to improve vaccination rates or reduce disease incidence in target populations.

Grant Award and Agreement: Successful applicants are notified and formal agreements are established to define the terms and conditions of the funding. These agreements typically include key elements such as disbursement schedules, reporting requirements, and milestones for progress. Grantmakers often release funds in phases, contingent upon the achievement of specified objectives or benchmarks. This phased disbursement ensures accountability and allows for course corrections if necessary.

Implementation: During the implementation phase, grantees execute their projects as outlined in their proposals. This phase involves mobilising resources, hiring personnel, and engaging with stakeholders. Regular communication with the grantmaker is essential to ensure alignment with project objectives. Challenges encountered during implementation, such as delays or unforeseen expenses, are addressed collaboratively to minimise disruptions. For instance, a non-profit implementing an educational programme might adjust timelines to accommodate local school schedules.

Monitoring and Reporting: Continuous monitoring is a critical component of the grant lifecycle. Grantees are required to submit periodic reports detailing their progress, financial expenditures, and preliminary outcomes. These reports are reviewed by the grantmaker to ensure compliance with the funding agreement and to assess the project's trajectory. Some grants incorporate third-party evaluations or site visits to validate reported data and gather qualitative insights. For example, a climate change mitigation grant might involve on-site inspections of renewable energy installations to verify their operational status.

Completion and Evaluation: Upon project completion, a comprehensive final report is submitted, summarising achievements, challenges, and lessons learned. This phase often includes an impact assessment to evaluate whether the project met its intended objectives and delivered measurable benefits. For instance, a grant aimed at improving literacy rates might use standardised test scores and enrolment figures to quantify its impact. Insights from this phase inform the grantmaker's future funding strategies and help refine programme designs.

Legacy and Sustainability: The final phase focuses on ensuring the longevity of the project's impact. Grantmakers often prioritise capacity-building initiatives that enable local communities or organisations to sustain the outcomes independently. For example, a healthcare grant might include training programmes for local medical staff to ensure continued service delivery after the funding period ends. Additionally, grantees may establish partnerships or seek follow-on funding to scale successful projects. The emphasis on sustainability ensures that the benefits of the grant extend beyond its formal duration.

Challenges of Leveraging Traditional Grants in Impact Investing

Traditional grants are a cornerstone of funding for initiatives addressing social, environmental, and cultural challenges. Despite their critical role in advancing public welfare, their utilisation in impact investing faces several challenges that span financial, operational, and strategic dimensions.

Limited Financial Sustainability: Traditional grants provide non-repayable funding, which, while beneficial for grantees in the short term, often fosters dependency rather than resilience. Organisations heavily reliant on grants may neglect the development of diversified revenue streams, leaving them vulnerable to funding discontinuation or shifts in donor priorities. This lack of financial sustainability limits the long-term scalability of grant-funded initiatives and inhibits innovation. Furthermore, grants frequently fail to incentivise grantees to explore self-sustaining models, such as integrating revenue-generating components or forming partnerships with private-sector actors. As a result, even highly impactful projects may falter once grant funding ceases, perpetuating a cycle of dependency rather than fostering independence and scalability.

Rigid Eligibility and Compliance Requirements: The eligibility criteria and compliance requirements tied to traditional grants are often rigid and exclusionary, favouring well-established organisations with significant administrative capacities. Smaller organisations, grassroots initiatives, and innovative startups—which often operate at the forefront of addressing systemic social and environmental issues—frequently lack the resources to meet these demands. Additionally, the bureaucratic nature of grant application processes diverts organisational focus from core programmatic activities, increasing administrative overhead and detracting from impact delivery. The prioritisation of compliance over flexibility can stifle creativity and deter organisations from pursuing transformative solutions that may not align neatly with predefined criteria or timelines.

Misalignment of Objectives: A significant challenge in leveraging traditional grants is the frequent misalignment between the objectives of grantmakers and grantees. Grantmakers typically operate within strategic frameworks that emphasise measurable outcomes tied to their specific mandates. While these outcomes may align with broader societal goals, they often fail to address the nuanced, context-specific needs of communities served by grantees. For example, a donor's focus on short-term deliverables may undermine grantees' efforts to achieve systemic, long-term change. This misalignment fosters a transactional relationship between donors and recipients, reducing the potential for collaborative innovation and limiting the transformative power of grant-funded interventions.

Fragmented Impact Measurement: The absence of standardised impact measurement frameworks creates significant inefficiencies in grant-funded initiatives. Grantees often face the burden of developing bespoke reporting mechanisms tailored to individual donors' preferences. This lack of consistency not only consumes resources but also impedes the aggregation of impact data across projects and organisations, limiting opportunities to derive insights or demonstrate sector-wide progress. Furthermore, qualitative aspects of impact, such as community empowerment or behavioural change, are frequently overlooked in favour of quantitative metrics that align with donor expectations. This narrow focus on measurable outputs can obscure the holistic value of grant-funded initiatives.

Inefficient Allocation of Resources: The competitive nature of traditional grants often results in an inefficient allocation of resources. Organisations invest substantial time and effort in crafting proposals, many of which are unsuccessful. This inefficiency favours entities with dedicated grant-writing teams over smaller or less-resourced organisations that may offer innovative solutions but lack administrative expertise. Additionally, grant funding is often concentrated on high-visibility issues or geographic regions, leaving less prominent but equally pressing challenges underfunded. This fragmented approach to resource allocation undermines the systemic impact of grant-making efforts.

Lack of Focus on Sustainability and Exit Strategies: Traditional grants are frequently structured to address immediate needs without adequately planning for the sustainability of outcomes. Many grants lack built-in mechanisms to ensure the continuation of project benefits once funding ends. This short-term approach can result in the collapse of initiatives, particularly in resource-constrained environments where grantees lack the capacity to sustain operations independently. The absence of clear exit strategies or transition plans—such as fostering partnerships, capacity-building, or integrating projects into local governance systems—further limits the long-term impact of grant-funded programmes.

Opportunities for Mitigating Challenges

Despite these challenges, traditional grants remain a vital tool in impact investing. This section discusses the innovative strategies that can foster collaboration and mitigate the barriers to enhance the effectiveness and sustainability of grant-funded initiatives.

Promoting Financial Sustainability through Innovative Models: Grantmakers should incorporate financial sustainability into grant design by incentivising grantees to adopt hybrid funding models. This could include combining grants with outcome-based funding mechanisms, such as social impact bonds or pay-for-success contracts, that encourage measurable, long-term outcomes. Additionally, grants can be paired with technical assistance programmes to help grantees build capacity for revenue diversification, such as launching fee-for-service models or social enterprises that generate sustainable income.

Rethinking Eligibility and Compliance Processes: Grantmakers can rethink traditional compliance processes by leveraging technology to simplify and streamline grant applications and reporting. AI-powered platforms can automate parts of the application process, allowing organisations to focus more on delivering impact. Furthermore, adopting flexible eligibility criteria that consider alternative credentials—such as track records of community engagement or innovative pilot programmes—can make grants accessible to smaller, high-potential initiatives.

Encouraging Co-Creation and Alignment of Objectives: To bridge the gap between donor priorities and community needs, grantmakers should engage grantees and beneficiaries in co-creating funding frameworks. Participatory grant-making, where stakeholders collaboratively design goals, metrics, and timelines, can ensure that funding aligns with both strategic objectives and local realities. Grantmakers can also allow adaptive programming, giving grantees the flexibility to pivot as circumstances evolve, ensuring relevance and responsiveness.

Standardising and Innovating Impact Measurement: Establishing globally recognised, sector-specific impact measurement frameworks would enable consistent reporting and comparative analysis. Digital dashboards with real-time data collection and visualisation capabilities can enhance transparency while reducing the reporting burden on grantees. Innovative metrics that integrate both quantitative outcomes (e.g., jobs created) and qualitative insights (e.g., community empowerment) can provide a more comprehensive view of impact.

Strategic Resource Allocation and Collaboration: Grantmakers can optimise resource allocation through pooled funding initiatives where multiple donors collaborate to address systemic issues. For example, creating regional or sector-focused grant pools can ensure funding reaches underserved areas or emerging challenges. Additionally, introducing a multi-stage funding approach—starting with smaller seed grants followed by larger scale-up grants for successful initiatives—can maximise the reach and effectiveness of resources.

Embedding Sustainability and Exit Strategies: Embedding sustainability requirements within grant agreements ensures grantees plan for long-term impact. This could involve tying funding to capacity-building milestones, such as training programmes for local staff or forming partnerships with private-sector entities to sustain projects post-grant. Grants can also be structured to include gradual funding reductions, preparing grantees for independence. Moreover, creating mechanisms for legacy impact—such as endowment funds or revolving grant pools—can ensure the longevity of successful projects.

Fostering Partnerships and Ecosystem Building: Grantmakers should actively foster ecosystems of collaboration by connecting grantees with private investors, government agencies, and other donors. Facilitating Public–Private Partnerships can amplify impact, while knowledge-sharing platforms can enable grantees to exchange best practices and replicate successful models. By building networks of aligned stakeholders, grantmakers can create environments where impact-driven initiatives thrive collectively.

Traditional grants provide critical support for initiatives addressing pressing social and environmental challenges. However, to maximise their potential, addressing challenges such as financial sustainability, alignment of objectives, and fragmented impact measurement is essential. Adopting innovative models, streamlining processes, embedding sustainability, and fostering collaboration enable grantmakers to enhance the effectiveness and scalability of grant-funded initiatives. Traditional grants can continue to catalyse transformative change through forward-thinking strategies, bridging the gap between societal needs and sustainable solutions.

(ii) **Program-Related Investments (PRIs) in Impact Investing**

Program-Related Investments (PRIs) are financial instruments used by philanthropic foundations to advance their charitable missions while potentially earning a financial return. Unlike traditional grants, PRIs are typically structured as investments, such as loans, equity stakes, or guarantees, that align with the foundation's mission-driven objectives. These investments are specifically designed to address societal challenges by funding ventures that may struggle to secure financing through conventional means. PRIs bridge the gap between philanthropic funding and traditional investments, blending social impact with financial sustainability. PRIs emerged as an innovative response to the limitations of traditional grant-making models. While grants remain a cornerstone of philanthropy, they often face challenges such as finite capital and lack of scalability. PRIs address these limitations by enabling foundations to recycle their capital—once loans are repaid or equity stakes are liquidated, the funds can be reinvested into new projects, ensuring a sustained impact. This approach has its roots in the broader evolution of impact investing, where financial tools are increasingly leveraged to achieve measurable social outcomes.

Traditional grants and PRIs share the same underlying mission: advancing societal goals. However, PRIs offer a dynamic mechanism to complement grants by supporting ventures that have the potential to become self-sustaining or scalable. For instance, a grant might fund the initial research and development of a clean water technology, while a PRI could support the commercialisation and scaling of that technology. This symbiotic relationship highlights how PRIs enrich the philanthropic toolkit by offering solutions that combine charitable intent with financial discipline. PRIs are unique because they are not driven primarily by the pursuit of profit. Instead, they prioritise advancing social, environmental, or cultural objectives. For example, a foundation might provide a low-interest loan to a renewable energy cooperative in a low-income community, enabling the development of sustainable energy solutions while ensuring the foundation's capital is eventually returned. PRIs are particularly useful for supporting innovative, high-risk ventures that are often overlooked by traditional investors.

In legal terms, PRIs must comply with specific regulations to qualify as a charitable activity. In the United States, for instance, the Internal Revenue Service (IRS) requires PRIs to significantly further the foundation's charitable purposes and not have profit generation as their primary goal. This distinction ensures that PRIs remain a tool for impact rather than a purely financial investment strategy.

Core Structure and Operational Model

Program-Related Investments operate within a framework that prioritises social impact while ensuring financial prudence. These investments are highly adaptable, allowing foundations to address a wide range of societal issues through tailored financial instruments. The structural elements of PRIs include loans, equity investments, and guarantees, each offering unique advantages for different contexts. Loans, often provided at below-market interest rates, are among the most common forms of PRIs. Foundations extend these loans to mission-aligned organisations, such as social enterprises or non-profits, to fund operations, scale initiatives, or address specific challenges. The terms of these loans are designed to reduce the financial burden on recipients, ensuring that repayment schedules align with their cash flow constraints. Equity investments enable foundations to take ownership stakes in mission-driven enterprises. This approach not only provides capital for growth but also allows foundations to participate in governance and influence strategic decisions. For instance, a foundation might invest in a startup developing affordable healthcare technologies, helping to guide its direction while fostering innovation. Guarantees are another pivotal component of PRIs, serving as a risk mitigation tool. By offering guarantees, foundations encourage third-party investors, such as banks or impact funds, to provide additional financing to high-impact projects. This de-risking mechanism is particularly valuable for ventures operating in underserved markets or addressing complex social challenges.

The operational model of PRIs revolves around alignment with the foundation's mission. Every PRI is carefully assessed to ensure that it furthers the foundation's charitable objectives. This assessment includes rigorous due diligence, evaluating the recipient's capacity to deliver impact and achieve financial sustainability. Foundations also adopt flexible structuring to meet the unique needs of each project, whether through customised repayment terms, co-investment arrangements, or shared decision-making processes. Monitoring and reporting are integral to the success of PRIs. Foundations establish robust systems to track both financial performance and social outcomes. Regular updates and impact assessments provide transparency, allowing foundations to make data-driven decisions and refine their strategies. Importantly, the recyclable nature of PRI capital sets it apart from traditional grants. Since the principal is often repaid or generates returns, foundations can reinvest these funds into new initiatives, creating a sustainable cycle of impact funding. Through this adaptable and mission-driven framework, PRIs enable foundations to address pressing societal issues while ensuring the longevity and sustainability of their philanthropic efforts.

Investment Lifecycle of Program-Related Investments

Initiation and Strategy Development: Foundations identify priority areas aligning with their mission and evaluate how PRIs can complement their grant-making activities. This phase includes defining specific impact objectives, determining the financial resources to allocate, and outlining criteria for eligible investments. For instance, a foundation focused on economic empowerment may decide to prioritise PRIs in Microfinance Institutions that target underserved populations. During this phase, foundations also consider how PRIs fit into their broader portfolio of philanthropic tools.

Sourcing and Screening: Opportunities for PRIs are identified through established networks, partnerships, or open calls for proposals. Foundations assess the alignment of these opportunities with their mission and perform initial screenings to evaluate the recipient organisation's operational capacity, business model, and potential for delivering meaningful social impact. A structured approach ensures that only the most viable and impactful opportunities proceed to the next stage.

Due Diligence: Due diligence is a comprehensive evaluation process aimed at ensuring the viability of the proposed investment. This phase typically involves an in-depth review of the recipient organisation's financial health, leadership team, governance practices, and operational track record. Foundations may conduct site visits to observe operations first-hand or engage independent consultants for expert insights. For example, a foundation considering an equity investment in a clean water initiative might investigate the scalability of the technology, its market readiness, and the organisation's ability to sustain operations over the long term.

Structuring the Investment: Once an opportunity is deemed viable, the foundation works closely with the recipient organisation to design a customised investment structure. This involves negotiating terms such as loan interest rates, equity ownership percentages, repayment schedules, and reporting requirements. The agreement also includes provisions to ensure accountability and alignment with the foundation's mission. For instance, a PRI structured as a loan may incorporate flexible repayment terms tied to the recipient's revenue cycles to reduce financial strain.

Approval and Disbursement: The proposed PRI is presented to the foundation's board or investment committee for approval. This stage ensures that the investment aligns with the foundation's strategic priorities and risk tolerance. Once approved, funds are disbursed according to the agreed terms, which may include phased disbursements tied to the achievement of specific milestones. For instance, a renewable energy project might receive funds in instalments as key construction phases are completed.

Implementation and Monitoring: During implementation, the recipient organisation uses the invested capital to advance its mission-aligned activities, such as expanding operations, launching new products, or delivering services to underserved populations. Foundations actively monitor progress through periodic updates, financial reports, and impact assessments. Regular communication ensures that challenges are identified early and adjustments are made as needed. Foundations may also provide non-financial support, such as strategic guidance or technical assistance, to enhance the likelihood of success.

Evaluation and Learning: As the PRI matures, foundations conduct evaluations to measure its effectiveness against the initial objectives. These evaluations focus on both financial performance and social outcomes. Metrics such as the number of beneficiaries reached, environmental improvements achieved, or economic opportunities created are analysed alongside financial returns. Insights gathered during this phase are documented to inform future investment strategies and improve overall decision-making processes.

Exit and Recycling of Capital: The exit phase marks the conclusion of the PRI, where the foundation recovers its invested capital through loan repayments, the sale of equity stakes, or the fulfilment of guarantee obligations. This phase is critical for enabling the foundation to recycle the recovered funds into new PRIs, thereby sustaining its philanthropic impact. For example, proceeds from a successful investment in a renewable energy startup might be reinvested into affordable housing projects, creating a multiplier effect for societal benefit.

Challenges of Leveraging Program-Related Investments (PRIs) in Impact Investing

Program-Related Investments (PRIs) represent a versatile tool for philanthropic foundations to advance their missions while generating potential financial returns. However, their effective utilisation in impact investing is constrained by several challenges that span financial, operational, and strategic dimensions.

Complex Structuring and Compliance Requirements: PRIs must adhere to stringent regulatory frameworks designed to ensure they align with philanthropic missions and charitable objectives. In jurisdictions like the United States, the IRS mandates that PRIs primarily advance charitable goals rather than generate profits. This dual requirement creates significant complexity in structuring PRIs. Foundations must navigate a labyrinth of legal stipulations, which often requires specialised expertise and increases transaction costs. Furthermore, the need to tailor each investment to comply with these regulations while addressing the unique needs of recipients makes the process resource-intensive and time-consuming. For smaller foundations, these challenges can act as a deterrent, limiting the use of PRIs despite their potential for impact.

Balancing Financial Returns and Social Impact: Striking the right balance between achieving social impact and ensuring financial returns is a persistent challenge. Foundations risk overemphasising financial sustainability, which could lead to mission drift, or focusing too heavily on impact, jeopardising the return of capital needed to recycle funds. This balancing act is particularly challenging in ventures operating in underserved markets, where achieving significant impact often necessitates accepting lower financial returns or extended payback periods. Moreover, foundations face pressure to demonstrate tangible results to stakeholders, which can incentivise short-term financial performance over long-term systemic change.

High Risk in Underserved Markets: PRIs are frequently deployed in high-risk markets—such as low-income communities or fragile economies—where systemic challenges like weak infrastructure, political instability, and limited access to skilled labour exacerbate investment risks. These ventures often lack the traditional financial benchmarks, such as credit ratings or audited financials, that are critical for underwriting. As a result, foundations must rely on alternative and often less reliable indicators to assess feasibility, making investment decisions more speculative. The heightened risk profile of these markets not only deters some foundations but can also result in higher-than-expected failure rates, diminishing confidence in PRIs as a viable tool for impact.

Resource-Intensive Due Diligence: The comprehensive due diligence required for PRIs involves evaluating an organisation's financial stability, governance practices, operational capacity, and potential for social impact. For smaller foundations or those new to PRIs, the lack of in-house expertise often necessitates hiring external advisors, increasing costs. The time and resources required to conduct thorough due diligence can also delay deployment, reducing the responsiveness of PRIs to urgent social or environmental challenges. Furthermore, the bespoke nature of each PRI means that lessons learned from one investment may not be easily transferable to others, limiting opportunities for efficiency gains.

Limited Internal Capacity and Expertise: Foundations accustomed to traditional grant-making often lack the financial expertise required to design, execute, and monitor PRIs effectively. Unlike grants, PRIs demand a sophisticated understanding of financial instruments, risk assessment, and investment management. This knowledge gap can lead to suboptimal structuring of investments or inadequate oversight, increasing the likelihood of underperformance. Additionally, reliance on external intermediaries to bridge this expertise gap can erode control and increase operational complexity, creating further barriers to scaling PRIs.

Challenges in Monitoring and Impact Measurement: PRIs necessitate dual monitoring of financial performance and social impact, which is inherently more complex than the reporting requirements for grants. Establishing robust and transparent systems to track outcomes is challenging, particularly for intangible or long-term impacts like community empowerment or systemic change. Many recipient organisations also lack the capacity to produce high-quality, consistent data, further complicating evaluation. This absence of reliable metrics not only impedes accountability but also limits the ability of foundations to iterate and improve their investment strategies.

Barriers to Scaling and Replication: The bespoke nature of PRIs limits their scalability. Unlike standardised financial products, PRIs are tailored to the specific needs of individual projects or organisations, making replication across different contexts challenging. The relatively small number of foundations engaged in PRIs also restricts opportunities for pooling resources, sharing best practices, or creating co-investment frameworks. This fragmentation of effort prevents PRIs from achieving the scale required to address systemic societal challenges effectively, reducing their transformative potential.

Opportunities for Mitigating Challenges

Unified Compliance and Simplified Regulatory Processes: To address regulatory complexity, foundations and policymakers should collaborate to create unified regulatory standards for PRIs. Streamlined compliance processes, including pre-approved templates for PRI structures, can reduce legal ambiguities and administrative burdens. This initiative would lower entry barriers for smaller foundations and enable more organisations to participate effectively.

Risk Pooling and De-Risking Mechanisms: Foundations can establish sector-specific risk-sharing funds that pool resources to mitigate investment risks. These funds could offer guarantees or first-loss capital to encourage co-investments from other foundations, development banks, and private investors. This approach ensures that high-risk, high-impact projects receive the necessary support while reducing individual foundation exposure.

Advanced Monitoring and Impact Analytics: Leveraging technology to integrate financial and impact data into dynamic dashboards can provide real-time monitoring of PRI outcomes. Tools like AI-driven analytics and blockchain-based reporting systems offer unparalleled transparency, enabling agile decision-making. These systems also foster accountability and allow foundations to evaluate both short-term results and long-term systemic changes.

Capacity-Building Through Strategic Partnerships: Foundations can collaborate with universities, think tanks, and industry experts to create tailored capacity-building programmes. These programmes should address gaps in financial literacy, governance, and impact measurement for both foundation staff and recipients. For example, offering training on structuring scalable business models can enhance recipient readiness and improve overall investment outcomes.

Collaborative Investment Platforms: Creating digital platforms that connect foundations, investors, and intermediaries can foster collaboration and resource sharing. Such platforms could feature due diligence repositories, co-investment opportunities, and a knowledge-sharing hub to disseminate best practices. By pooling efforts, foundations can reduce duplication and amplify the impact of their investments.

Embedding Flexibility in PRI Design: Introducing adaptive structures like revenue-contingent loans or tiered repayment schemes allows PRIs to accommodate fluctuating market conditions. These flexible designs align repayment obligations with the financial realities of recipients, ensuring stability and fostering long-term success. This adaptability is particularly crucial for ventures operating in volatile or underserved markets.

Scalable Public–Private Partnerships: PRIs should be integrated into broader public–private initiatives that leverage government funding, corporate expertise, and philanthropic resources. For instance, renewable energy PRIs can be paired with government subsidies and private-sector innovation to accelerate the transition to sustainable energy. Such partnerships can achieve a systemic impact while maximising resource efficiency.

PRIs are influential for advancing philanthropic missions while fostering financial sustainability. Innovative approaches such as unified regulatory frameworks, advanced monitoring tools, and scalable Public–Private Partnerships, can enable foundations overcome existing barriers and unlock the full potential of PRIs. These strategies ensure that PRIs bridge the gap between traditional philanthropy and market-based solutions and drive systemic and transformative societal impact.

2.5 Legal and Regulatory Landscape for Impact Investing

The prevailing legal and regulatory frameworks deeply influence impact investing's effectiveness and scalability. Impact investing operates at the intersection of financial markets and social or environmental objectives, necessitating clear legal and regulatory guidance.

2.5.1 Legal Recognition and Frameworks

Legal recognition of impact investing encompasses the formal structures, policies, and regulations that enable and govern investments aimed at achieving both financial and non-financial outcomes. Across jurisdictions, these frameworks vary in maturity, complexity, and comprehensiveness, reflecting differing legal traditions, market priorities, and government support levels. Legal entities and corporate forms represent foundational mechanisms through which impact investing is operationalised. They can be broadly categorised into those explicitly designed for impact-driven purposes and traditional profit-driven entities integrating impact objectives.

Impact-Driven Entities Impact-Driven Entities
These organisations prioritise generating positive social and environmental impact alongside financial returns. They intentionally design their business models, operations, and strategies to address critical challenges and contribute to a more sustainable and equitable world. Such entities embed accountability for their stated missions, ensuring alignment with impact goals.

Key examples include:

- **Social Enterprises**: Social enterprises primarily aim to address social or environmental challenges while maintaining financial sustainability. Their structure often combines elements of for-profit and non-profit models. These entities reinvest significant profits into their social missions, ensuring continuity and scale of impact. Requirements typically include transparent impact reporting, governance mechanisms to prevent mission drift, and community involvement in decision-making processes.
- **Cooperatives**: Cooperatives are member-owned organisations that prioritise shared value and community benefits over maximising profits. Their democratic governance structures ensure that decisions align with the collective goals of members. In impact investing, cooperatives often focus on areas like renewable energy, fair trade, and financial inclusion. Impact is implemented through equitable profit distribution, reinvestment in community projects, and adherence to cooperative principles such as sustainability and social responsibility.
- **Non-profit Organisations Engaging in Impact Investing**: Many non-profits use impact investing to advance their mission, often through Program-Related Investments (PRIs). Unlike grants, PRIs are designed to generate financial returns that can be recycled into future mission-driven activities. Non-profits engaging in impact investing must ensure their investments align with their charitable purposes, maintain tax-exempt status, and comply with strict reporting requirements. Their focus on measurable impact is

achieved through rigorous due diligence, monitoring, and alignment with global standards such as the Sustainable Development Goals (SDGs).

- **Benefit Corporations**: Benefit Corporations are for-profit entities with a legally defined mission to generate public benefits alongside financial returns. Their governance structures require directors to consider the impact of their decisions on stakeholders, including workers, communities, and the environment. To maintain transparency, they must produce annual reports assessed against third-party standards detailing their social and environmental performance.
- **Community Interest Companies (CICs)**: CICs are specialised entities that operate to benefit the community. Their legal framework includes asset locks and dividend caps to ensure profits are primarily reinvested into their social missions. CICs must adhere to stringent reporting requirements to demonstrate how their activities serve the public interest.

Traditional Profit-Driven Entities

Traditional corporations, while historically focused on shareholder value, are increasingly integrating impact objectives into their operations, with some aligning closely with impact investing principles. While CSR programmes, ESG integration, and sustainability-linked instruments provide context, true impact investing within traditional entities involves deliberate efforts to achieve measurable social or environmental outcomes alongside financial returns. Examples include:

- **Corporate Impact Initiatives**: Traditional entities may establish impact investment funds or direct portions of their corporate capital to projects with clearly defined and measurable impact outcomes, often aligning with global frameworks like the Sustainable Development Goals (SDGs).
- **Strategic Partnerships**: Collaborations with impact-driven organisations or Development Finance Institutions to co-finance or co-develop projects that deliver both commercial and societal benefits.

Traditional profit-driven entities pursuing impact investing must navigate a complex landscape of legal and governance considerations. These considerations highlight the need for directors to ensure accountability and transparency when aligning impact outcomes with corporate strategies. The integration of impact objectives into investment decisions often hinges on fiduciary duties, which traditionally prioritise financial returns. Scholarly discourse has extensively examined the evolving nature of fiduciary duties in impact investing. Freshfields Bruckhaus Deringer (2005), which laid the groundwork for subsequent regulatory changes, argued that integrating ESG considerations into investment decisions is consistent with fiduciary duties, provided they align with the best interests of beneficiaries. Also, the UNEP FI & PRI (2015) *Fiduciary Duty in the 21st Century* report underscored integrating ESG factors into fiduciary practices to manage risks and identify opportunities, advocating for regulatory reforms to eliminate ambiguity. Legal scholars have increasingly explored the concept of double materiality, emphasising that fiduciaries should consider both the financial impacts of ESG risks on investments and the broader impacts of investments on society and the environment. This paradigm shift challenges traditional interpretations of fiduciary duty and supports a more holistic approach to impact investing. Furthermore, several jurisdictions have clarified or expanded the scope of fiduciary duties to accommodate ESG considerations. In particular, the European Union's Sustainable Finance Disclosure Regulation (SFDR) and other legislative initiatives explicitly require financial market participants to assess and disclose sustainability risks. These regulations embed ESG considerations into fiduciary responsibilities, marking a shift from voluntary to mandatory practices. The US Department of Labour (DOL) issued guidance clarifying that fiduciaries of private pension plans governed by the Employee Retirement Income Security Act (ERISA) can incorporate ESG factors when they are material to investment performance. Recent developments further emphasise that fiduciaries may pursue ESG objectives if they align with the financial interests of beneficiaries. The United Kingdom's Pensions Regulator provided similar guidance, encouraging pension trustees to incorporate climate risks and other ESG factors into their investment strategies, consistent with their fiduciary duties.

2.5.2 *Regulatory Support for Impact Investing*

Governmental and multilateral efforts to regulate and incentivise impact investing have created enabling environments that foster growth in this sector. Regulatory support encompasses a variety of mechanisms, including fiscal incentives, market regulations, and capacity-building initiatives.

Tax Incentives: Tax incentives remain among the most impactful tools for promoting impact investing. Jurisdictions worldwide use targeted tax relief schemes to attract capital into socially beneficial ventures. For example, the United Kingdom's Social Investment Tax Relief (SITR) provides individual investors with income tax relief for investments in qualifying social enterprises, reducing the financial risk for investors and encouraging participation. Similarly, the United States offers tax credits for investments in Opportunity Zones, designed to spur economic development in distressed communities by lowering capital gains tax liabilities for investors who commit their resources to these areas over a specified time frame. These measures effectively reduce the cost of capital for socially impactful projects, incentivising private-sector involvement.

Disclosure Regulations: Transparency fosters trust and accountability in impact investing. Disclosure regulations ensure that investors and companies articulate their sustainability and impact objectives clearly. For instance, the European Union's Sustainable Finance Disclosure Regulation (SFDR) mandates institutional investors and asset managers to disclose how sustainability risks are integrated into their investment decisions. The regulation also requires reporting on the adverse impacts of these investments on environmental and social factors. Such frameworks aim to prevent greenwashing, where entities overstate or misrepresent their environmental and social contributions. Disclosure require ments provide market participants with verifiable and comparable data, enhancing decision-making processes and strengthening overall market integrity.

Development of Green and Social Bond Markets: The expansion of Expanding bond markets represents a transformative development in impact investing. Regulatory frameworks like the Green Bond Principles (GBP) and Social Bond Principles (SBP) have provided issuers with standardised guidelines on using proceeds, project selection, and impact reporting.[12] These bonds allow investors to allocate capital towards projects with explicit environmental or social objectives, such as renewable energy, affordable housing, or community healthcare. Sovereign green bonds issued by governments further legitimise the market and set benchmarks for private-sector participation. Regulatory oversight ensures that bond proceeds are allocated to projects that align with declared objectives, building investor confidence and driving market growth.

Regulatory Sandboxes: Regulatory sandboxes offer a controlled environment for testing innovative financial instruments and business models under the oversight of regulatory authorities. These frameworks encourage experimentation with new impact-oriented investment vehicles while ensuring compliance with essential legal standards. Sandboxes have been particularly valuable in fostering financial inclusion and technology-driven solutions in emerging markets, where traditional financing mechanisms often fall short (Jenik & Lauer, 2017). By reducing regulatory uncertainty for innovators, sandboxes accelerate the development and scaling of impactful solutions.

Capacity-Building Programmes: Capacity-building initiatives aim to strengthen the ecosystem for impact investing by providing technical assistance, training, and resources to stakeholders. Governments, multilateral organisations, and non-profits collaborate to enhance the operational and financial capabilities of social enterprises, intermediaries, and investors. For example, the Global Impact Investing Network (GIIN) offers educational programmes and tools that help organisations integrate impact measurement and management into their operations. These programmes are crucial in reducing information asymmetries and equipping stakeholders to navigate the complexities of impact investing effectively.

Public–Private Partnerships (PPPs): Public–Private Partnerships (PPPs) have emerged as a key mechanism for financing large-scale impact projects. Leveraging public funds to attract private investment enables PPPs to address funding gaps in critical sectors such as urban infrastructure, renewable energy, and affordable housing. Governments often provide guarantees or co-financing arrangements to de-risk these projects, ensuring they align with public policy objectives and investor expectations. Successful examples include renewable energy projects where public funds cover initial development costs, enabling private investors to step in during later stages.

2.5.3 *Role of Multilateral and International Organisations*

International Organisations like the United Nations and Multilateral institutions like the World Bank and Development Finance Institutions (DFIs) have been instrumental in shaping the legal and regulatory environment for impact investing by providing technical assistance, capacity-building, and policy guidance. Their efforts have created frameworks that align private investments with public objectives, facilitating the growth of impact investing on a global scale. The collective efforts of multilateral organisations underscore the importance of global cooperation in advancing impact investing. These organisations operate with a dual structure encompassing normative and operational functions. Normative functions involve establishing standards, frameworks, and best practices to guide the development of impact investing globally. On the other hand, operational functions focus on implementing these standards through direct interventions, technical assistance, and capacity-building initiatives.

The normative role of International Organisations and Multilateral institutions includes defining global priorities for sustainability and impact, often through collaborative platforms and agreements. For instance, international organisations such as the United Nations (UN) leverage their convening power to set ambitious targets like the Sustainable Development Goals (SDGs), which serve as a universal framework for aligning investments with societal and environmental objectives. The UN also spearheads initiatives such as the Principles for Responsible Investment (PRI), which encourage institutional investors to integrate Environmental, Social, and Governance (ESG) factors into their investment processes. These standards have been critical in mainstreaming sustainability within financial systems and fostering stakeholder accountability. On the operational side, multilateral organisations engage directly with national governments, financial institutions, and private-sector entities to implement impact-focused programmes. Development Finance Institutions (DFIs), such as the International Finance Corporation (IFC) and the African Development Bank (AfDB), exemplify this role by providing financial instruments that mitigate risks and mobilise private capital for high-impact projects. DFIs address the barriers that often deter private investment in underserved markets through guarantees, concessional loans, and blended finance mechanisms. The operational role also extends to capacity-building efforts, which aim to strengthen the institutional and technical capabilities of stakeholders in the impact investing ecosystem.

These interventions are complemented by training programmes and knowledge-sharing platforms that disseminate best practices in impact measurement, reporting, and project management.

Multilateral synergies ensure that impact investing frameworks are globally consistent and locally actionable by integrating normative and operational functions. Their ability to harmonise efforts across regions enhances the scalability of impact investing initiatives, addressing systemic barriers and creating enabling environments for sustainable finance. However, challenges persist, particularly in harmonising ESG standards and addressing the unique needs of politically or economically unstable regions. Continued collaboration and adaptability will be essential to fully realise the transformative potential of multilateral synergies in impact investing. The interventions by multilateral and international organisations have significantly enhanced the ecosystem for impact investing. These entities have de-risked high-impact projects by providing financial and technical support, enabling private capital to flow into underserved markets. They have also set global benchmarks for sustainability, ensuring consistency and transparency in impact investing practices.

Multilateral organisations also facilitate global dialogue and collaboration, creating synergies among diverse stakeholders. Platforms such as the G20 Sustainable Finance Working Group and the Organisation for Economic Co-operation and Development (OECD) Centre on Green Finance and Investment provide forums for member countries, financial institutions, and regulators to harmonise standards and mobilise resources. These collaborative efforts aim to address fragmentation in regulatory frameworks and promote the alignment of financial systems with sustainability goals. In addition to their normative and operational functions, multilateral organisations play a catalytic role in mobilising private capital for impact investing. These entities attract co-investment from private-sector players, amplifying their initiatives' scale and reach by leveraging their financial and technical expertise. However, the effectiveness of multilateral organisations is not without challenges. While these entities have successfully set global standards and mobilised resources, the harmonisation of ESG frameworks across jurisdictions remains incomplete. Furthermore, political instability and weak governance in certain regions hinder the implementation of impact-focused initiatives. Addressing these challenges will require strengthened partnerships with local governments, increased coordination among multilateral entities, and sustained advocacy for regulatory reforms.

2.6 When Impact Investing Goes Wrong

Despite its noble intentions, impact investing is not immune to failure. When it goes wrong, it not only fails to deliver on its promises but can also exacerbate the very issues it seeks to solve, leaving behind a trail of disillusionment and unintended harm. At the heart of the conundrum lies the intricate balancing act between impact and profit. The tension of whether investors should prioritise financial returns to satisfy stakeholders or focus solely on maximising social and environmental benefits means that they often navigate a complex web of competing priorities. This tension is compounded by the diverse and sometimes conflicting expectations of beneficiaries, policymakers, and financial backers. The very promise of "doing well by doing good" can quickly unravel if these priorities clash, revealing deep systemic and operational flaws. The literature is replete with cautionary tales. Scholars have highlighted instances where investments, though well-intentioned, failed to account for local contexts, leading to outcomes that ranged from ineffectual to harmful. For example, a microfinance initiative in South Asia intended to empower women through small loans instead spiralled into a crisis of over-indebtedness, triggering social unrest and, tragically, a series of borrower suicides (Ashta et al., 2011; Engel & Pedersen, 2019). Such evidence underscores the critical need for a nuanced understanding of local dynamics and a commitment to ethical, context-sensitive practices (Bengo et al., 2021). Moreover, the rise of impact washing has further muddied the waters. As the popularity of impact investing grows, so does the temptation for firms to exaggerate their contributions to social and environmental causes. This phenomenon undermines trust in the sector and diverts crucial resources away from genuinely impactful initiatives. The challenge, therefore, is not merely to invest but to invest responsibly, with a clear-eyed awareness of the potential pitfalls and a steadfast commitment to accountability. The failure of impact investing is a stark reminder that good intentions alone are not enough. Even the most promising ventures can falter without rigorous planning, transparent measurement, and genuine stakeholder engagement. This section delves into the circumstances under which impact investing goes wrong, offering an analysis of key challenges, illustrative case studies, and potential solutions.

2.6.1 *Misalignment of Goals*

One of the most pervasive challenges in impact investing arises from the misalignment of goals between investors and the communities or sectors they aim to support. While impact investing seeks to balance social and environmental benefits with financial returns, this balance can be delicate and, at times, contentious. When financial priorities overshadow impact objectives, the essence of impact investing is compromised, leading to suboptimal or even harmful outcomes. Daggers (2019) argues that the tension between financial returns and impact objectives is at the heart of many misaligned investments. Investors, often under pressure to deliver competitive returns, may prioritise profit-making ventures at the expense of long-term social and environmental benefits. This misalignment not only dilutes the intended impact but can also exacerbate the very issues that impact investing seeks to address (Daggers, 2019).

The Case of the Addax Bioenergy Project
The Addax Bioenergy Project, launched in 2010, is a cautionary tale of impact investing gone wrong. The project was marketed as an innovative solution to produce bioethanol from sugarcane, create employment opportunities, and stimulate economic development in Sierra Leone, one of the world's poorest nations.[13] However, the project's implementation revealed significant flaws in aligning its stated impact goals with on-the-ground realities. The Addax Bioenergy Project acquired 54,000 hectares of land in northern Sierra Leone, intending to cultivate sugarcane for bioethanol production and electricity generation. Promises were made to create thousands of jobs, enhance local livelihoods, and contribute to sustainable energy solutions. The initiative attracted funding from various international development institutions, including the African Development Bank and other European financiers, further underscoring its appeal as a model impact investment. However, the actual outcomes fell far short of these aspirations (Fielding et al., 2015).

The Problems

- **Displacement and Land Rights Issues**: The large-scale land acquisition disrupted the lives of local farmers who relied on the land for subsistence farming. Despite an Environmental, Social, and Health Impact Assessment (ESHIA) conducted prior to the project's initiation, local communities reported being inadequately informed about the land leasing process. Compensation was insufficient, with landowners receiving just USD 12.35 per hectare per year, a rate widely considered inadequate given the loss of livelihood. Additionally, informal land users were excluded from compensation, further deepening social inequities. Over 25,000 people were affected by the project's land acquisition, yet only a small fraction of the promised benefits materialised for these communities.

- **Environmental Degradation**: The project's reliance on monoculture sugarcane farming resulted in significant environmental harm. Intensive irrigation from local water sources reduced water availability for nearby communities and degraded aquatic ecosystems. Soil erosion, loss of biodiversity, and the overuse of chemical fertilisers further compounded these issues. Despite being marketed as a sustainable project, the environmental costs were substantial and undermined the project's long-term viability.
- **Unfulfilled Promises of Employment**: Although the project initially created jobs, most positions were seasonal, temporary, or low-paying. Women, in particular, were marginalised, with only 10 per cent of the workforce comprising female employees, often relegated to the lowest-paying roles. Furthermore, skilled roles were disproportionately allocated to expatriates, leaving local workers without opportunities for meaningful capacity-building. The lack of investment in local skill development further limited the project's potential to deliver long-term economic benefits.
- **Economic Fallout and Project Abandonment**: By 2015, the Addax Bioenergy Project scaled down its operations due to financial difficulties, including low sugarcane yields and the impact of the Ebola outbreak. The sudden cessation of operations left local communities without access to leased land and without the promised economic opportunities. Workers faced layoffs, salary cuts, and reduced economic security, while local food security deteriorated due to restricted land access and unfulfilled agricultural development promises. The project's assets were eventually sold to new investors, but this transition did little to address the grievances of affected communities.

Lessons from Addax Bioenergy Project

The Addax Bioenergy Project illustrates how impact investing can fail when insufficient attention is paid to local contexts, stakeholder engagement, and environmental sustainability. Key insights from this case include:

- **Comprehensive Stakeholder Analysis**: Effective impact investing requires meaningful engagement with local communities, ensuring that their needs, rights, and priorities are fully integrated into project design. This entails moving beyond tokenistic consultations to genuinely collaborative processes that empower local voices in decision-making. The lack of transparent and inclusive consultation processes in the Addax project fostered mistrust and resistance, highlighting the critical importance of stakeholder participation.
- **Robust Environmental and Social Safeguards**: Impact investments must prioritise thorough environmental impact assessments beyond compliance checkboxes. The reliance on monoculture farming in the Addax project underscores the need for agroecological approaches that protect biodiversity and ensure the sustainable use of natural resources. These safeguards should include ongoing monitoring and the flexibility to adapt to evolving challenges.
- **Equitable Compensation Mechanisms**: Compensation frameworks must account for the full spectrum of land users, including informal and marginalised groups. Transparent and fair compensation mechanisms are essential to avoid exacerbating existing inequalities. Lessons from Addax highlight the need for compensatory mechanisms that reflect not only the economic value of the land but also its social and cultural significance to the affected communities.
- **Sustainable Financial Models**: Impact investments must be underpinned by robust financial planning that balances profitability with social and environmental objectives. This includes realistic revenue projections, diversification of funding sources, and contingency plans to address unforeseen challenges. The financial collapse of the Addax project highlights the dangers of overambitious projections and insufficient risk mitigation, serving as a cautionary tale for future investments.
- **Accountability and Exit Strategies**: Development Finance Institutions (DFIs) and other stakeholders must adopt clear accountability

mechanisms to address grievances and ensure the continuity of benefits. In the case of Addax, the DFIs failed to provide adequate oversight and compensation when the project was scaled down, leaving communities to bear the brunt of its failures. Exit strategies should include robust measures to safeguard local livelihoods and minimise disruptions.
- **Gender Inclusion and Social Equity**: A critical oversight in the Addax project was the marginalisation of women in employment and decision-making processes. Future impact investments must prioritise gender-inclusive approaches, ensuring women have equitable access to jobs, training, and leadership opportunities. Addressing social equity is essential for fostering more resilient and cohesive communities.
- **Policy and Institutional Support**: Governments and DFIs must strengthen regulatory frameworks and institutional capacities to ensure that large-scale investments align with national development goals. In Sierra Leone, weak governance structures exacerbated the challenges faced by the Addax project. Policymakers should establish clear guidelines for investor accountability and create enabling environments for sustainable development.

The Addax Bioenergy Project compromised its social and environmental objectives by prioritising financial returns and scaling operations aggressively, failing the very communities it sought to help. This case underscores the importance of aligning investment practices with local needs, sustainability, and a commitment to accountability, providing invaluable lessons for the future of impact investing.

2.6.2 *Superficial Measurement*

One of the critical pitfalls of impact investing lies in the over-reliance on metrics that emphasise outputs rather than outcomes (Dadush, 2012). While metrics are indispensable for tracking progress and demonstrating accountability, an overemphasis on quantifiable indicators often leads to superficial assessments that fail to capture the nuanced, long-term impacts of investments. This reductionist approach undermines the credibility of impact investing and perpetuates systemic challenges rather than addressing them. Metrics that focus on outputs—such as the number of individuals receiving loans or the total households gaining access to electricity—are often prioritised because they are easier to measure, report, and communicate (Daggers, 2019). However, these metrics often fail to evaluate whether interventions achieve meaningful, sustainable change. For example, a programme that distributes microloans to 10,000 individuals may look successful on paper, but if these loans do not improve financial stability or empower recipients to escape poverty, the intervention's impact is superficial at best. This tendency to overemphasise outputs over outcomes stems from several interrelated factors:

- **Ease of Communication**: Quantifiable metrics simplify the narrative for investors and stakeholders. Statements such as "50,000 people served" are digestible and compelling, but they often obscure the complexity and multidimensional nature of the impact.
- **Investor and Donor Pressures**: Impact investors and donors often operate under tight timelines and face pressure to demonstrate quick results. As a result, programmes may prioritise easily achievable outputs rather than designing for sustained and transformative outcomes.
- **Challenges in Measuring Outcomes**: Evaluating qualitative changes—such as improvements in well-being, empowerment, or long-term resilience—requires sophisticated methodologies, longitudinal studies, and significant resources. Many projects lack the capacity or willingness to invest in such rigorous assessments.
- **Perverse Incentives**: When organisations are incentivised to achieve numerical targets, they may prioritise quantity over quality. This can lead to unintended consequences, such as compromising the welfare of beneficiaries or exacerbating existing inequities.

The Case of Andhra Pradesh—Microfinance Crisis

The Andhra Pradesh microfinance crisis of 2010 offers a sobering example of how superficial measurement and aggressive practices can lead to adverse outcomes. Initially, microfinance in India was heralded as a groundbreaking tool for poverty alleviation, with millions of small loans disbursed to underserved communities. However, this rapid expansion of Microfinance Institutions (MFIs) was driven by an overemphasis on metrics like "loans disbursed" and "borrowers served", masking deeper structural issues within the sector. Andhra Pradesh became the epicentre of microfinance activity in India, with MFIs leveraging the state's well-established Self-Help Group (SHG) networks to scale rapidly. By 2010, approximately one-third of India's microfinance borrowers were concentrated in the state. This rapid expansion, however, was underpinned by aggressive lending practices and inadequate regulatory oversight. Borrowers were often encouraged to take multiple loans from different MFIs without sufficient checks on their repayment capacity, leading to over-indebtedness. The situation escalated as MFIs adopted coercive recovery methods to ensure repayment. Reports surfaced of field agents employing public shaming, intimidation, and harassment to recover loans, disproportionately affecting women who constituted the majority of borrowers. These practices created significant social and psychological distress, contributing to a wave of borrower suicides that captured national attention. Media reports documented over 75 suicides linked to microfinance debt in Andhra Pradesh, exposing the darker side of what was marketed as a tool for empowerment (Mader, 2013; Priyadarshee & Ghalib, 2011).

Structural Issues

- **Over-Indebtedness and Loan Saturation**: Andhra Pradesh households faced an unsustainable debt burden due to the proliferation of MFIs and overlapping loans. A 2010 study revealed that 84% of borrower households in the state held multiple loans, with many using new loans to repay existing debts. The lack of coordinated oversight among MFIs allowed borrowers to be trapped in a cycle of high-interest loans, creating systemic financial instability.
- **Weak Regulatory Frameworks**: The absence of robust regulatory mechanisms allowed MFIs to operate unchecked, prioritising profit over social impact. The Andhra Pradesh Microfinance Institutions (Regulation of Money Lending) Ordinance, introduced in October 2010, was a reactive measure aimed at curbing abusive practices. However, it failed to address root causes such as over-saturation, unethical lending, and the lack of borrower protections. Regulatory gaps prior to the crisis enabled predatory practices to flourish.
- **Exploitation of Aspiration Paradoxes**: MFIs capitalised on the aspiration paradox, where borrowers' aspirations to improve their quality of life outpaced their financial realities. Loans were often marketed for non-productive purposes like consumer goods, further

exacerbating borrowers' financial vulnerability. Without mechanisms to assess the capacity for repayment or to channel loans into productive uses, borrowers were pushed deeper into cycles of debt.

- **Social and Cultural Impacts**: The crisis disproportionately affected women, who comprised most of the microfinance borrowers. Beyond financial stress, the coercive recovery tactics employed by some MFIs caused social stigma, fractured familial relationships, and undermined community trust. This highlights the intersection of financial practices with broader socio-cultural dynamics that were poorly understood or ignored by MFIs.

Lessons from Andhra Pradesh

The Andhra Pradesh crisis underscores the dangers of superficial measurement in impact investing. It highlights the urgent need for structural reforms and strategic realignment within the microfinance sector. Key lessons include:

- **Contextual Understanding**: One of the critical errors leading to the Andhra Pradesh crisis was the failure to adapt microfinance interventions to local socio-economic realities. Andhra Pradesh had become oversaturated with microfinance due to its historically well-organised Self-Help Groups (SHGs), which private Microfinance Institutions (MFIs) exploited to rapidly scale operations. However, these models often ignored the nuanced socio-economic contexts of borrowers, resulting in widespread over-indebtedness. Standardised models are ill-suited to regions with diverse socio-economic structures. For example, SHGs in Andhra Pradesh were initially state-sponsored, with strong ties to local governance, but private MFIs leveraged these structures primarily for profit without strengthening borrowers' capacity for sustainable credit use. Future interventions must involve participatory planning and community-driven needs assessments to avoid creating systems that exacerbate inequality.
- **Beyond Superficial Metrics: Outcome-Based Measurement**: The crisis revealed the inadequacy of traditional metrics, such as the number of loans disbursed or repayment rates, which masked underlying issues like coercive repayment practices and borrower

distress. High repayment rates were often achieved through mechanisms of social pressure, which increased borrowers' vulnerability to psychosocial harm, including shame and stigmatisation Success in microfinance must prioritise transformative outcomes—such as financial resilience, poverty alleviation, and economic empowerment—over transactional outputs. Integrating qualitative data into evaluations (e.g., changes in quality of life or community dynamics) provides a more comprehensive understanding of impact.

- **Strengthened Regulatory Oversight**: The absence of robust regulatory frameworks enabled private MFIs to prioritise profit over ethical practices. MFIs often disbursed loans irresponsibly, targeting multiple loans at the same households without assessing repayment capacity. This laissez-faire environment allowed exploitative practices, such as aggressive recovery measures, to flourish, culminating in borrower suicides and widespread public backlash. Regulatory mechanisms must balance financial innovation with borrower protection. Recommendations include capping interest rates, mandating transparency in loan terms, and establishing independent grievance redressal systems. Further, routine audits by regulatory bodies can preempt unethical practices and improve sectoral accountability.
- **Empowering Borrowers**: Many borrowers in Andhra Pradesh lacked financial literacy, which compounded their vulnerability to over-indebtedness. MFIs often exploited the aspiration paradox, encouraging borrowers to take loans for non-productive expenditures like consumer goods, which deepened their financial precarity. Financial empowerment must go beyond access to credit. Borrowers require training in financial literacy, budgeting, and investment. Such empowerment not only mitigates risks of over-indebtedness but also ensures that credit serves as a tool for development rather than dependency.
- **Integration of Holistic Support System**: The Andhra Pradesh crisis highlighted the inadequacy of microfinance as a standalone solution. While credit access was widespread, it lacked integration with capacity-building, market linkages, and social safety nets. Borrowers often used loans to address immediate consumption needs, perpetuating a cycle of debt. MFIs should collaborate with local governments, NGOs, and community organisations to develop integrated frameworks that combine financial services with education, health, and livelihood support. Such an ecosystem approach

ensures that financial inclusion translates into meaningful socio-economic development.

- **Ethical Business Practices**: The profit-driven orientation of many MFIs in Andhra Pradesh fundamentally conflicted with the ethos of microfinance as a development tool. Credit commodification led to practices prioritising institutional gain over borrower welfare. A sustainable microfinance industry requires a dual focus on profitability and social impact. Establishing industry-wide ethical standards, supported by third-party monitoring bodies, can foster transparency and rebuild trust in the sector. Encouraging social enterprises and mission-driven MFIs can also create models that align financial sustainability with borrower welfare.
- **Crisis Mitigation and Risk Management**: The lack of early warning systems and crisis mitigation frameworks exacerbated the fallout in Andhra Pradesh. MFIs lacked mechanisms to track repayment capacities and systemic risks, allowing localised issues to spiral into a state-wide crisis. Proactive crisis management strategies are essential for safeguarding the microfinance sector. MFIs should invest in data-driven tools to monitor borrower health and repayment trends in real time. Furthermore, regulatory bodies should establish contingency plans to address potential sectoral crises preemptively.

2.6.3 Market Failures

Market failures present significant obstacles to the effectiveness of impact investing, particularly in the provision of public goods and the internalisation of externalities. Systemic inefficiencies in sectors like public goods provision or externality mitigation can hinder the optimal allocation of resources (Schmidt, 2023). Public goods are defined by their accessibility to all (non-excludability) and their ability to be consumed by one individual without diminishing availability to others (non-rivalrous consumption). Examples include resources like clean air, national defence, and public parks. These characteristics make it difficult for private investors to exclusively capture the benefits of such goods, thereby reducing their incentive to fund these projects. This often results in market failures, as essential services and resources remain underfunded despite their societal importance. In impact investing, systemic inefficiencies arise when markets fail to allocate resources efficiently for the provision of public goods. This inefficiency is primarily due to the free-rider problem, where individuals or entities benefit from resources without paying for their cost, leading to underinvestment in essential services. For instance, in sectors like education, healthcare, and clean energy, the societal benefits often far exceed the direct financial returns to investors. As a result, these areas may suffer from chronic underfunding.

Schmidt (2023) discussed how markets tend to undervalue positive externalities—benefits experienced by third parties—such as improved public health or reduced greenhouse gas emissions. This undervaluation leads to a misalignment between private incentives and social welfare, causing essential services to be underprovided. Businesses and investors have an ethical obligation to address these market failures by actively contributing to the provision of public goods and mitigating negative externalities. A pertinent example is the renewable energy sector. Despite its critical role in combating climate change, investments in renewable energy often struggle to compete financially with fossil fuels. This challenge is exacerbated by inadequate carbon pricing mechanisms that fail to internalise the environmental costs of fossil fuel consumption (Sterner et al., 2024). Consequently, the market continues to favour investments in polluting energy sources, perpetuating an inefficient allocation of resources and hindering the transition to sustainable energy solutions. Addressing these systemic inefficiencies requires a multifaceted approach, including the implementation of effective carbon pricing, the development of policies that incentivise investment in public goods, and the promotion of ethical business practices that prioritise social and environmental outcomes alongside financial returns. By aligning private incentives with public welfare, and done right, impact investing is an essential tool for correcting market failures and fostering a more equitable and sustainable economy.

The Case of Enbridge and Sustainability-Linked Loans—A Lesson in Market Failure

Enbridge Inc., a Canadian multinational energy transportation company, is one of North America's largest pipeline operators. Enbridge historically faced significant criticism over its environmental impact, particularly due to its role in the transportation of fossil fuels and its controversial Line 3 replacement project.[14] In an effort to position itself as a sustainability-conscious corporation, Enbridge utilised sustainability-linked loans. In

February 2021, the company secured a three-year, syndicated Sustainability Linked Credit Facility for US$1.0 billion, facilitated by a consortium including CIBC, Scotiabank, BMO, RBC, and TD Securities.[15] The credit facility was the first in the North American midstream energy sector. The instrument tied Enbridge's borrowing costs to its performance on predetermined sustainability metrics, including commitments to reduce greenhouse gas (GHG) emissions intensity and increase diversity within its workforce. Investors and stakeholders initially perceived this as a sign that even traditionally "dirty" industries were taking meaningful strides towards sustainability. Enbridge's loan, structured by leading global banks, validated the potential for financial instruments to drive corporate ESG improvements. Despite the optimistic framing, cracks in the narrative began to emerge almost immediately.

Soon after it was announced issues with adopting the financial instrument were brought to light. Organisations focused on environmental advocacy and corporate accountability,[16] closely examined the details of the loan and raised concerns about the weakness of the sustainability targets and the exclusion of Scope 3 emissions.[17] The analyses revealed critical flaws in the impact investing ecosystem. Some underlying issues with Enbridge financing became clear, including:

1. **Lack of Ambitious Targets**: Enbridge's emissions targets were based on emissions intensity—a measure relative to output—rather than absolute reductions. This allowed the company to technically meet its targets while continuing to expand its pipeline operations and overall emissions footprint.
2. **Opaque Metrics and Reporting**: Transparency regarding the measurement and verification of progress towards these targets was minimal. Stakeholders questioned the methodologies used to calculate emissions intensity reductions and workforce diversity improvements. This lack of clarity weakened investor confidence in the authenticity of the commitments.
3. **Misaligned Incentives**: The financial penalties embedded in the SLL—an incremental increase in loan interest rates for failing to meet targets—were negligible compared to Enbridge's overall revenues. This diluted the financial incentive to prioritise meaningful ESG performance.

4. **Conflict with Core Business Model**: Enbridge's reliance on fossil fuel transportation fundamentally contradicted its sustainability narrative. Even as it committed to ESG goals, the company continued to invest in pipeline expansions, including the controversial Line 3 project. This project faced widespread opposition from Indigenous communities, environmental activists, and legal experts, who argued that it perpetuated dependence on fossil fuels and posed significant environmental risks.

Market failure, in the context of Enbridge's sustainability-linked loans, highlights how systemic inefficiencies and weak governance within the impact investing ecosystem contributed to unfulfilled promises of genuine sustainability. The failures stemmed from the inability of markets to allocate resources efficiently towards impactful outcomes, leading to misaligned incentives and unintended consequences. The first major issue lies in the absence of robust regulatory frameworks. Without stringent standards, Enbridge was able to establish its own sustainability targets with limited external scrutiny. This allowed the company to define metrics that were either insufficiently ambitious or inherently flawed, such as relying on emissions intensity instead of absolute reductions. Consequently, the broader market failed to discern between meaningful commitments and superficial efforts, undermining the credibility of sustainability-linked instruments. Another critical element of market failure was investor complicity. Financial institutions and investors, eager to support ESG initiatives, failed to conduct thorough due diligence on Enbridge's commitments. By structuring and endorsing the sustainability-linked loan, these institutions effectively validated Enbridge's sustainability narrative without demanding accountability or substantive action. This not only perpetuated greenwashing but also diluted the broader impact of ESG financing by rewarding superficial compliance rather than transformative change. Market failure was further exacerbated by the misalignment of incentives. The penalties tied to Enbridge's sustainability-linked loans were insignificant compared to its revenues, diminishing the financial motivation to achieve its ESG targets. This lack of meaningful consequences revealed the inadequacy of current market mechanisms in aligning corporate behaviour with sustainability goals. The result was a system where financial products appeared to promote ESG priorities but lacked the necessary accountability to ensure real progress.

Lessons Learned

Enbridge's SLL case is a stark reminder of the limitations of market-driven sustainability initiatives. Here are some of the key lessons:

- **The Need for Absolute Metrics**: One of the most critical flaws in Enbridge's sustainability strategy was its reliance on intensity-based metrics rather than absolute reductions. Intensity-based metrics measure emissions relative to output, allowing companies to meet targets while expanding operations and increasing total emissions. To drive genuine environmental progress, sustainability targets must prioritise absolute reductions, forcing companies to rethink their overall environmental footprint rather than relying on incremental efficiency gains. Absolute metrics also enhance comparability across industries and align with global climate goals, such as the Paris Agreement, which focuses on total emissions reductions to limit global warming.
- **Greater Transparency**: Transparency is foundational to building trust in sustainability-linked instruments. Enbridge's lack of clarity regarding how it measured progress against its ESG targets raised doubts about the authenticity of its commitments. To ensure accountability, companies must adopt standardised reporting frameworks, such as the Global Reporting Initiative (GRI) or the Sustainability Accounting Standards Board (SASB). Third-party verification is equally critical, as it provides an independent assessment of a company's performance and prevents manipulation or misrepresentation of data. Transparency not only strengthens stakeholder confidence but also ensures that financial instruments genuinely reflect a company's sustainability efforts.
- **Stronger Penalties**: The financial penalties embedded in Enbridge's SLL were negligible compared to the company's overall revenues, resulting in weak incentives to prioritise ESG goals. Penalties and rewards tied to sustainability performance must be proportional to a company's financial scale to drive meaningful change. For example, a failure to meet ESG targets could result in higher interest rates, reduced access to capital, or reputational consequences that impact stock valuations. Stronger penalties ensure that sustainability commitments are more than superficial promises and compel companies to integrate ESG priorities into their core operations.

- **Holistic Stakeholder Engagement**: Sustainability requires companies to go beyond corporate boardrooms and engage with a diverse range of stakeholders, including Indigenous communities, environmental advocates, and affected local populations. Enbridge's exclusion of key stakeholders in its ESG decision-making process highlighted a significant gap in its approach. Effective stakeholder engagement involves meaningful consultation, where all voices are heard and integrated into corporate strategies. This ensures that sustainability initiatives address real-world challenges and mitigates risks of opposition and strengthens the social licence to operate. Companies must recognise that sustainability is as much about social equity as it is about environmental stewardship.

By exploiting weak frameworks and superficial commitments, Enbridge capitalised on the growing demand for ESG-focused financial products while continuing to perpetuate environmental harm. This example underscores the urgent need for stricter oversight, stronger accountability mechanisms, and a more critical approach to evaluating corporate sustainability claims. Only then can impact investing fulfil its promise as a force for genuine social and environmental progress.

2.6.4 Systemic Risks in Impact Investing

Systemic risks in impact investing represent overarching challenges inherent to the broader market or sector in which these investments operate. These risks are particularly pronounced in cross-border contexts, where investments frequently target emerging markets characterised by underdeveloped infrastructure, volatile political climates, and weak institutional capacity. Such risks can affect entire regions or sectors, amplifying the potential for project failures, financial losses, and reputational damage, ultimately undermining the intended social and environmental outcomes. A nuanced understanding of systemic risks in impact investing requires a multidisciplinary perspective, drawing on insights from economics, political science, and sociology. These risks often mirror those observed in traditional investing, including governance challenges, market inefficiencies, and macroeconomic instability. However, the dual mandate of impact investing introduces additional complexities. By examining systemic risks through this lens, it is possible to assess their implications for investment performance and impact outcomes more comprehensively. The risk factors include:

- **Institutional Weaknesses**: Institutional weaknesses are a recurring theme in failed impact investments. Fragile governance, regulatory loopholes, and a lack of accountability mechanisms often create environments where projects fail to meet their objectives. For instance, renewable energy projects in countries with inconsistent enforcement of policies frequently encounter delays or are abandoned altogether. Bureaucratic inefficiencies and corruption further compound these risks, leading to cost overruns and eroded trust among stakeholders. Hence inclusive institutions are crucial for sustainable development (Robinson & Acemoglu, 2012). In impact investing, the absence of such institutions often leads to a misalignment between project design and implementation capabilities, resulting in suboptimal outcomes or outright failures.
- **Market Inefficiencies**: Market inefficiencies are a key driver of failures in impact investing. Information asymmetry, for example, often prevents investors from fully understanding the risks associated with projects, leading to poorly informed decisions. Inconsistent or non-existent reporting frameworks further exacerbate this problem, allowing impact washing to proliferate—where claims of positive outcomes are exaggerated or false. The fragmentation of financial infrastructure also contributes to systemic failures. In regions where credit markets and payment systems are underdeveloped, investors face heightened transaction costs and limited scalability. Such inefficiencies often render smaller projects unviable, perpetuating cycles of underdevelopment and limiting the reach of impact investments.
- **Cultural and Contextual Misalignment**: Cultural and contextual misalignment represents another systemic risk that often leads to failed impact investment projects. When investment strategies fail to account for the socio-cultural dynamics of target communities, they risk alienating stakeholders and undermining project adoption. External solutions imposed without recognising local knowledge and practices frequently result in resistance and suboptimal outcomes (Escobar, 2011). For example, the introduction of advanced agricultural technologies without involving local farmers in the design and implementation processes often leads to low adoption rates and

diminished impact. Participatory approaches that integrate community perspectives are critical for mitigating these risks and ensuring alignment with local contexts.

- **Macroeconomic Instability**: Macroeconomic instability, including inflation, currency fluctuations, and economic recessions, poses systemic risks that directly affect the viability of impact investments. Projects reliant on long-term financing, such as renewable energy infrastructure or social housing initiatives, are particularly vulnerable to economic volatility. In their historical analyses, Rogoff and Reinhart (2009) demonstrate how macroeconomic instability undermines investment climates, leading to increased risks of default and reduced investor confidence. Mitigating these risks often requires sophisticated financial instruments, such as hedging mechanisms and blended finance models, which add complexity and cost to investment structures.
- **Political Uncertainty**: Political uncertainty, encompassing abrupt policy shifts, geopolitical tensions, and governance instability, represents a systemic risk that can derail impact investment projects. For example, changes in subsidy policies for renewable energy or sudden nationalisation of assets can render investments financially unviable. World Bank (2017) highlights the importance of policy predictability in fostering private-sector confidence. Collaborative engagement with policymakers, alongside the establishment of stable regulatory frameworks, is essential for mitigating political risks and ensuring the long-term sustainability of impact investments.

Systemic risks are central to understanding why impact investments often fail to deliver on their dual mandates of financial returns and measurable social or environmental impact. By critically examining the structural, institutional, and contextual dimensions of these risks, this discussion highlights the urgent need for robust risk management strategies. Future research should prioritise the development of innovative tools and methodologies to address systemic risks, including technology-driven solutions for real-time monitoring and risk assessment. Additionally, fostering cross-sectoral collaborations and investing in institutional capacity-building are critical for creating an enabling environment for impact investing. Addressing systemic risks is not merely a technical exercise but a foundational step towards ensuring that impact investing realises its transformative potential.

Case of Honduras Renewable Energy Development
The Honduras Renewable Energy Development initiative of the Patuca III hydroelectric dam project, sought to diversify the country's energy matrix by expanding renewable energy generation. While the project aimed to meet the growing electricity demand and reduce reliance on fossil fuels, it faced significant challenges that derailed its objectives. It was an ambitious renewable energy development plan to achieve an 80 per cent renewable energy share in its energy mix by 2038. Patuca III, a 104 MW hydroelectric dam, was central to this vision. The government envisioned the project as a means to improve energy access, stimulate economic growth, and reduce greenhouse gas emissions in alignment with international commitments such as the Paris Agreement. Despite its promise, the project faced immediate resistance from local communities and indigenous groups. Concerns about displacement, environmental degradation, and inadequate consultation led to protests and legal challenges. These disruptions caused significant delays in construction and strained relationships between stakeholders. Navigating Honduras' complex regulatory environment proved to be a major obstacle. The approval process for environmental and construction permits was slow, exacerbating delays. Overlapping mandates among the National Electric Energy Company (ENEE), Electric Energy Regulatory Commission (CREE), and Energy Secretariat (SEN) further complicated project execution. The lack of clear guidelines for hydropower development highlighted systemic governance issues. Cost overruns became a significant issue as delays and unforeseen technical challenges inflated the project's budget. The National Electric Energy Company (ENEE), already burdened by substantial debt and operational inefficiencies, struggled to secure additional financing. Moreover, optimistic revenue projections based on high energy prices and demand proved unrealistic, leading to financial uncertainty. Honduras' vulnerability to extreme weather events, including hurricanes and droughts, further complicated the project's progress. Hydropower's dependency on stable water levels made it particularly susceptible to climate variability. At the same time, the country's outdated and underdeveloped electricity transmission and distribution networks resulted in high energy losses, limiting the project's efficiency and effectiveness (IRENA, 2023). The systemic factors that led to the failure of the Patuca III hydroelectric dam project include:

- **Governance and Institutional Fragmentation**: The fragmented governance structure within the energy sector. The overlapping mandates and lack of coordination among key institutions—including ENEE, CREE, and SEN—created inefficiencies in policy implementation and decision-making. These institutional weaknesses hindered the ability to execute large-scale renewable energy projects effectively, resulting in delays and increased costs.
- **Social and Environmental Vulnerabilities**: The project's failure to address deep-rooted social and environmental vulnerabilities further exacerbated its issues. Honduras' history of deforestation, land degradation, and marginalisation of indigenous communities heightened resistance to the Patuca III project. The lack of a comprehensive stakeholder engagement strategy undermined efforts to build trust and consensus, leading to protests, legal challenges, and project delays. These vulnerabilities reflect broader systemic issues of inequality and environmental mismanagement that were not adequately factored into the project's design.
- **Climate and Infrastructure Resilience**: Honduras is one of the countries most affected by extreme weather events, such as hurricanes and droughts. These climate risks posed significant challenges to the viability of renewable energy infrastructure, particularly hydropower projects like Patuca III. The project's inability to incorporate resilience measures against such events underscores a critical oversight in addressing systemic risks associated with climate variability. Additionally, the country's outdated and underdeveloped transmission and distribution networks contributed to high electricity losses, further undermining the project's efficiency and financial viability.
- **Economic and Financial Instability**: Systemic economic challenges, including ENEE's financial instability and the high cost of energy subsidies, compounded the risks associated with the project. ENEE's significant debt burden and operational inefficiencies limited its capacity to finance and manage large-scale infrastructure investments. Furthermore, the country's reliance on cross-subsidies and public funds to offset energy costs created a precarious financial environment, which discouraged private-sector investment and strained public resources.

Lessons Learned

The Patuca III project offers several key lessons for future renewable energy and impact investment initiatives:

- **Holistic Community Engagement**: Engaging communities early and inclusively is critical. A robust stakeholder engagement strategy must extend beyond token consultations to involve local communities as genuine partners. Understanding their cultural, social, and economic contexts ensures that projects align with community needs and mitigates resistance, thereby reducing costly delays and fostering trust. This approach should include mechanisms for continuous dialogue and conflict resolution.
- **Strengthening Institutional Coordination**: Fragmented governance undermines project execution. Establishing a cohesive framework that clarifies institutional roles and responsibilities is paramount. Institutions like ENEE, CREE, and SEN must collaborate under well-defined mandates to avoid redundancies and conflicts. Enhanced capacity-building programmes for these institutions can help streamline permitting processes and improve operational efficiency.
- **Incorporating Climate Resilience**: Climate risks must be central to the design and planning of renewable energy projects. For hydropower projects, integrating adaptive technologies such as variable water flow management systems can mitigate the impact of climate variability. Furthermore, investing in climate risk assessments and scenario planning will ensure that infrastructure remains robust against extreme weather events, enhancing long-term sustainability.
- **Improving Financial Stability**: Financial instability in institutions like ENEE poses systemic risks to project viability. Comprehensive financial reforms, including debt restructuring and transparent accounting practices, are essential. Establishing Public–Private Partnerships (PPPs) with clear risk-sharing mechanisms can attract private investment while reducing the financial burden on public institutions.
- **Transparent Financial Modelling**: Overly optimistic revenue projections undermine credibility and financial sustainability. Employing conservative estimates, stress testing revenue assumptions, and accounting for potential fluctuations in market and environmental conditions can help create realistic financial models. These

models should also include contingencies for unexpected delays or cost overruns.

- **Focus on Infrastructure Modernisation**: Modern and efficient transmission and distribution networks are critical for reducing energy losses and maximising the impact of renewable energy projects. Investment in smart grid technologies and energy storage solutions can enhance grid stability and support the integration of variable renewable energy sources such as solar and wind.
- **Aligning Policy with Action**: Ambitious energy and climate policies must be accompanied by actionable implementation strategies. Policymakers must ensure that sufficient funding, human resources, and technical capacity are available to achieve stated goals. Regular monitoring and evaluation mechanisms can track progress and make necessary adjustments to align policies with ground realities.
- **Promoting Inclusive Development**: Renewable energy projects should not only focus on environmental and economic objectives but also aim to promote social equity. Ensuring that project benefits, such as job creation and improved energy access, are distributed equitably can address systemic inequalities and build broader support for such initiatives. Special attention should be given to empowering marginalised groups, including women and indigenous communities, through targeted capacity-building programmes.

2.7 The Importance of Recognising and Addressing Failures in Impact Investing

Recognising and addressing failures in impact investing is paramount, as these failures offer critical insights that can shape the future of the sector. Impact investing, with its ambitious promise to align financial returns with social and environmental benefits, operates in a complex landscape. The challenges range from systemic market inefficiencies to operational missteps, often compounded by the dual mandate of generating profits while effecting transformative change. When these investments fail, the repercussions are not confined to financial losses but extend to the erosion of trust, exacerbating social inequities and sometimes irreparable harm to communities and ecosystems. Thus, a careful and transparent analysis of these failures is essential to refine strategies, build resilience, and foster accountability within the sector. One of the foremost reasons for recognising and addressing failures lies in the lessons they unveil. Every failure, whether due to misalignment of goals, inadequate stakeholder engagement, or superficial impact measurement, underscores specific gaps in planning, execution, or oversight. For instance, the Addax Bioenergy Project in Sierra Leone illustrates how insufficient stakeholder consultation and unrealistic financial projections can lead to widespread displacement, environmental degradation, and economic instability. Analysing such cases will enable impact investors and policymakers to understand better the importance of aligning project objectives with the needs of local communities, implementing robust environmental and social safeguards, and adopting realistic financial models. These lessons, when applied effectively, can prevent the recurrence of similar mistakes and contribute to the development of more resilient and equitable investment frameworks.

Addressing failures also demands a commitment to accountability and transparency, which are foundational to restoring and maintaining trust in the impact investing ecosystem. As seen in cases of "impact washing", where firms exaggerate their contributions to social or environmental causes, the lack of stringent accountability mechanisms can erode confidence among stakeholders. Transparent reporting, third-party evaluations, and inclusive decision-making processes are not merely procedural requirements but vital tools to ensure that investments are genuinely impactful and aligned with their stated objectives. Acknowledging failures openly and implementing corrective measures enable the sector to enhance its credibility and demonstrate a genuine commitment to social and environmental progress. Furthermore, recognising failures provides a unique opportunity to address the structural and systemic challenges that impede the effectiveness of impact investing. Systemic risks, such as weak regulatory frameworks, macroeconomic instability, and cultural misalignment, often underlie project failures. For instance, the challenges faced by the Patuca III hydropower project in Honduras highlight how fragmented governance, inadequate financial planning, and the absence of climate resilience measures can derail even well-intentioned initiatives. Tackling these systemic risks requires collaborative efforts among governments, investors, and development institutions to create enabling environments that support sustainable and equitable investments. By addressing these structural issues, the sector can move beyond reactive measures and adopt proactive strategies that anticipate and mitigate potential risks.

Lastly, addressing failures is essential for fostering innovation and adaptability within the impact investing sector. Each setback provides a critical impetus to re-evaluate existing practices, explore alternative approaches, and experiment with innovative solutions. For example, the inadequacies of traditional metrics that focus on outputs rather than outcomes have prompted a shift towards more holistic evaluation frameworks that capture the long-term social and environmental impacts of investments. Similarly, failures in financial sustainability have led to the exploration of blended finance models that combine public, private, and philanthropic capital to de-risk investments and enhance scalability. The sector can evolve to meet the complex and dynamic challenges of sustainable development by embracing a culture of learning and innovation. Recognising and addressing failures in impact investing is not merely an exercise in retrospection but a critical step towards building a more robust, accountable, and transformative investment ecosystem. Failures, when analysed thoughtfully, serve as invaluable guides that illuminate pathways for improvement, foster stakeholder trust, and inspire systemic change. As the sector continues to grow, its success will increasingly depend on its ability to learn from its shortcomings and adapt to the evolving needs of the communities and ecosystems it seeks to serve. Through rigorous reflection and decisive action, impact investing can realise its potential as a powerful tool for driving inclusive and sustainable development.

2.8 Conclusion

Impact investing challenges the long-standing dichotomy between profit and purpose, seeking social good with expectation of financial gain. This dual focus presents unique opportunities, as well as challenges, for investors, enterprises, and regulatory bodies navigating this evolving landscape. The concept itself has not always been clearly defined. In its early years, the boundaries of impact investing were blurred, encompassing a broad spectrum of investments with varying degrees of commitment to social or environmental change. Some feared that strict definitions would alienate potential investors, while others worried that ambiguity would dilute the field's credibility. Over time, however, consensus has emerged that impact investing requires more than good intentions; it necessitates clear, measurable objectives and robust strategies for tracking outcomes. This shift has given rise to the importance of intentionality. As impact investing has matured, so too have the structures through which it is implemented. The field has moved beyond simple equity investments into a diverse range of financial instruments designed to balance risk, return, and impact. However, as the field has expanded, so too have its challenges. One of the most pressing concerns is the regulatory and legal landscape, which remains fragmented across different jurisdictions. Without standardised impact measurement frameworks and disclosure requirements, investors struggle to compare investments, assess risks, and verify claims. Some regulatory bodies have begun to address this gap, but inconsistencies remain, creating opportunities for misrepresentation and reducing investor confidence. Additionally, the influx of mainstream institutional investors into the impact space has raised concerns about mission drift, where financial returns gradually overshadow impact objectives. The pressure to meet shareholder expectations can lead firms to prioritise profitable projects over those that generate the most meaningful social or environmental change.

Yet, perhaps the most important lessons in impact investing come not from its successes, but from its failures. While the field has produced remarkable success stories, it has also encountered projects that fell short of their intended impact. Despite their best intentions, some investments failed to deliver the anticipated outcomes due to flawed assumptions, misaligned incentives, or external market shocks. Others struggled with scalability, as impact-focused enterprises faced difficulties in transitioning from grant-dependent models to commercially viable ones. These setbacks, while frustrating, serve as valuable learning opportunities. Recognising and addressing failures is essential for the field's long-term credibility. Each challenge presents a chance to refine measurement frameworks, adjust investment strategies, and strengthen accountability mechanisms, ensuring that impact investing continues to evolve into a more rigorous, results-driven discipline. Impact investing is more than a financial trend; it is a movement that challenges traditional notions of capital and responsibility. It asks whether markets can be designed to serve investors and society at large. While obstacles remain, the field's continued refinement signals a shift in the way capital is mobilised, structured, and evaluated. As impact investing gains momentum, the challenge will be to preserve its integrity while ensuring that financial markets remain a powerful force for positive, lasting change.

Notes

1. For example, see Arosio (2011) and Harji and Jackson (2012).
2. See Agrawal and Hockerts (2021) for a review of the definitions of impact investing.
3. See Mudaliar and Bass (2017) and Kilmurray et al. (2021) for review of studies and evidence on competitive returns on impact investment.
4. See a bibliometric analysis by Shome et al. (2023).
5. The Impact Management Project (IMP) is a forum that used insight from over one thousand stakeholders in the industry to understand how to measure and manage impact. They developed the five dimensions modelled by IRIS+.
6. See Ely and Hearn (2021), Holtslag et al. (2021) and Nachyła and Justo (2024).
7. Brest (2015) discusses impact investing in countries with weak legal institutions and suggests work-arounds.
8. Such networks can leverage economies of scale and collective power, compared to the relatively minimal impact of solo individual practices (Perez et al., 2015).
9. Where funds are marketed as sustainable without delivering genuine environmental benefits.
10. GPs receive a management fee (typically 1.5 per cent–2.5 per cent annually) and a performance fee (commonly 20 per cent of profits above a predefined hurdle rate).
11. See empirical studies like Guerin et al. (2018) and Mjuza and Ntsalaze (2019).
12. See reports at https://www.icmagroup.org/.
13. With an average GDP per capita of USD 650 and about 60 per cent of the population living below national poverty line.
14. See Orenstein and Schneider (2019), Simpson (2023).
15. See Enbridge (2021), Reuters (2021).
16. See Banktrack (2021).
17. Scope 3 emissions also known as value chain emissions are indirect greenhouse gas (GHG) emissions that occur from activities outside of an organisation's direct control (Environment Protection Agency, 2025).

References

Addison, T., & Roe, A. (2018). *Extractive industries: The management of resources as a driver of sustainable development*. Oxford University Press.

Adeoye, O. B., Chigozie, A. E., Nwakamma, N.-E., Danny, J. M., Usman, F. O., & Olu-Lawal, K. A. (2024). A conceptual framework for data-driven sustainable finance in green energy transition. *World Journal of Advanced Research and Reviews, 21*(2), 1791–1801.

Aggarwal, D., & Elembilassery, V. (2018). Sustainable finance in emerging markets: A venture capital investment decision dilemma. *South Asian Journal of Business and Management Cases, 7*(2), 131–143. https://doi.org/10.1177/2277977918774651

Agrawal, A., & Hockerts, K. (2021, April 03). Impact investing: Review and research agenda. *Journal of Small Business & Entrepreneurship, 33*(2), 153–181. https://doi.org/10.1080/08276331.2018.1551457

Arosio, M. (2011). Impact Investing in Emerging Markets: Issues for Responsible Investors. *Responsible Research*. https://www.responsibleresearch.com

Ashta, A., Khan, S., & Otto, P. (2011, March 03). Does microfinance cause or reduce suicides? Policy recommendations for reducing borrower stress. *Strategic Change, 24*. https://doi.org/10.2139/ssrn.1715442

Banktrack. (2021). Enbridge's "sustainable" credit: Tar sands and rights violations. https://www.banktrack.org/article/enbridge_s_sustainable_credit_tar_sands_and_rights_violations

Bengo, I., Borrello, A., & Chiodo, V. (2021, March 06). Preserving the integrity of social impact investing: Towards a distinctive implementation strategy. *Sustainability, 13*, 2852. https://doi.org/10.3390/su13052852

Bolton, R., & Foxon, T. J. (2015). Infrastructure transformation as a socio-technical process—Implications for the governance of energy distribution networks in the UK. *Technological Forecasting and Social Change*, *90*, 538–550. https://doi.org/10.1016/j.techfore.2014.02.017

Bouri, A., Mudaliar, A., Schiff, H., Bass, R., & Dithrich, H. (2018). *Roadmap for the future of impact investing: Reshaping financial markets*. GIIN.

Brest, P. (2015). *Impact investing in developing countries: Legal institutions and work-arounds*. Stanford Law and Policy Lab.

Brest, P., & Born, K. (2013). Unpacking the impact in impact investing. *Stanford Social Innovation Review*, *11*(4), 22–31.

Bugg-Levine, A., & Emerson, J. (2011). *Impact investing: Transforming how we make money while making a difference*. Wiley. https://books.google.co.uk/books?id=LhxIwPYYhfEC

Casasnovas, G., & Jones, J. (2022, September 01). Who has a seat at the table in impact investing? Addressing inequality by giving voice. *Journal of Business Ethics, 179*(4), 951–969. https://doi.org/10.1007/s10551-022-05154-6

Castellas, E. I.-P., Ormiston, J., & Findlay, S. (2018). Financing social entrepreneurship: The role of impact investment in shaping social enterprise in Australia. *Social Enterprise Journal, 14*(2), 130–155.

Chen, S., Song, Y., & Gao, P. (2023, November 01). Environmental, social, and governance (ESG) performance and financial outcomes: Analyzing the impact of ESG on financial performance. *Journal of Environmental Management, 345*, 118829. https://doi.org/10.1016/j.jenvman.2023.118829

Clark, C., Emerson, J., & Thornley, B. (2014). *The impact investor: Lessons in leadership and strategy for collaborative capitalism*. Wiley.

Cuccia, A. (2023). Venture philanthropy. In *Encyclopedia of sustainable management* (pp. 3925–3930). Springer.

Dadush, S. (2012). Impact investment indicators: A critical assessment. In Davis et al. (Eds.), *Governance by indicators: Global power through quantification and rankings*. Oxford University Press.

Daggers, J. (2019). *'Solving' social problems with markets and measurement: A critical study of social impact investing*. Birkbeck, University of London.

Ebrahim, A., & Rangan, V. K. (2014). What impact? A framework for measuring the scale and scope of social performance. *California Management Review, 56*(3), 118–141.

Eccles, R. G., Serafeim, G., Seth, D., & Ming, C. C. Y. (2013). The performance frontier: Innovating for a sustainable strategy: Interaction. *Harvard Business Review, 91*(7), 17–18.

EIB. (2017). In Impact Investing: Do good and make money. https://www.eib.org/en/podcasts/impact-investing

Ely, A., & Hearn, D. (2021). Impact investors need to share power, not just capital. *Stanford Social Innovation Review*. https://ssir.org/articles/entry/impact_investors_need_to_share_power_not_just_capital

Enbridge. (2021). *Enbridge publishes sustainability-linked bond framework*. Enbridge. Retrieved November 11, from, https://www.enbridge.com/media-center/news/details?id=123680&lang=en

Engel, S., & Pedersen, D. (2019). Microfinance as poverty-shame debt. *Emotions and Society, 1*(2), 181–196.

Environmental Protection Agency. (2025). Scope 3 Inventory Guidance. https://www.epa.gov/climateleadership/scope-3-inventory-guidance

Escobar, A. (2011). *Encountering development: The making and unmaking of the third world* (Vol. 1). Princeton University Press.

Fielding, M., Davis, M., Weitz, N., Cummings-John, I., Hickey, A., Johnson, F., Senyagwa, J., Martinez, L., & Sun, M. (2015). *Agricultural investment and rural transformation: A case study of the Makeni bioenergy project in Sierra Leone*. JSTOR.

Financial Conduct Authority. (2023). *Sustainability disclosure requirements and investment labels*.

Findlay, S., & Moran, M. (2019). Purpose-washing of impact investing funds: Motivations, occurrence and prevention. *Social Responsibility Journal, 15*(7), 853–873. https://doi.org/10.1108/SRJ-11-2017-0260

Frede, J. (2021). *Impact value added (IVA)*. Available at SSRN 3868250.

Freshfields Bruckhaus Deringer. (2005). *A legal framework for the integration of environmental, social and governance issues into institutional investment*. European Company Law.

Gasper, D. (2019, May 04). The road to the Sustainable Development Goals: Building global alliances and norms. *Journal of Global Ethics, 15*(2), 118–137. https://doi.org/10.1080/17449626.2019.1639532

Geczy, C., Jeffers, J. S., Musto, D. K., & Tucker, A. M. (2021, November 01). Contracts with (Social) benefits: The implementation of impact investing. *Journal of Financial Economics, 142*(2), 697–718. https://doi.org/10.1016/j.jfineco.2021.01.006

Gianoncelli, A., & Gaggiotti, G. (2021). Managing for impact: The role of investors' networks in enhancing appropriate impact measurement and management practices. In A. Richards & J. Nicholls (Eds.), *Generation impact* (pp. 219–233). Emerald Publishing Limited. https://doi.org/10.1108/978-1-78973-929-920200022

GIIN. (2019). *The core characteristics of impact investing*. https://thegiin.org/publication/post/core-characteristics-of-impact-investing/

GIIN. (2021). *The state of impact measurement and management practice*.

GIIN. (2023). *Impact investing: A guide to this dynamic market*.

Guerin, I., Labie, M., & Morvant-Roux, S. (2018). Inadequate growth, over-indebtedness, and crises in microcredit: What have we learned? *Enterprise Development & Microfinance, 29*(2), 118–132.

Gutterman, A. (2024). *Impact investment tools, structures and instruments*.

Hand, D., Sunderji, S., Ulanow, M., Remsberg, R., & Xiao, K. (2024). *State of the market 2024: Trends, performance and allocations*. https://thegiin.org/publication/research/state-of-the-market-2024-trends-performance-and-allocations/

Harji, K., & Jackson, E. (2011). Accelerating impact: Achievements, challenges and what's next in building the impact investing industry. *Monitor Group*. https://www.monitorinstitute.com

Hockerts, K., Hehenberger, L., Schaltegger, S., & Farber, V. (2022, September 01). Defining and conceptualizing impact investing: Attractive nuisance or catalyst. *Journal of Business Ethics, 179*(4), 937–950. https://doi.org/10.1007/s10551-022-05157-3

Holtslag, M., Chevrollier, N., & Nijhof, A. (2021). Impact investing and sustainable market transformations: The role of venture capital funds. *Business Ethics, the Environment & Responsibility, 30*(4), 522–537.

Hummels, H., & de Leede, M. (2014). The emergence of impact investments: The case of microfinance. In *Socially responsible investment in the 21st century: Does it make a difference for society?* (Vol. 7, pp. 91–115). Emerald Group Publishing Limited. https://doi.org/10.1108/S2043-905920140000007004

Kilmurray, D., Melin, L., & Mercereau, B. (2021). Integrating Impact Funds into Mainstream Portfolios. *The Journal of Impact and ESG Investing*, *1*(4), 103–119.

Koh, H., Karamchandani, A., & Katz, R. (2019). From Blueprint to Scale: the case for philanthropy in impact investing. *Gates open research*, *3*(1357), 1357.

Impact Alliance. (2020). *Classifiying a fund's impacts using the IMP+ACT Classification System: Case Study Acumen.*

International Finance Corporation. (2019). *Creating impact: The promise of impact investing*. World Bank.

IRENA. (2023). *Renewables readiness assessment honduras.*

Jenik, I., & Lauer, K. (2017). *Regulatory sandboxes and financial inclusion.* CGAP

John, R. (2006). *Venture philanthropy: The evolution of high engagement philanthropy in Europe.*

Lydenberg, S., Burckart, W., & Ziegler, J. (2017). *Effective Investing for the Long Term: Intentionality at systems levels.*

Mader, P. (2013). Rise and fall of microfinance in India: The Andhra Pradesh crisis in perspective. *Strategic Change, 22*(1–2), 47–66.

Martin, M. (2013). *Making impact investible.*

Mjuza, N., & Ntsalaze, L. (2019). Over-indebtedness of microfinance borrowers in South Africa. *African Review of Economics and Finance, 11*(2), 98–128.

Moretto, L., & Scola, B. (2017). *Development finance institutions and financial inclusion: From institution-building to market development.* CGAP Focus Note. https://www.cgap.org/sites/default/files/Focus-Note-DFIs-and-Financial-Inclusion-Mar-2017-rev_0.pdf

Mudaliar, A., & Bass, R. (2017). GIIN perspectives: Evidence on the financial performance of impact investments. *Global impact investing network.* https://thegiin.org/research/publication/financial-performance

Nachyła, P., & Justo, R. (2024, June 01). How do impact investors leverage non-financial strategies to create value? An impact-oriented value framework. *Journal of Business Venturing Insights, 21*, e00435. https://doi.org/10.1016/j.jbvi.2023.e00435

Nicholls, A. (2021). Impact measurement and management in sustainable finance. *Asian Development Outlook*, 1–54.

O'Reilly, C., Walsh, L., & Mottiar, Z. (2023, 05/17). Considerations for scaling a social enterprise: Key factors and elements. *The Irish Journal of Management, 42*. https://doi.org/10.2478/ijm-2023-0008

OECD. (2017). *Responsible business conduct for institutional investors: Key considerations for due diligence under the OECD Guidelines for Multinational Enterprises.*

OECD. (2019). *Social impact investment 2019: The impact imperative for sustainable development*. O. Publishing. https://doi.org/10.1787/9789264311299-en

OECD netFWD. (2014). *Venture philanthropy in development: Dynamics, challenges and lessons in the search for greater impact*. O. D. Centre.

Orenstein, W., & Schneider, G. (2019). Enbridge's Line 3 project is becoming an issue in the 2020 presidential race. *MINNPOST*. https://www.minnpost.com/environment/2019/07/enbridges-line-3-project-is-becoming-an-issue-in-the-2020-presidential-race/

Perez, M. M., Patel, D. M., & Cuff, P. A. (2015). *Empowering women and strengthening health systems and services through investing in nursing and midwifery enterprise: Lessons from lower-income countries: Workshop summary.* National Academies Press.

Priyadarshee, A., & Ghalib, A. K. (2011). *The Andhra Pradesh microfinance crisis in India: Manifestation, causal analysis, and regulatory response* (Working Paper 157). Brooks World Poverty Institute.

Reisman, J., & Olazabal, V. (2021). Impact measurement and management techniques to achieve powerful results. In E. D. M. Sarmento & R. P. Herman (Eds.), *Global handbook of impact investing-solving global problems via smarter capital markets towards a more sustainable society* (pp. 667–696). Wiley.

Reuters. (2021). Enbridge raises funds from sustainability bonds in climate-goal push. *Reuters*. https://www.reuters.com/business/sustainable-business/enbridge-raises-funds-sustainability-bonds-climate-goal-push-2021-06-29/

Robinson, J. A., & Acemoglu, D. (2012). *Why nations fail: The origins of power, prosperity and poverty*. Profile.

Rogoff, K. S., & Reinhart, C. (2009). *This time is different*. Princeton University Press.

Roth, B. (2021). Impact investing: A theory of financing social enterprises. *Harvard Business School Entrepreneurial Management Working Paper* (20-078).

Ruch, F. U. (2020). *Prospects, risks, and vulnerabilities in emerging and developing economies: lessons from the past decade* (Working Paper 9181) World Bank Policy Research.

Sarmento, E. D. M., & Herman, R. P. (2020). Measuring and evaluating social impact in impact investing: an overview of the main available standards and methods. In E. D. M. Sarmento & R. P. Herman (Eds.), *Global handbook of impact investing: Solving global problems via smarter capital markets towards a more sustainable society* (pp. 555–666). Wiley.

Schmidt, R. (2023, May 01). Are business ethics effective? A market failures approach to impact investing. *Journal of Business Ethics, 184*(2), 505–524. https://doi.org/10.1007/s10551-022-05133-x

Shome, S., Hassan, M. K., Verma, S., & Panigrahi, T. R. (2023, March 01). Impact investment for sustainable development: A bibliometric analysis. *International Review of Economics & Finance, 84*, 770–800. https://doi.org/10.1016/j.iref.2022.12.001

Simpson, N. (2023). *Line 3: An unthinkable pipeline powered by our cash*. bank.green. https://bank.green/blog/an-unthinkable-pipeline-powered-by-our-cash

Sterner, T., Ewald, J., & Sterner, E. (2024, April 01). Economists and the climate. *Journal of Behavioral and Experimental Economics, 109*, 102158. https://doi.org/10.1016/j.socec.2023.102158

Trelstad, B. (2016). Impact investing: A brief history. *Capitalism & Society, 11*(2), 1–14.

UNEP FI & PRI. (2015). *Fiduciary duty in the 21st century*. SSRN 2724866. https://www.unpri.org/fiduciary-duty/fiduciary-duty-in-the-21st-century-final-report/4998.article

Urban, B., & Mapula-e Lehasa, O. (2024). Investing in a social venture to generate social impact or financial return. *Business Perspectives and Research, 12*(4), 633–646. https://doi.org/10.1177/22785337221132614

World Bank. (2017). *World development report 2017: Governance and the law*. The World Bank.

Yelnikova, Y. V., & Kwilinski, A. (2020). Impact-investing in the healthcare in terms of the new socially responsible state investment policy. *Business Ethics and Leadership, 4*(3), 57–64.

PART II

Financialization and Growth

CHAPTER 3

The Interplay of Financialisation and Sustainable Growth

3.1 Introduction

The growing entanglement of financialisation with sustainable growth has reshaped contemporary economic systems, raising critical questions about the role of financial actors, instruments, and motives in shaping development outcomes. Financialisation, broadly understood as the increasing dominance of financial markets, institutions, and logic in economic and social life, has permeated sustainable finance, influencing policy frameworks and investment strategies. This chapter interrogates how financialisation facilitates or constrains sustainable growth, assessing its structural implications and the tensions it introduces. The discussion explores the concept of financialisation, tracing its historical trajectory and expansion beyond traditional financial markets. The transformation of economies through financial deregulation, the proliferation of financial instruments, and the prioritisation of shareholder value has reconfigured corporate and governmental behaviour. It delves into the role of the state and market in financing sustainable growth. The chapter further explores the complex reality of financing sustainable growth, addressing the inherent trade-offs embedded within financialised approaches to development. The discussion then turns to the tensions between financialisation and sustainable growth, particularly the conflict between profit-driven incentives and

N. Udohaya, *Impact Investing and Financial Inclusion*, Sustainable Development Goals Series,
https://doi.org/10.1007/978-3-031-87841-1_3

long-term sustainability objectives. Lastly, the chapter considers the structural inequalities reinforced by financialisation, particularly how capital flows prioritise profitable markets while marginalising vulnerable communities. This section critically examines whether financialisation inherently reinforces existing global inequities or whether it can be restructured to foster more equitable and inclusive development. Through this exploration, the chapter aims to provide a nuanced understanding of financialisation's dual role in sustainable growth—both as a catalyst for mobilising capital and as a potential source of systemic inefficiencies and inequalities, urging a balanced approach that ensures financial mechanisms support, rather than undermine, equitable and sustainable development outcomes. By critically engaging with these themes, this chapter sets the foundation for evaluating the broader implications of financialisation in shaping the future of sustainable finance.

3.2 The Concept of Financialisation

The concept of financialisation has a long and intricate history, characterised by its expansion from traditional economic sectors into broader areas such as development and sustainability. Financialisation can be broadly understood as the increasing dominance of financial motives, financial institutions, and financial actors in shaping economic and social dynamics (Aalbers & Haila, 2018). Particularly in sustainable finance, financialisation entails the influence of these elements on development policies and outcomes. It leverages private capital and market-based mechanisms to address development challenges, often emphasising financial returns alongside, or sometimes over, social objectives (Bernards, 2024). It can be traced back to the post-war era, particularly gaining momentum in the 1970s as economies underwent significant shifts. The collapse of the Bretton Woods system in 1971 marked a turning point, as countries moved away from fixed exchange rates and financial markets began to play a more central role in shaping economic outcomes. In the ensuing decades, financial deregulation, technological advances, and the liberalisation of capital markets paved the way for rapid growth in financial activities. Financial institutions expanded their influence, and the proliferation of financial products, from derivatives to complex credit instruments, became integral to economic functioning. This period also saw the increasing separation of financial and productive activities, with financial markets and institutions exerting an ever-growing influence over real sectors of the economy. As financial markets expanded, the emphasis shifted towards maximising shareholder value, often at the expense of other stakeholders, which significantly shaped corporate behaviour.

The proliferation of financialisation into the sustainable investment space, however, is a more recent development. In the late twentieth century, as neoliberal economic policies gained prominence, there was a growing focus on the role of private capital in addressing social and environmental issues. The limitations of public funding in solving these complex challenges led to a search for innovative financing mechanisms, resulting in the blending of financial and social objectives. The emergence of microfinance in the 1980s and 1990s is often cited as an early example of financialisation within the social investment space. By the early 2000s the formalisation of impact investing as a distinct sector, which sought to mobilise private capital to address social and environmental challenges, positioned financial markets as a solution to global issues. Financialisation's influence became evident with the development of innovative financial products like social impact bonds (SIBs) and green bonds (Knoll & Fraser, 2024; Sinclair et al., 2021), which aim to attract investors by blending financial returns with social or environmental outcomes.

3.3 Market and the State

The theoretical underpinnings in financing sustainable growth are primarily market-oriented, drawing on neoliberalism and public choice theory. Neoliberalism advocates for minimal state intervention and posits that market mechanisms are the most efficient resource allocation. This perspective has heavily influenced the design and implementation of financialised development tools, reinforcing the belief that the private sector is better suited to drive economic growth and deliver public goods. Public choice theory, which views government actors as inherently self-interested and prone to inefficiency, further strengthens the argument for private-sector-led development. Nonetheless, neoliberal theories have faced substantial criticism for failing to account for the social dimensions of development and the frequent occurrence of market failures. Its critics argue that approaches rooted in such theories often prioritise efficiency and profitability, overlooking the importance of equity, social justice, and the structural determinants of poverty (Fine, 2019). In a market-oriented system, the state's role in addressing such systemic issues is marginalised. Rather than the provider and the driver, the role of the state shifts to that of a facilitator of private investment. Their role is operationalised through regulatory frameworks, credit enhancements, and risk-sharing mechanisms designed to attract private capital. While proponents of market-oriented systems argue that this enables the state to leverage private-sector efficiency, critics contend that it reduces public control over essential services and subjects such systems to the volatility of market dynamics. The retreat of the state from direct service provision has, in many cases, resulted in reduced access and increased costs, particularly for marginalised communities. In sectors such as water and healthcare, privatisation has often led to outcomes that contradict the equity and inclusiveness goals of development, highlighting the inherent limitations of relying on private capital to deliver public goods (Hall, 2015). Financialisation then introduces novel financial instruments and expanded funding sources for addressing social issues, but not without its challenges. The argument is that its emphasis on financial returns can undermine the social objectives that underpin impact investments, raising important questions about the consequences for equitable and sustainable development.[1]

3.4 The Complex Reality of Financing Sustainable Growth

The concern in development finance literature is that financialisation risks sidelining core sustainability goals outcomes like poverty reduction, social equity, and climate change. Chiapello (2020, p. 81) describes financialisation as "...a specific process of transforming the world, objects, organisations and the problems we encounter, by the introduction of 'financialised' practices, theories and instruments". This introduction of financialised practices has profound implications for how development objectives are conceived, financed, and implemented. Financialised practices encompass a wide array of financial tools and strategies, many of which are designed to attract private investment by mitigating risks and ensuring profitability. The tools and strategies seek to reshape the traditional landscape of development interventions and are often justified by theories that extol the virtues of market discipline, efficiency, and the ability of private actors to innovate in addressing development challenges. However, they present a complex reality for growth and can sometimes lead to unintended consequences that may undermine the very goals they aim to achieve. On one hand, the tools and strategies have introduced innovative ways to mobilise capital and align incentives for development outcomes. They pool private-sector resources and expertise, expand funding for critical development projects, and introduce performance-based accountability. On the other hand, they introduce the risk of financialisation, which brings about significant trade-offs. Such risks include the prioritisation of financial returns and risk mitigation for investors, which can undermine social objectives, leading to an imbalance where the most profitable projects receive funding while those with the most significant social need are neglected. The theoretical assumptions underpinning financialisation—such as the belief in market efficiency and the primacy of private-sector solutions—do not always hold true in the context of sustainable growth and development. Markets are often imperfect, especially in low-income settings where information asymmetry, lack of infrastructure, and systemic inequalities can limit the effectiveness of market-based interventions. Financialisation tends to favour projects that can generate measurable financial returns, which may not align with the complex, multidimensional nature of development challenges. Issues such as poverty, inequality, and environmental degradation require holistic solutions that address both immediate needs and structural causes, something that purely financial approaches are often ill-equipped to achieve.

Chiapello's (2020) framework of financialisation offers a critical lens to understand its profound influence on sustainable development. It highlights financialisation as a technical mechanism and a paradigm that reframes development challenges through the logics of capital markets. The author identifies three processes integral to financialisation, namely problematisation, tangibilisation, and financial structuring. Problematisation involves redefining complex social and environmental challenges—such as poverty, climate change, and inequality—as financial problems. This reframing aligns development priorities with market imperatives, privileging solutions that can be expressed in financial terms over those grounded in social or ecological contexts. Tangibilisation then converts these reframed problems into financial metrics, products, or instruments, such as carbon credits, green bonds, and sustainability-linked loans. These instruments often reduce multidimensional challenges into narrowly defined targets, such as emission reductions or profit margins, potentially obscuring the broader systemic issues they aim to address. Financial structuring completes this process by embedding these instruments into global financial markets, effectively transforming development interventions into commodities traded for profit. While these mechanisms mobilise substantial capital for sustainability initiatives, they risk subordinating developmental goals to the demands of financial profitability. This commodification can marginalise critical but less monetisable objectives, such as community empowerment, cultural preservation, or equitable resource distribution. Moreover, the reliance on financial instruments often introduces market volatility into development financing, destabilising long-term efforts to address systemic issues.

Furthermore, the assumption that aligning development goals with market mechanisms inherently enhances efficiency and effectiveness is called into question as critics have argued such alignment often depoliticises development by framing socio-economic inequalities and environmental degradation as technical issues solvable through market-based solutions rather than as manifestations of power imbalances and structural inequities. This depoliticisation diverts attention from systemic reforms needed to address the root causes of unsustainable practices. Integrating financial logic into development also poses risks of reproducing and exacerbating global inequalities. Financial markets prioritise regions and projects offering higher returns, often sidelining low-income and high-risk areas that require the most support. The resulting capital flows reinforce existing disparities, channelling resources to already-developed regions while leaving marginalised communities further behind. Furthermore, the focus on financial metrics can incentivise greenwashing, thereby undermining the credibility of sustainable finance (Sun, 2024).

3.5 The Tensions Between Financialisation and Sustainable Growth

Profit vs. Purpose: One of the fundamental tensions between financialisation and sustainable growth lies in the conflict between profit maximisation and the pursuit of broader social and environmental objectives. Financialisation, emphasising financial returns and shareholder value, often prioritises short-term gains over long-term, purpose-driven initiatives. This profit-driven mindset can divert resources and attention away from addressing pressing societal and environmental challenges. For instance, impact investing—initially conceived as a way to align capital with positive social outcomes—can sometimes be co-opted by financial actors whose primary focus remains on financial returns, thus compromising the integrity of social and environmental objectives. When profit takes precedence, projects that yield high financial returns but do not align with sustainability goals are often prioritised, leaving purpose-driven initiatives underfunded or neglected. This divergence creates a systemic bias against projects with longer payback periods or those whose returns are difficult to quantify in purely financial terms, thereby undermining the potential for genuinely sustainable growth.

Risk of Exclusion: Another significant tension is the risk of exclusion, where financialised models driven by profitability concerns can sideline underserved populations. Financialisation tends to favour sectors, projects, and individuals that promise higher returns, often at the expense of marginalised groups whose needs may be less profitable to address. In sustainable finance, this exclusionary tendency can manifest in several ways. For example, microfinance—once celebrated as a financial innovation that empowered underserved communities—has in some instances become commercialised to the point where its focus on profitability overrides its original mission of social impact. High interest rates and stringent repayment conditions have, at times, exacerbated the financial vulnerabilities of borrowers rather than alleviating them. Additionally, large-scale green infrastructure projects funded through financialised mechanisms may overlook the needs and voices of local communities, particularly in the Global South, where access to decision-making processes is often limited. The result is a form of sustainable development that does not equitably benefit all stakeholders, further entrenching social inequalities.

Metrics and Accountability: The reliance on financial metrics to assess success presents another challenge to achieving sustainable growth. Financialised models often emphasise metrics such as return on investment (ROI) or internal rate of return (IRR), which are inherently limited in capturing the full spectrum of social and environmental impacts. This over-reliance on financial metrics can lead to a skewed understanding of growth, where the true costs and benefits to society and the environment are overlooked. For instance, a renewable energy project might be deemed successful based on its financial returns. Yet, the broader impacts—such as disruptions to local ecosystems or social dislocation—may be disregarded if they do not translate into financial terms. The challenge, therefore, is to develop more holistic impact assessments that prioritise long-term, systemic change over short-term financial gains, ensuring that accountability extends beyond financial performance to include genuine social and environmental contributions.

As a consequence of financialisation, the commodification of social and environmental issues—where these issues are viewed primarily through the lens of market opportunities—can detract from the core mission of development, By prioritising financial returns, financialisation may perpetuate existing inequalities, as capital flows tend to favour more developed regions and sectors with established revenue potential, rather than the poorest and most marginalised communities who need support the most.

3.6 Structural Inequalities Reinforced by Financialisation

Financialisation has significant implications for social equity. This discussion explores how financialisation reinforces structural inequalities through capital flow dynamics, imbalanced power and representation, and its impact on widening socio-economic disparities.

Capital Flow Dynamics: The dynamics of capital flow under financialisation reveal a tendency for financial resources to be directed towards "safe" and profitable markets, often at the expense of truly underserved regions. Financialisation tends to prioritise investments that promise immediate returns while minimising risk. As a result, capital is disproportionately directed to regions and sectors perceived as stable and profitable, such as real estate in established urban centres or infrastructure in developed economies. This capital flow bypasses regions in greater need of investment but deemed risky, such as those with weaker financial institutions or political instability, often found in the Global South. This dynamic exacerbates existing inequalities as underserved regions struggle to attract the capital necessary for sustainable development. Financial institutions, driven by a focus on risk-adjusted returns, are reluctant to allocate resources to projects that do not offer apparent, short-term profitability, even if those projects have substantial potential to generate long-term social and economic benefits. For instance, many regions in Sub-Saharan Africa and parts of South Asia remain marginalised in capital investment due to the perception of high risk. Consequently, the regions most in need of infrastructure, healthcare, and educational development are sidelined, reinforcing existing disparities in economic opportunity and growth.

Power and Representation: The mechanisms of financialisation also play a significant role in shaping power dynamics and representation in decision-making processes. Financialised systems are often characterised by centralised decision-making structures dominated by institutional investors, multinational corporations, and powerful financial intermediaries. These actors, primarily concerned with maximising financial returns, frequently operate with little accountability to local communities or the individuals directly impacted by their investment decisions. Local voices and communities, particularly those in underserved or marginalised areas, are often excluded from critical decisions that affect their lives. This exclusion is evident in large-scale infrastructure and development projects, where financial backers and multinational corporations make decisions without meaningful consultation or input from local stakeholders. For example, in many large renewable energy projects, such as hydropower dams or wind farms, local communities have reported inadequate involvement in the planning process, leading to social displacement and environmental degradation without adequate compensation or support. The concentration of decision-making power in the hands of a few financial actors thus perpetuates an imbalance in representation, where those with the most significant stake in development outcomes have the least influence over them.

Impact on Inequality: Financialisation's impact on inequality is profound, as it perpetuates and even exacerbates existing disparities. At its core, financialisation is concerned with the commodification of assets and the maximisation of financial returns. This profit-driven approach often prioritises wealth accumulation for investors and shareholders rather than addressing the needs of broader society. The disproportionate benefits accruing to those who already possess capital—such as institutional investors and wealthy individuals—create a feedback loop that reinforces wealth concentration and deepens socio-economic divides. One illustrative example of this dynamic is the housing market. The financialisation of housing, where properties are treated as financial assets rather than basic human necessities, has led to rising property values and rent prices, particularly in urban areas. Institutional investors have increasingly entered the housing market, buying up large portfolios of residential properties and driving up costs. This trend has made housing less affordable for low-income populations, effectively displacing vulnerable communities and exacerbating inequality in access to shelter. Similarly, in agriculture, financial speculation on commodity prices has contributed to increased volatility, which disproportionately impacts smallholder farmers in developing countries who lack the resources to absorb price shocks, further entrenching poverty and inequality. The unequal distribution of benefits and risks inherent in financialised systems has far-reaching implications for social equity. While financialisation has contributed to economic growth and created new opportunities for wealth generation, it has done so in a way that often privileges those with existing financial power. Addressing these entrenched inequalities requires a rethinking of how financial systems are structured, emphasising ensuring that capital is deployed in ways that promote inclusive development and empower marginalised communities.

3.7 Manifestations and Dualities of Financialisation in Sustainable Finance

The manifestation of financialisation in sustainable finance unfolds through several interconnected dimensions that fundamentally reshape the goals, structures, and practices of development finance and sustainability. Drawing on academic discussions and critical analyses, this section explores the nuanced ways financialisation materialises in sustainable finance.

Commodification of Development Goals: Financialisation transforms complex, multidimensional social and environmental challenges into tradable financial assets. This commodification is deeply problematic as it often narrows the scope of sustainability, prioritising measurable, monetisable outcomes over systemic change. For example, carbon markets—a flagship mechanism—frequently encourage offset purchases rather than fostering fundamental shifts in production processes or consumption patterns (Knox-Hayes, 2013). Moreover, by translating public goods like biodiversity and clean air into financial products, commodification risks sidelining intangible values such as community well-being or cultural preservation by translating public goods like biodiversity and clean air into financial products. Despite these critiques, aspects of commodification remain indispensable for aligning sustainability objectives with global financial markets (Hahn et al., 2015). By embedding development goals into tradable mechanisms, commodification attracts vast private capital flows that public funding alone cannot achieve. This alignment allows for market-driven scalability and the development of innovative financial instruments, such as biodiversity credits and green bonds, which make sustainability initiatives financially viable and competitive in market-driven economies.

Institutional Restructuring: As an element of financialisation, institutional restructuring introduces profound implications for development and sustainability. The shift by Development Finance Institutions (DFIs) and multilateral organisations towards financialised models reorient priorities to align with market logic. This reorientation frequently sidelines grassroots initiatives and community-driven projects that lack immediate financial appeal or quantifiable outcomes demanded by private investors. Such sidelining exacerbates systemic inequalities, leaving marginalised communities and locally grounded solutions without adequate support. Additionally, institutional restructuring often results in a concentration of funding on large-scale, market-friendly projects that align with the profitability expectations of private capital. While impactful in terms of scale, these projects may neglect the nuanced, context-specific needs of local populations. The top-down nature of these investments risks creating solutions that are misaligned with community priorities, leading to inefficiencies and potential resistance. Furthermore, the drive to integrate private capital frequently prioritises financial metrics over developmental objectives, undermining the transformative potential of sustainability-focused initiatives. Despite these drawbacks, institutional restructuring remains crucial for addressing the immense financing gaps in global sustainability efforts. DFIs and multilateral organisations can attract significant private-sector participation by leveraging public funds to de-risk private investments. This collaboration enables large-scale funding for infrastructure, renewable energy, and other high-impact projects that would otherwise remain underfunded. When executed with care, institutional restructuring can bridge public-sector priorities and private-sector efficiency, fostering innovation and scalability. Relevant institutions must incorporate safeguards to ensure that grassroots initiatives and marginalised communities are not excluded. Policymakers should develop frameworks that prioritise inclusivity, local engagement, and long-term impact over short-term financial returns. A balanced approach that integrates the strengths of private capital while preserving the developmental focus of public institutions is essential for achieving equitable and sustainable outcomes.

Proliferation of ESG Metrics and Financial Instruments: The integration of ESG factors into financial markets has driven the creation of sophisticated metrics and investment products, such as green bonds, sustainability-linked loans, and development impact bonds. These tools align financial returns with ESG outcomes, facilitating the incorporation of sustainability into mainstream finance. However, the proliferation of ESG metrics has introduced significant inconsistencies. Divergent methodologies among rating agencies undermine their reliability. Also, the emphasis on measurable outcomes can neglect less quantifiable but equally critical aspects, such as social cohesion or cultural resilience. The focus on financial products often prioritises short-term gains over transformative, long-term investments. For instance, renewable energy projects with predictable cash flows are more likely to attract funding than community-led initiatives addressing systemic social inequalities. This short-termism reflects the inherent tension between market imperatives and the temporal horizons required for meaningful sustainability transitions. The complexity of the financial instruments often excludes smaller actors and community-based organisations that lack the resources or expertise to navigate these financialised landscapes. Financialised tools risk privileging capital-intensive solutions while neglecting grassroots innovation and equity-driven approaches (Mader et al., 2020). Additionally, complex financial instruments amplify market volatility, as smaller actors and developing regions face exclusion due to their inability to navigate intricate compliance requirements. The implication of this concentration of investments in regions and sectors deemed commercially viable at the expense of initiatives targeting social equity or grassroots empowerment is the reshaping of development priorities around profitability rather than impact (Chiapello, 2020). Despite these drawbacks, financial instruments are indispensable for scaling sustainability efforts and fostering alignment between capital markets and developmental goals. They international capital flows facilitated by these instruments and standards are critical for addressing global challenges such as climate change and biodiversity loss.

Dependency on Market Mechanisms: Financialisation embeds market mechanisms as the central pillar of sustainability financing, this brings both significant opportunities and critical challenges. On one hand, market mechanisms inherently prioritise financial profitability, often at the expense of broader developmental goals. Projects that generate high financial returns, such as large-scale renewable energy installations, attract disproportionate investments. Meanwhile, initiatives targeting systemic issues like social equity, grassroots empowerment, or decentralised community solutions often struggle to secure funding due to their limited short-term profitability. This imbalance creates a resource allocation problem, sidelining projects that could deliver long-term societal benefits. Reliance on market mechanisms exacerbates systemic inequalities. High-income regions with stable regulatory environments naturally attract more investment, leaving marginalised and politically unstable regions underserved. This dynamic reinforces global disparities, as low-income communities are deprived of the resources necessary for sustainable development. Market-driven approaches also tend to favour short-term financial returns over the long-term systemic changes required for addressing complex global challenges like climate adaptation and poverty alleviation. Yet on the other hand, market mechanisms are necessary for mobilising the vast private capital needed to achieve global sustainability goals. They create incentives for innovation and efficiency by aligning sustainability objectives with the profit motives of private actors, thereby engaging a broader pool of resources. For instance, market mechanisms have catalysed the growth of green bond markets, attracting institutional investors and unlocking billions for renewable energy and sustainable infrastructure projects. In addition, market mechanisms provide a framework for scalability, allowing successful projects to be replicated and expanded across regions and sectors. They also encourage the development of financial products that integrate sustainability into mainstream investment practices, promoting a culture of accountability and measurable impact. A robust regulatory framework and targeted incentives that prioritise inclusivity, equity, and long-term sustainability outcomes can induce less dependence on market mechanisms. Policymakers and multilateral organisations play a vital role in ensuring that market mechanisms are designed to balance profitability with the broader social and environmental imperatives of sustainable development.

Influence of Financial Actors: The argument against the dominance of financial institutions in sustainable finance is that it centralised decision-making processes around profitability and risk aversion. Asset managers, institutional investors, and banks act as gatekeepers, determining the allocation of resources based on financial viability rather than social impact. This centralisation exacerbates global inequalities, channelling funds towards high-income regions with stable regulatory environments while excluding low-income areas perceived as high-risk. Rizzi et al. (2018) observed that the concentration of power aligns resource allocation with the risk-return preferences of financial actors. Grassroots initiatives, community-led projects, and programmes with long-term impact but limited immediate profitability frequently struggle to secure funding. This imbalance marginalises efforts aimed at addressing systemic inequities and amplifies existing disparities within the global financial ecosystem. Also, the prioritisation of shareholder value inherent in financial actors' strategies can conflict with sustainability objectives. For instance, large-scale infrastructure projects backed by institutional investors may prioritise financial returns over community involvement or environmental integrity. Such profit-driven orientation fosters a transactional approach to sustainability, reducing complex developmental challenges to financial opportunities. However, financial actors are indispensable for development and sustainability. They mainstream sustainability within global financial systems, and their expertise in managing complex portfolios and structuring innovative instruments such as green bonds and sustainability-linked loans has been critical in mobilising capital at unprecedented scales. Financial actors bring technical knowledge and market access that enable the development and execution of large-scale sustainability initiatives. While their influence requires careful regulation to prevent the prioritisation of profits over impact, their role is integral to achieving the scale and efficiency needed to address global sustainability challenges.

Fragmentation of Regulatory Frameworks: Financialisation intensifies the complexity and fragmentation of regulatory frameworks in sustainable finance. The proliferation of ESG standards, taxonomies, and reporting requirements reflects financial markets' response to the demand for sustainability metrics. While these frameworks aim to enhance transparency and accountability, they often lack coordination, leading to a patchwork of overlapping and inconsistent standards. This fragmentation increases compliance costs, disproportionately affecting smaller actors and entities in emerging markets, and creates barriers to equitable participation in sustainable finance. Moreover, the focus on financialised metrics such as ESG ratings can lead to discrepancies between different rating agencies, further complicating the landscape. These inconsistencies undermine trust and transparency, essential pillars of sustainable finance. Embedding financialisation into regulatory structures often shift the emphasis from substantive sustainability outcomes to the ability to meet specific, narrowly defined financialised criteria. Nonetheless, regulatory frameworks remain indispensable for fostering accountability and ensuring market integrity and, therefore, sustainability. Efforts to harmonise fragmented standards are critical to addressing the inefficiencies introduced by financialisation. A coordinated regulatory approach can balance the need for transparency with inclusivity, enabling diverse actors to participate meaningfully in sustainable finance markets while maintaining robust accountability mechanisms.

Securitisation and Microfinance: The application of securitisation within development finance is another manifestation of financialisation. Securitisation involves the bundling of financial assets, such as microloans, into securities that are subsequently sold to investors. This process introduces several challenges. First, the financial imperatives of securitisation shift the focus from social objectives to profitability. In the case of microfinance, securitisation has led to higher interest rates and aggressive lending practices, contributing to over-indebtedness among vulnerable borrowers (Bateman & Chang, 2012). These practices undermine the original mission of microfinance, which aims to empower underserved communities, highlighting the tension between financial returns and developmental goals. Gabor (2019) cautions that the structural ambition of organising development finance around securitisation, may prioritise market mechanisms and profitability over genuine developmental and environmental outcomes. Furthermore, securitisation mirrors the speculative tendencies of mainstream financial markets, introducing volatility and systemic risks into the development ecosystem. Despite these risks, securitisation is not inherently detrimental. It enhances the liquidity of financial institutions, expands access to credit and facilitates the mobilisation of private capital for development projects. The mechanism fosters collaboration among diverse stakeholders, enabling the pooling of resources to address large-scale sustainability challenges (Gabor, 2019). When carefully regulated, securitisation can balance financial imperatives with developmental objectives, providing a powerful tool for achieving systemic impact.

3.8 Balancing Market Efficiency with Transformative Equity

Financialisation has played a pivotal role in shaping capital allocation, expanding access to new financial instruments, and facilitating private-sector participation in sustainability. A market-efficient approach to sustainability finance typically seeks to optimise capital allocation, minimise risks, and generate predictable returns. While these principles have enabled rapid growth in financialised sustainability instruments—such as green bonds, sustainability-linked loans, and impact investing—they also introduce systemic risks. Market-based financialisation tends to concentrate investments in sectors, regions, and projects that offer the highest financial returns, often at the expense of communities and sectors that may be critical for equitable development but are perceived as high risk or low yield. The result is a misalignment between investment priorities and the urgent need for inclusive and transformative development. Addressing this misalignment can help balance the efficiency of the market with transformative equity. It requires recalibrating financialisation for social and environmental impact. Financialisation must be complemented by frameworks that prioritise transformative equity, ensuring that financial innovations do not merely reinforce existing inequalities but actively contribute to social justice and environmental restoration. This requires mechanisms that explicitly account for fair resource distribution, participatory governance, and long-term impact assessment.

A key challenge in balancing financialisation and transformative equity lies in capital access and distribution. Financialised markets tend to favour large-scale, market-friendly initiatives, sidelining grassroots and community-led projects that often lack financial leverage but are essential for sustainable development. A more balanced approach requires impact-driven financial mechanisms that channel resources into underfunded regions, including blended finance models, concessional lending, and outcome-based financing structures. These mechanisms help mitigate investment risks while ensuring that marginalised communities benefit from financial innovation. Another crucial factor is inclusive financial governance. Decision-making in financialised markets is often concentrated among institutional investors, banks, and multinational corporations, limiting the representation of local communities and civil society in financial governance structures. Strengthening multi-stakeholder decision-making and regulatory oversight can help ensure that financialised instruments align with broader developmental goals. Governments and regulatory bodies must adopt policies that incentivise equity-driven investment practices, requiring financial actors to integrate social and environmental impact metrics into their investment decisions. Balancing efficiency with transformative equity requires recalibrating financialisation for social and environmental impact.

The recalibration of financialisation requires a shift from a profit-first model to an impact-first model, where financial returns are pursued in conjunction with systemic social and environmental improvements through:

- **Enhanced impact measurement frameworks** that prioritise long-term transformation over short-term financial gains.
- **Progressive regulatory frameworks** that enforce accountability in sustainability-linked financial instruments, preventing greenwashing and speculative distortions.
- **Incentivising investments in high-impact sectors** such as climate adaptation, affordable housing, and financial inclusion, even when these projects may not yield immediate financial returns.
- **Integrating participatory finance models** that allow communities to have greater agency over capital flows and investment priorities.

Financialisation must evolve beyond efficiency metrics and short-term risk management to embrace a broader vision of sustainability—one that places human well-being and planetary resilience at the core of financial decision-making. A more balanced approach will ensure the continued growth of sustainable finance and make it a truly transformative force in addressing global inequality and environmental challenges.

3.9 Conclusion

While financialisation has unlocked significant capital flows for sustainability initiatives, it has also introduced structural inefficiencies and ethical dilemmas that must be critically examined. The growing reliance on market-based mechanisms and financial instruments to drive sustainability has reshaped the development landscape, raising essential questions about inclusivity, accountability, and long-term impact. One of the central insights from this discussion is that financialisation has facilitated the mobilisation of private capital through instruments such as green bonds and social impact bonds. These mechanisms have expanded the financing available for critical environmental and social initiatives, bridging gaps left by traditional public funding. However, the financial imperatives embedded within these instruments often prioritise profitability over developmental impact, leading to potential distortions where investments are channelled into commercially viable projects rather than those with the greatest social need. Another key tension highlighted in this chapter is the reallocation of economic power. Financialisation has reinforced the dominance of financial actors who wield considerable influence over sustainability agendas. This concentration of decision-making power can marginalise grassroots initiatives and communities, particularly in the Global South, where access to finance remains constrained. The risks of exclusion and deepening inequalities underscore the need for financial governance frameworks that prioritise equitable resource distribution and participatory decision-making. Moreover, the commodification of sustainability has raised concerns about the efficacy of market-driven solutions. While financial markets can offer scale and efficiency, they are also susceptible to short-termism, speculative behaviour, and greenwashing. This raises fundamental questions about whether financialisation is genuinely advancing sustainability objectives or merely repackaging them in ways that cater to investor interests without delivering meaningful structural change. To mitigate these risks, a balanced approach to financialisation is necessary—one that aligns financial innovation with sustainability imperatives while maintaining regulatory oversight and accountability. Financialisation harnessed effectively, can be a powerful catalyst for sustainable growth. However, without critical interventions, it risks reinforcing existing inequalities and prioritising financial returns over long-term societal well-being. The future of sustainable finance must, therefore, be guided by a commitment to recalibrating financial systems to serve both people and the planet, ensuring that financialisation evolves as a means of achieving—not obstructing—genuine sustainability transformation.

Note

1. For instance, the high interest rates and profit-driven practices of microfinance, and the appropriateness of applying market-based

approaches to complex social problems as it relates to social impact bonds. Also see discussion by Kuokkanen (2024).

References

Aalbers, M., & Haila, A.-K. E. (2018). The financialization of housing: A political economy approach. *Urban Studies, 55*(15), 1821–1835.

Bernards, N. (2024, January 02). Where is finance in the financialization of development? *Globalizations, 21*(1), 88–102. https://doi.org/10.1080/14747731.2023.2222481

Bateman, M., & Chang, H.-J. (2012). Microfinance and the illusion of development: From hubris to nemesis in thirty years. *World Economic Review, 1*, 13–36. Retrieved from https://ssrn.com/abstract=2385482

Chiapello, E. (2020). Financialization as a socio-technical process. In *The Routledge international Handbook of financialization* (pp. 81–91). Routledge.

Fine, B. (2019). Post-apartheid South Africa: It's neoliberalism, stupid! In J. Reynolds, B. Fine & R. van Niekerk (Eds.), *Race, class and the post-apartheid democratic state*. University of KwaZulu-Natal Press.

Gabor, D. (2019). Securitization for sustainability. *Does it help achieve the sustainability development goals*. Heinrich Boll Foundation.

Hahn, T., McDermott, C., Ituarte-Lima, C., Schultz, M., Green, T., & Tuvendal, M. (2015, December 01). Purposes and degrees of commodification: Economic instruments for biodiversity and ecosystem services need not rely on markets or monetary valuation. *Ecosystem Services, 16*, 74–82. https://doi.org/10.1016/j.ecoser.2015.10.012

Hall, D. (2015). *Reclaiming public services. How cities and citizens are turning back privatisation*. Transnational Institute (TNI).

Knoll, L., & Fraser, A. (2024, January 02). Social impact bond assetization struggles: A comparative case study of the United Kingdom and Germany. *Economy and Society, 53*(1), 92–111. https://doi.org/10.1080/03085147.2024.2309802

Knox-Hayes, J. (2013, December 01). The spatial and temporal dynamics of value in financialization: Analysis of the infrastructure of carbon markets. *Geoforum, 50*, 117–128. https://doi.org/10.1016/j.geoforum.2013.08.012

Kuokkanen, N. (2024, March 01). A problematizing review of the financialization of living beings. *Critical Perspectives on Accounting, 99*, 102739. https://doi.org/10.1016/j.cpa.2024.102739

Mader, P., Mertens, D., & Van Der Zwan, N. (2020). Financialization: An introduction. In *The Routledge international Handbook of financialization* (pp. 1–16). Routledge.

Rizzi, F., Pellegrini, C., & Battaglia, M. (2018, January 01). The structuring of social finance: Emerging approaches for supporting environmentally and socially impactful projects. *Journal of Cleaner Production, 170*, 805–817. https://doi.org/10.1016/j.jclepro.2017.09.167

Sinclair, S., McHugh, N., & Roy, M. J. (2021). Social innovation, financialisation and commodification: A critique of social impact bonds. *Journal of Economic Policy Reform, 24*(1), 11–27.

Sun, W. (2024). Unmasking the hidden risks in ESG reporting: How egoism distorts transparency and what can be done about it. *Research in Economics and Management Science, 1*(1).

CHAPTER 4

Rethinking Financialisation

4.1 Introduction

This chapter explores the imperative of rethinking value creation within sustainable finance, advocating for an approach that moves beyond financial maximisation towards long-term resilience, equitable growth, and ecological restoration. At the heart of this discussion is the argument that rejecting financialisation outright is neither practical nor desirable. Instead, the focus must be on reshaping financial mechanisms to align with sustainability imperatives. The chapter critically examines how financial instruments can be redesigned to support holistic outcomes, embedding ethical and regulatory safeguards that ensure capital is channelled towards projects that generate multidimensional impact. Through the lens of alternative theoretical frameworks such as stakeholder value and shared value creation, this chapter challenges conventional notions of financial success, proposing a more inclusive and regenerative model of economic progress. A central component of this re-evaluation is the development of dynamic and adaptive methodologies for assessing sustainability outcomes. The chapter introduces the concept of regenerative metrics, which go beyond conventional ESG indicators by actively measuring restoration efforts in ecosystems, climate resilience, and community well-being. Rather than focusing solely on minimising harm, these metrics emphasise proactive interventions that enhance environmental and social

N. Udohaya, *Impact Investing and Financial Inclusion*, Sustainable Development Goals Series,
https://doi.org/10.1007/978-3-031-87841-1_4

conditions over time. Further, the discussion delves into the role of real-time adaptability in financial sustainability, enabled by emerging technologies such as blockchain and artificial intelligence (AI). These tools enhance transparency, accountability, and the capacity for responsive financial decision-making. For instance, blockchain can track supply chain sustainability in real time, while AI-powered analytics provide predictive insights on climate risks and investment impact.

Recognising the importance of collaborative co-creation, the chapter highlights the necessity of engaging diverse stakeholders—governments, private-sector actors, civil society, and Indigenous communities—in shaping adaptive sustainability metrics. In addition, the chapter underscores the role of circular economic principles, where financial instruments are structured to promote resource efficiency, waste reduction, and closed-loop production systems in societies. Finally, the chapter examines the challenges of universal standardisation and regulatory alignment in sustainable finance. While numerous sustainability frameworks exist, fragmentation often leads to inefficiencies and greenwashing. Harmonising regulatory approaches and impact measurement standards is essential for ensuring accountability and comparability in global financial markets. Through a comprehensive exploration of these themes, this chapter sets the stage for a transformative rethinking of value creation in sustainable finance. It argues that by embedding adaptive, inclusive, and regenerative financial models, the global economy can transition towards sustainability in a manner that balances financial viability with long-term societal and planetary health.

4.2 Redefining Value Creation for Sustainable Growth

The complexity of sustainable finance and the pervasiveness of financialisation mean that attempting to avoid financialised approaches entirely is neither practical nor desirable. Instead, the focus should be on leveraging financialisation to achieve holistic outcomes that encompass environmental, social, and financial dimensions. By carefully designing financial instruments, embedding social and environmental safeguards in partnerships, and shifting the concept of value beyond mere financial returns, sustainable finance can harness the power of financial markets for transformative development. The challenge lies not in rejecting financialisation but in reshaping it to align with the imperatives of sustainable and equitable growth. Rethinking value creation within the context of financing for growth requires a fundamental paradigm shift—one that moves away from a narrow focus on maximising financial returns to a broader conceptualisation of value that encompasses social equity, environmental health, and long-term resilience. Financialisation, when guided by appropriate ethical and regulatory frameworks, can serve as a mechanism for channelling capital towards projects that create positive, multidimensional impacts. This requires embracing alternative theoretical frameworks such as stakeholder value and shared value creation, which recognise that value is created not only for shareholders but for a diverse array of stakeholders, including communities, employees, and the environment. The concept of value creation in sustainable finance requires a fundamental re-evaluation within the context of financialisation. Financialised practices have predominantly framed value through a narrow lens of financial returns, often sidelining broader societal and environmental objectives. Although significant progress has been made through initiatives like impact-weighted accounting, green bonds, and blended finance models, the journey to fully redefine value creation for sustainability is ongoing. It requires continuous effort, innovation, and systemic integration across global financial systems. It is crucial to focus efforts on deepening existing frameworks, ensuring they are universally adopted, and pushing the boundaries of what sustainability metrics can achieve. Value creation must evolve to incorporate dynamic, adaptive metrics that not only address biodiversity loss, climate change, and social equity disparities but also anticipate and integrate solutions for emerging challenges, fostering regeneration, resilience, and long-term planetary health. That measures the outputs of sustainability initiatives and provides a framework for anticipating, mitigating, and solving emergent challenges. Additionally, fostering a culture of forward-thinking innovation—from leveraging digital technologies like blockchain to advancing community-driven financial models—will play a pivotal role in shaping the future of sustainable finance.

4.2.1 Dynamic and Adaptive Methodology

To achieve an evolution of value creation in sustainable finance, dynamic and adaptive metrics must be developed and integrated into global financial systems. These metrics need to go beyond traditional performance indicators and reflect a holistic understanding of interconnected Environmental, Social, and Governance (ESG) challenges. They should not only measure the outputs of sustainability initiatives but also provide a framework for anticipating, mitigating, and solving emergent challenges. Achieving this transformation requires a multifaceted approach rooted in innovation, collaboration, and systemic integration. Adaptive metrics can enable financial systems to move towards a paradigm prioritising regenerative outcomes, equity, and resilience.

(a) **Regenerative Metrics**

The development of regenerative metrics represents a transformative shift in the approach to sustainability. Unlike traditional measures that focus on minimising harm, regenerative metrics emphasise the active restoration and enhancement of ecosystems and communities. For example, biodiversity metrics could include ecosystem health indices that assess the active restoration of degraded landscapes through reforestation efforts, the resurgence of native species populations, and tangible improvements in water quality, highlighting regenerative processes rather than passive measurement. On the other hand, climate action metrics should assess long-term carbon sequestration potential, integrating factors such as reforestation efforts and soil health restoration rather than focusing solely on immediate emissions reductions. Regenerative metrics require interdisciplinary collaboration to create comprehensive frameworks. Scientific methodologies must be robust, scalable, and adaptable to regional contexts. For instance, metrics for marine ecosystems could evaluate coral reef regeneration, ocean acidification levels, and the recovery of fish stocks, demonstrating how regenerative efforts restore ecological balance and contribute to broader environmental resilience. Regenerative metrics encourage a shift from short-term sustainability goals to long-term, resilient growth by focusing on systemic improvements.

(b) **Real-Time Adaptability**

Building real-time adaptability into sustainability metrics is crucial for addressing rapidly evolving global challenges. Technologies such as blockchain and artificial intelligence (AI) are pivotal in enabling this adaptability. Blockchain can ensure transparency, accountability, and traceability in sustainability data, creating trust among stakeholders. For instance, blockchain could track the provenance of raw materials in real time, ensuring transparency and enabling financial systems to adapt to disruptions while maintaining ethical and environmental standards, which directly supports real-time adaptability in dynamic contexts. AI-powered analytics provide the capability to predict trends, simulate scenarios, and generate actionable insights. For example, satellite data analysed by AI can monitor deforestation, track carbon emissions, and assess the impact of climate change on vulnerable regions. Such real-time data enables dynamic responses, such as deploying resources to protect threatened ecosystems or adjusting investment strategies to mitigate emerging risks. Integrating these technologies into adaptive metrics ensures financial systems remain proactive and resilient, driving impactful decision-making.

(c) **Collaborative Co-Creation**

The co-creation of adaptive metrics with diverse stakeholders ensures that these frameworks are inclusive, equitable, and contextually relevant. Governments, businesses, academic institutions, and local communities each contribute unique perspectives and expertise to the process. Indigenous knowledge, often undervalued in traditional financial systems, offers invaluable insights into sustainable resource management and ecosystem resilience. For instance, indigenous practices such as rotational farming and water conservation can directly inform metrics by providing measurable outcomes, including increased soil fertility, improved water retention, and enhanced agricultural yield, thus making them integral to the design of contextually adaptive sustainability frameworks. These practices provide quantifiable data on soil health, water retention, and agricultural yield improvements, creating a blueprint for developing adaptive metrics that align with environmental sustainability and local livelihoods. Metrics that reflect cultural and ecological interdependencies align financial objectives with the aspirations of local populations. For example, community-driven approaches to renewable energy projects can incorporate metrics that track the extent of local employment generated, equitable access to energy, and community satisfaction. Engaging diverse perspectives can enable adaptive metrics to become tools for global alignment and local empowerment, ensuring sustainability efforts resonate at every level.

(d) **Circular Economic Principles**

The integration of circular economic principles into financial metrics has the potential to revolutionise how value is measured and created. Circularity emphasises reducing waste, extending product life cycles, and designing closed-loop systems. Metrics aligned with these principles should track resource efficiency, product durability, and waste reduction across supply chains. For example, in manufacturing, metrics could assess the percentage of recycled materials, product lifespan, and the carbon footprint per product life cycle, ensuring these principles are theoretical and also actively drive industries towards circular practices that reduce waste and resource dependency. Incorporating circular economic principles into adaptive metrics encourages businesses to adopt sustainable practices while driving systemic shifts across industries. For instance, metrics for the construction sector could evaluate the use of recycled materials, energy efficiency of buildings, and end-of-life recyclability of construction components. Such metrics incentivise innovation, reduce environmental impacts, and promote a regenerative economy.

(e) **Universal Standardisation and Regulatory Alignment**

Universal standardisation and regulatory alignment are essential to ensure the adoption and effectiveness of adaptive metrics. The fragmentation of sustainability standards often leads to inconsistencies, inefficiencies, and greenwashing. Regulatory bodies, industry consortia, and international organisations must collaborate to harmonise metrics, establish robust reporting frameworks, and enforce accountability. Initiatives such as the Task Force on Climate-related Financial Disclosures (TCFD) provide a valuable foundation for aligning financial decision-making with sustainability objectives. Expanding these frameworks to include adaptive metrics for biodiversity, social equity, and climate resilience will strengthen their relevance and impact. Harmonised standards also facilitate comparability across industries and regions, fostering transparency and trust among investors and stakeholders. Regulatory alignment ensures that sustainability metrics are not just aspirational but integral to global financial practices.

(f) **Continuous Learning and Innovation**

Fostering a culture of continuous learning and innovation within the financial sector is critical for the evolution of adaptive metrics. Financial institutions must prioritise research and development to explore new methodologies, test emerging technologies, and refine metrics to meet evolving challenges. Pilot projects and experimental approaches should be encouraged, enabling stakeholders to evaluate the effectiveness of adaptive metrics in diverse contexts. Collaboration with academic institutions and think tanks accelerates knowledge-sharing and capacity-building. For instance, partnerships with universities can advance technologies for real-time data analysis, allowing metrics to evolve iteratively and remain effective in addressing dynamic sustainability challenges, while ensuring alignment with emerging global priorities. Similarly, think tanks can provide forward-looking policy insights, ensuring that metrics remain aligned with both global sustainability goals and emerging issues, thereby fostering a culture of continuous innovation and adaptability. Continuous innovation ensures that metrics remain relevant, impactful, and capable of addressing the complexities of.

4.2.2 *Integrating Stakeholder-Centric Models*

Integrating stakeholder-centric models into sustainable finance represents a pivotal mechanism for creating inclusive, equitable, and enduring value. Unlike traditional frameworks that prioritise shareholder returns, stakeholder-centric models expand the focus to include the interests of diverse groups such as employees, customers, communities, and the environment. These models aim to align financial objectives with the well-being of all stakeholders, fostering trust, resilience, and long-term sustainability. Achieving this shift requires rethinking governance structures, promoting transparency, and embedding participatory decision-making processes into financial systems. Prioritising these principles allows stakeholder-centric models to address systemic inequalities, drive holistic impact, and contribute to a sustainable global economy.

(a) **Rethinking Governance Structures**

Effective stakeholder-centric models begin with reimagining governance frameworks to reflect the interests of all stakeholders. Traditional governance models often prioritise the needs of shareholders, sidelining other voices in critical decision-making processes. This has historically led to decisions that maximise short-term financial returns at the expense of long-term social and environmental stability. In contrast, stakeholder-centric approaches adopt governance structures that ensure equitable representation and accountability to all affected parties. Multi-stakeholder governance frameworks are a cornerstone of this approach. For instance, cooperatives and benefit corporations provide governance structures where decision-making authority is distributed among stakeholders, ensuring their diverse interests are represented. In financial institutions, the inclusion of multi-stakeholder advisory boards can guide investment strategies, ensuring that projects align with community needs, environmental goals, and ethical standards. Additionally, mechanisms such as weighted voting rights and stakeholder councils can institutionalise stakeholder representation, preventing the dominance of a single group and fostering inclusive decision-making. Governance structures that embed stakeholder perspectives enhance trust and credibility. By involving diverse voices in decision-making, organisations can ensure that their strategies are socially legitimate and contextually relevant. This not only reduces the risk of social opposition and regulatory scrutiny but also strengthens the long-term viability of sustainable investments.

(b) **Transparency and Accountability**

Transparency is a cornerstone of stakeholder-centric models, as it ensures that all stakeholders have access to relevant information about the social, environmental, and financial impacts of investments. Transparency builds trust, facilitates informed participation, and acts as a safeguard against unethical practices such as greenwashing. Technological advancements, particularly in digital tools, have the potential to revolutionise transparency. Blockchain technology, for example, can enable real-time tracking of supply chain activities, ensuring that investments adhere to ethical and sustainability standards. Blockchain's decentralised nature provides an immutable ledger of transactions, enhancing accountability and reducing the risk of manipulation. For instance, investors can use blockchain to verify whether the raw materials used in a project are sourced responsibly and align with sustainability goals. In addition to technological tools, robust reporting frameworks such as the Global Reporting Initiative (GRI) and the Sustainability Accounting Standards

Board (SASB) play a critical role in promoting transparency. Expanding these frameworks to include metrics that directly measure stakeholder well-being—such as community satisfaction, worker safety, and biodiversity conservation—can further enhance accountability. Transparent reporting ensures that stakeholders are equipped to hold organisations accountable for their actions, fostering a culture of ethical behaviour and continuous improvement.

(c) **Participatory Decision-Making**

A stakeholder-centric approach requires embedding participatory mechanisms into decision-making processes. This involves actively engaging stakeholders at every stage of the investment lifecycle, from project design to implementation and evaluation. Such engagement ensures that investments are aligned with stakeholder needs and reflect local contexts. Participatory decision-making can take various forms, including consultations, workshops, and collaborative planning sessions. Community-driven development models are particularly effective in empowering local stakeholders to define priorities and allocate resources. For example, in renewable energy projects, participatory processes could involve local communities in site selection, ensuring equitable access to energy and sharing economic benefits through revenue-sharing agreements. This enhances the legitimacy of investments and also improves their social and environmental outcomes. Effective participatory processes also require mechanisms to address power imbalances among stakeholders. Marginalised groups, such as indigenous communities and low-income populations, often lack the resources and influence to participate meaningfully in decision-making. Providing capacity-building support, such as training and access to information, can empower these groups to engage effectively. Participatory decision-making enhances the inclusivity, accountability, and impact of sustainable investments through meaningful stakeholder engagement.

(d) **Aligning Value Creation with Stakeholder Needs**

Stakeholder-centric models redefine value creation to include not just financial returns but also social and environmental outcomes that directly benefit stakeholders. This holistic approach requires developing metrics and evaluation frameworks that capture these multidimensional impacts.

For example, Social Return on Investment (SROI) is a powerful tool for quantifying the broader benefits of investments. SROI considers factors such as improved community health, reduced inequalities, and enhanced ecosystem services, translating these outcomes into monetary values. Participatory impact assessments, which involve stakeholders in evaluating project outcomes, are another effective method for aligning value creation with stakeholder needs. These assessments provide insights into how investments affect various stakeholder groups, ensuring their perspectives are integrated into decision-making processes. These models foster long-term resilience and shared prosperity by aligning value creation with stakeholder needs. Organisations that prioritise stakeholder well-being are more likely to build strong relationships with communities, enhance employee satisfaction, and gain customer loyalty, all of which contribute to sustained success.

(e) **Building Resilience Through Collaboration**

Stakeholder-centric models thrive on collaboration. Fostering partnerships among businesses, governments, non-profits, and communities enables stakeholder-centric models to create synergies that amplify impact and build resilience. Collaborative efforts enable the pooling of resources, sharing risks, and leveraging expertise to address complex sustainability challenges. Public–Private Partnerships (PPPs) and multi-stakeholder platforms are effective mechanisms for collaboration. For instance, a PPP in agriculture could involve private investors funding sustainable farming practices, governments providing technical support, and local communities contributing traditional knowledge. Such initiatives ensure that investments address the needs of multiple stakeholders while distributing benefits equitably. Additionally, cross-sector collaborations can drive innovation by integrating diverse perspectives and expertise, resulting in more effective and scalable solutions. Collaboration also plays a critical role in addressing systemic risks and uncertainties. Working together enables stakeholders to develop coordinated responses to challenges such as climate change, biodiversity loss, and economic inequality. Collaborative governance structures like regional sustainability councils can facilitate dialogue and joint action, fostering a shared commitment to sustainable development.

4.2.3 Expanding the Role of Public and Development Institutions

Expanding the role of public and development institutions in sustainable finance is imperative for addressing pressing global challenges such as climate change, social inequality, and biodiversity loss. These institutions must transcend conventional mechanisms and adopt paradigms rooted in sustainable value creation—a concept that aligns financial ecosystems with long-term resilience, equitable growth, and environmental stewardship. They can catalyse systemic shifts in global finance by integrating innovative frameworks, fostering collaborative partnerships, and enacting transformative policies. Such an approach requires bold leadership and a nuanced understanding of the complex interplay between economic, social, and environmental factors.

(a) **Deploying Patient and Concessional Capital**

Public and development institutions need to rethink how they structure and deploy these resources to make patient and concessional capital more impactful. A key innovation would be to transition from merely funding projects to actively shaping markets and driving systemic change. Institutions should prioritise projects that have multiplier effects on local economies, promote innovation, and address deep-rooted structural challenges. For example, concessional financing can be tied to innovation challenges, where recipients are incentivised to develop scalable solutions to complex sustainability problems, such as carbon capture or water conservation technologies. Instead of solely providing loans, institutions could explore outcome-based financing mechanisms, where disbursements are linked to measurable impacts, such as greenhouse gas reductions or job creation in marginalised communities. This approach ensures accountability and aligns funding with tangible results. Blended finance models can also be re-engineered to include dynamic risk-sharing mechanisms. For instance, public institutions could offer adaptive guarantees that adjust based on project performance or evolving risks, thereby reducing the cost of capital for private investors while maintaining fiscal prudence. Another potential strategy is to embed circular economy principles within financing terms, encouraging projects to incorporate resource efficiency and waste minimisation into their operations. Additionally, digital platforms powered by blockchain could enhance transparency and traceability in the use of concessional funds. Real-time data on how funds are allocated and utilised can promote trust for these platforms among stakeholders and attract additional private-sector co-investment. Institutions like the Green Climate Fund could pilot such platforms to demonstrate their effectiveness in improving governance and impact.

(b) Establishing Regulatory and Policy Frameworks

To enable sustainable value creation, regulatory and policy frameworks must evolve to incentivise transformative change and foster accountability. Public and development institutions should go beyond conventional compliance measures to design systems that drive sustainable innovation and align global financial systems with climate and social goals. For instance, carbon pricing can be enhanced by integrating artificial intelligence and real-time data analytics to dynamically adjust pricing structures based on emerging climate trends and emissions levels. Similarly, sustainability reporting mandates could evolve into continuous monitoring frameworks that provide stakeholders with actionable insights into a company's progress on ESG goals, rather than static annual reports. At the international level, regulatory frameworks should aim for standardisation, but with adaptive flexibility to account for local and regional variations. Institutions can champion the creation of a global taxonomy for sustainable finance, supported by digital platforms that allow investors to assess real-time compliance and performance metrics. These harmonised frameworks could unlock cross-border investments while ensuring consistent accountability across jurisdictions.

(c) Fostering Innovation and Capacity-Building

Public and development institutions are uniquely positioned to drive innovation in sustainable finance by incubating disruptive technologies and models. Instead of incremental improvements, these institutions can enable breakthroughs by funding high-risk, high-reward ventures that address the root causes of unsustainability. For example, blockchain can be used to tokenise environmental assets such as biodiversity credits or carbon offsets, democratising access to sustainable investments. Artificial intelligence could enhance project evaluation processes, identifying the most impactful investments through predictive modelling and big data analysis. Institutions can channel resources into projects with transformative potential by exploring innovative financial instruments such as catastrophe bonds linked to climate adaptation or community-driven impact funds. Capacity-building should also extend beyond technical skills to foster a culture of sustainability across all stakeholders. Institutions could establish global knowledge-sharing networks that leverage virtual platforms to connect policymakers, investors, and researchers. Regional hubs could focus on tailoring solutions to specific local challenges, such as urban resilience in coastal cities or regenerative agriculture in arid regions. Embedding sustainability principles into every layer of capacity-building will enable public institutions can create a knowledge base that drives systemic change.

(d) **Catalysing Partnerships and Collaborative Action**

Collaboration is central to sustainable value creation, and public and development institutions must take the lead in fostering partnerships that break down silos and amplify impact. These partnerships should be structured around shared goals, mutual accountability, and co-creation of solutions. Public–Private Partnerships (PPPs) must evolve into innovation ecosystems. For instance, renewable energy PPPs could integrate cutting-edge energy storage solutions, digital smart grids, and decentralised energy systems, creating comprehensive models that scale effectively. Similarly, cross-sectoral initiatives—such as combining public financing with grassroots expertise—could address challenges like deforestation through sustainable land-use planning. Platforms for collaboration must prioritise inclusivity and ensure that diverse voices are represented. Digital tools, such as AI-driven stakeholder mapping, can identify gaps and opportunities for collaboration, ensuring that partnerships are not only strategic but also equitable. Institutions could also facilitate cross-industry alliances, bringing together sectors such as agriculture, technology, and finance to co-develop scalable solutions for shared challenges.

(e) **Ensuring Equity and Inclusivity**

Equity and inclusivity are essential for ensuring sustainable finance addresses systemic injustices and promotes shared prosperity. Public and development institutions must design financial ecosystems that prioritise historically marginalised communities and ensure their active participation in decision-making. Participatory financing models, such as community-driven impact funds, can enable local ownership of sustainability initiatives. These models could be further enhanced by digital platforms that allow individuals to directly invest in projects aligned with their values, creating a more democratic approach to finance. Institutions should also develop financing solutions tailored to underrepresented groups, such as women entrepreneurs, Indigenous communities, and youth-led ventures. Capacity-building programmes should focus on bridging knowledge and resource gaps in marginalised communities. Training modules on financial literacy, climate resilience, and digital tools can empower these groups to actively engage with and benefit from sustainable finance ecosystems. Embedding equity into every stage of project design and implementation can enable public institutions to create inclusive systems that leave no one behind.

4.2.4 Innovative Financial Instruments for Holistic Value

Innovative financial instruments hold immense potential to redefine the parameters of value creation, enabling a shift from traditional financial returns to a multidimensional approach that integrates social, environmental, and governance outcomes. These instruments are uniquely positioned to address the growing complexities of global challenges, such as climate change, inequality, and resource scarcity, by embedding mechanisms that align investor incentives with holistic value creation. Achieving this transformation requires a paradigm shift in financial design, integrating advanced technologies, and fostering collaborative ecosystems that prioritise inclusivity and sustainability. Advancing this agenda can support innovative financial instruments as a critical catalyst for systemic change, ensuring that finance drives equitable and sustainable development.

(a) **Redefining Green Bonds and Sustainability-Linked Instruments**

Green bonds and sustainability-linked instruments have emerged as vital tools in financing sustainability. However, these instruments must evolve beyond their current design and application to realise their full transformative potential. A critical area for improvement lies in expanding their scope to address interconnected sustainability challenges while incorporating adaptive mechanisms to respond to emerging global priorities. For instance, green bonds could adopt a systems-thinking approach by targeting cross-sectoral impacts. Rather than funding individual renewable energy projects, these bonds could prioritise integrated solutions that combine renewable energy generation with energy efficiency upgrades, ecosystem restoration, and community resilience initiatives. Sustainability-linked bonds, on the other hand, could broaden their impact by embedding multidimensional performance indicators that simultaneously address Environmental, Social, and Governance (ESG) dimensions. These indicators could include gender equity, access to essential services, and climate adaptation measures, enabling the instruments to drive comprehensive progress on the Sustainable Development Goals (SDGs). Furthermore, integrating blockchain technology into these instruments can enhance transparency and accountability. Blockchain enables real-time tracking of fund allocation and impact outcomes, ensuring that resources are utilised effectively and in alignment with stated objectives. The incorporation of data analytics and AI-driven monitoring systems can further bolster the credibility and scalability of these instruments, providing investors with actionable insights and fostering confidence in sustainable finance markets.

(b) **Tokenisation of Environmental Assets**

The tokenisation of environmental assets represents a groundbreaking innovation that can democratise access to sustainability markets while enhancing market efficiency. Environmental assets such as carbon credits, biodiversity offsets, and water rights can be tokenised to enable fractional ownership and broader participation by various stakeholders, including retail investors. Tokenisation offers numerous benefits, including increased liquidity, improved price discovery, and enhanced traceability of asset transactions. For example, a decentralised marketplace for carbon credits could facilitate peer-to-peer trading, enabling smaller enterprises and individual investors to participate in carbon offsetting initiatives. This approach expands market access and promotes a more equitable distribution of sustainability benefits. Moreover, tokenisation can address issues of double-counting and fraud by ensuring that each asset is uniquely identifiable and linked to verified environmental outcomes. The potential of tokenisation extends beyond individual assets to include bundled solutions that address multiple sustainability objectives. For instance, tokenised platforms could combine biodiversity credits with carbon offsets, creating integrated products that incentivise comprehensive ecosystem management. Embedding advanced analytics and predictive modelling, can enable the platforms to provide stakeholders with dynamic tools for decision-making and impact assessment, paving the way for more effective and inclusive sustainability markets.

(c) **Blended Finance and Outcome-Based Structures**

Blended finance mechanisms have long been recognised for their ability to mobilise private capital to support public goals. However, these mechanisms must incorporate innovative structures that prioritise measurable outcomes and adaptive risk-sharing to maximise their effectiveness. Blended finance can unlock significant investment in high-impact sectors often overlooked by traditional markets by aligning financial incentives with long-term sustainability objectives. Outcome-based structures represent an auspicious approach. They link financial returns to the achievement of predefined impact metrics, such as reductions in greenhouse gas emissions, improvements in education access, or enhancements in community health. These models ensure that financial flows deliver tangible and lasting benefits by shifting the focus from inputs to outcomes. Public institutions can enhance the scalability of blended finance by introducing dynamic guarantees that adjust based on project performance. For instance, a blended finance vehicle targeting regenerative agriculture could offer higher risk coverage for projects that demonstrate innovation and scalability. Additionally, the integration of data-driven impact assessment tools can enable real-time monitoring and course correction, ensuring that resources are directed towards the most effective interventions.

(d) **Development of Climate Resilience Bonds**

Climate resilience bonds are a critical innovation for addressing the escalating impacts of climate change, particularly in vulnerable regions. These bonds are uniquely suited to fund adaptive infrastructure projects, such as flood-resistant housing, climate-resilient transportation networks, and drought-tolerant agricultural systems. Incorporating insurance-like mechanisms can enable resilience bonds to provide timely financial support in response to climate events, mitigating economic and social disruptions. To enhance their effectiveness, climate resilience bonds should prioritise participatory design processes that involve local communities and stakeholders. This approach ensures that funded projects are context-specific and address the unique vulnerabilities of target regions. For example, bonds designed to support coastal resilience could incorporate traditional ecological knowledge alongside advanced engineering solutions, creating hybrid models that are both culturally relevant and technically robust. Moreover, partnerships with technology providers can enhance the predictive accuracy of climate risk assessments, enabling more precise targeting of resources. Advanced modelling techniques, such as machine learning and geospatial analysis, can identify areas of highest risk and inform the prioritisation of investments. By combining financial innovation with evidence-based decision-making, climate resilience bonds can contribute to both immediate and long-term climate adaptation efforts.

(e) **Integrating Artificial Intelligence in Financial Design**

Artificial intelligence (AI) offers transformative potential in designing, implementing, and monitoring innovative financial instruments. Harnessing the analytical power of AI can enable stakeholders to optimise investment decisions, forecast long-term outcomes, and enhance the efficiency of sustainability-focused financial markets. For instance, AI-driven algorithms can analyse vast datasets to identify high-potential projects that align with multidimensional impact objectives. These algorithms can assess factors such as climate risk, social vulnerability, and economic feasibility, enabling more informed decision-making. Additionally, AI can be integrated into sustainability-linked instruments to monitor issuer performance against impact metrics continuously. This real-time feedback loop allows for dynamic adjustments in financial terms, such as interest rates or repayment schedules, ensuring that instruments remain aligned with their intended goals. AI-powered platforms can also facilitate the aggregation and analysis of impact data, providing stakeholders with comprehensive insights into market trends and investment outcomes. The platforms can build trust among investors and attract more significant participation in sustainable finance markets by enhancing transparency and accountability.

(f) **Ensuring Equity and Inclusivity**

The design and deployment of innovative financial instruments must prioritise equity and inclusivity to ensure that marginalised communities are not left behind. Participatory models, such as community-driven impact funds and cooperative green bonds, can empower local stakeholders to co-own and benefit from sustainability initiatives. For example, bonds designed to finance renewable energy projects in underserved regions could incorporate revenue-sharing mechanisms that directly benefit local communities. These mechanisms enhance social equity and foster local support and engagement, increasing the likelihood of project success. Additionally, Microfinance Institutions can leverage innovative instruments to expand access to capital for women entrepreneurs, smallholder farmers, and other underrepresented groups. Capacity-building initiatives should complement these efforts, equipping marginalised communities with the knowledge and tools needed to engage with and benefit from sustainable finance systems. Training programmes on financial literacy, climate resilience, and digital technologies can empower individuals and communities to participate actively in the transition to sustainability.

4.2.5 Building Inclusive Financial Systems

Building inclusive financial systems is a cornerstone of sustainable development and equitable economic growth. Inclusive financial systems extend access to financial services to underserved populations, enabling their participation in economic activities and fostering social and economic resilience. However, achieving true inclusivity requires moving beyond basic access to create systems that address structural inequities, leverage technological innovation, and integrate multidimensional impact objectives. Such an approach necessitates a reimagining of financial systems as catalysts for systemic transformation, fostering both individual empowerment and societal resilience. Embedding inclusivity into financial architecture can help stakeholders to unlock pathways for shared prosperity and long-term sustainability.

(a) **Leveraging Technology for Financial Inclusion**

Technological innovation is a critical enabler of financial inclusion, but its transformative potential is realised only when applied strategically and equitably. Digital technologies such as mobile banking, digital wallets, and fintech platforms have expanded access to financial services for millions, particularly in remote and underserved areas. However, to fully harness their potential, these technologies must address the specific needs and barriers marginalised populations face. Blockchain technology, for example, offers a decentralised and transparent framework for enhancing trust and accessibility in financial transactions. They reduce transaction costs, making financial services more affordable for low-income users. Smart contracts can further streamline disbursement processes, enabling conditional cash transfers to be executed automatically upon meeting predefined milestones, such as school attendance or medical check-ups. Similarly, artificial intelligence (AI) and machine learning (ML) hold significant promise for expanding access to credit. By utilising alternative data sources—such as mobile phone usage, utility payments, and social media activity—AI-driven credit-scoring models can evaluate the creditworthiness of individuals without formal financial histories. This innovation can democratise credit access, enabling smallholder farmers, micro-entrepreneurs, and low-income households to secure financing. However, ensuring these technologies are free from biases and designed with ethical considerations is critical to preventing unintended exclusions.

(b) **Addressing Structural Inequities**

Structural inequities remain one of the most significant barriers to financial inclusion. These inequities often arise from historical, social, and economic injustices that disproportionately affect women, Indigenous populations, and other marginalised groups. Addressing these deep-rooted challenges requires a multifaceted approach involving policy reforms, institutional innovation, and community empowerment. Policy interventions must prioritise inclusive design. For instance, governments can implement gender-responsive financial policies that incentivise financial institutions to extend services to women-led enterprises. These policies could include subsidised interest rates, reduced collateral requirements, and mentorship programmes aimed at fostering entrepreneurship among women. Similarly, regulatory frameworks can mandate financial inclusion targets for banks, ensuring that a portion of their portfolios is directed towards underserved regions and communities. Culturally tailored financial products also play a critical role in addressing inequities. For example, Islamic finance—which prohibits interest-based transactions—can be leveraged to provide Sharia-compliant financial solutions in Muslim-majority areas. Cooperative finance models, where members pool resources to access collective credit, can empower rural communities to overcome capital constraints and build resilience. These approaches demonstrate the importance of designing financial systems that respect cultural contexts while promoting inclusivity.

(c) **Participatory Approaches to Financial System Design**

True inclusivity requires the active participation of underserved populations in the design and implementation of financial solutions. Participatory approaches ensure that financial products and services align with the lived experiences, aspirations, and priorities of target communities. This co-creation process fosters a sense of ownership and sustainability, increasing the likelihood of success. Community-driven financial models, such as village savings and loan associations (VSLAs), exemplify participatory approaches. These grassroots initiatives enable communities to pool savings and extend microloans, fostering collective decision-making and resource mobilisation. Formal financial institutions can partner with these groups to provide additional capital, training, and digital tools, creating a bridge between informal and formal financial ecosystems. Digital platforms can further enhance participatory engagement by providing mechanisms for real-time feedback and consultation. For example, mobile-based survey tools can capture user insights on financial services, enabling continuous refinement and adaptation. These platforms can also deliver financial literacy training, equipping individuals with the knowledge and skills needed to navigate financial systems effectively.

(d) **Integrating Impact Measurement and Accountability**

Inclusive financial systems must deliver measurable social and economic benefits to stakeholders. Impact measurement frameworks are essential for tracking progress, identifying gaps, and ensuring accountability. These frameworks should encompass multidimensional metrics, covering well-being, inclusion, and sustainability.[1] Advanced analytics and digital tools can significantly enhance the effectiveness of impact measurement. Geospatial mapping, for example, can identify financial service gaps by visualising underserved areas, while big data analytics can track the usage patterns and outcomes of financial products. Integrating these tools into reporting frameworks can help stakeholders to gain actionable insights and make data-driven decisions to improve inclusivity. Accountability mechanisms are equally important. Independent audits, participatory monitoring committees, and multi-stakeholder review processes can ensure that financial institutions remain aligned with inclusivity objectives. By fostering transparency and trust, these mechanisms can create a culture of accountability that drives continuous improvement.

(e) **Building Resilience Through Inclusive Finance**

Inclusive financial systems play a pivotal role in enhancing economic and social resilience, particularly in the face of global challenges such as climate change, pandemics, and economic shocks. Access to financial services enables individuals and communities to build buffers, recover from disruptions, and adapt to changing circumstances. Micro-insurance products, for instance, provide low-income households with protection against climate-related risks, such as crop failure, flooding, or property damage. Similarly, flexible savings accounts can help individuals manage financial emergencies without resorting to high-interest loans or informal borrowing. Financial institutions can also design credit products that support climate adaptation measures, such as investments in drought-resistant crops, renewable energy, or disaster-resilient infrastructure. Moreover, inclusive finance can facilitate collective resilience. For example, community-based insurance schemes can pool risks and resources, providing a safety net for vulnerable populations. Aligning financial systems with resilience-building objectives can help stakeholders to enhance both individual and collective capacity to navigate uncertainty and build sustainable futures.

4.3 Achieving Balanced Outcomes in Sustainable Growth

The increasing complexity of development finance makes it challenging to escape the influence of financialisation. Given this reality, the focus should shift from mitigating financialisation to finding ways to achieve both environmental and social outcomes while also generating financial returns. Financialisation, rather than being inherently problematic, can be a powerful catalyst for sustainable development if its mechanisms are properly aligned with broader societal goals. This perspective necessitates a nuanced understanding of financialisation as a tool that, when governed effectively, can mobilise capital towards addressing some of the most pressing challenges of our time, such as climate change, social inequality, and economic resilience. To harness the potential of financialisation, it is essential to put in place appropriate frameworks, regulations, and incentives that prioritise long-term value creation over short-term profit maximisation. This means reconceptualising the purpose of finance beyond mere wealth accumulation, towards fostering inclusive growth, environmental stewardship, and community well-being. Sustainable or development finance must thus operate on principles that integrate financial, environmental, and social objectives, recognising that true value lies not just in financial returns but also in the positive impacts generated for society and the planet. Such an approach requires collaborative efforts from governments, private-sector actors, and civil society to build financial ecosystems that support sustainable outcomes and are accountable to a broader set of stakeholders.

Finance for growth operates at the intersection of multiple economic, environmental, and social objectives, making it inherently difficult to navigate without engaging financialised tools and markets. The key to ensuring that, in spite the risk of financialisation, finance contributes positively to sustainable growth and development lies in the deliberate structuring and regulation of financial mechanisms to prioritise non-financial outcomes alongside profit. This balance requires a reimagining of value creation, where financial returns are not considered the sole metric of success but are integrated with broader social and environmental metrics, thereby creating a more holistic and impactful approach.

Blended Finance as a Lever for Impact: Blended finance is one of the most effective tools for aligning financialisation with sustainable outcomes, particularly when it integrates social and environmental goals into its foundational framework. By leveraging public or concessional capital to absorb risks, blended finance incentivises private investors to participate in sectors they might otherwise avoid due to perceived financial risks. However, a critical challenge lies in ensuring that public capital is utilised in ways that genuinely enhance social impact rather than simply subsidising private profit. For blended finance to serve its intended purpose, it must be strategically deployed to focus on projects that yield significant social and environmental benefits, such as renewable energy, climate adaptation, education, and equitable healthcare access. Additionally, blended finance should be subject to rigorous impact assessments that ensure that public investments translate into meaningful outcomes, thereby safeguarding against the risk of public funds being misappropriated for private gain.

Public–Private Partnerships with Social Safeguards: Public–Private Partnerships (PPPs) are another financial mechanism that, when structured thoughtfully, can achieve balanced and sustainable outcomes. The success of PPPs in sustainable finance hinges on embedding enforceable social and environmental standards within contractual agreements. These standards must include provisions for equitable access, community empowerment, and environmental protection, thereby ensuring that the benefits of infrastructure and service projects extend beyond mere financial returns to deliver tangible social value. Furthermore, transparency in risk-sharing arrangements is critical to prevent the undue burden of financial risk from falling on the public sector—a common pitfall of financialised projects. For instance, infrastructure projects financed through PPPs should be designed to meet profitability criteria and enhance local communities' quality of life through job creation, improved access to essential services, and ecological preservation. To this end, integrating social impact assessments and public accountability mechanisms is crucial for fostering trust and ensuring that PPPs contribute effectively to sustainable development.

Impact Investing and Results-Based Financing: Impact investing and results-based financing, including development impact bonds (DIBs), represent financialised approaches that can be leveraged to achieve sustainable outcomes if carefully calibrated. The allure of impact investing lies in its dual focus on generating financial returns while achieving measurable social or environmental benefits. However, for impact investing to fulfil its potential, robust metrics must be developed that capture immediate, quantifiable outcomes and long-term systemic changes. This requires moving beyond simplistic indicators and developing comprehensive frameworks that assess multidimensional impacts, such as improvements in social equity, biodiversity conservation, and community resilience. Similarly, results-based financing should reward projects that contribute to transformative change rather than focusing narrowly on outputs that are easy to quantify. Results-based financing can be pivotal in aligning financial incentives with Sustainable Development Goals by fostering a culture of accountability and long-term impact orientation.

4.4 Case Studies: Financialisation in Action

4.4.1 Accelerated Growth in Certain Sectors

In spite of the challenges posed by financialised tools and strategies, certain sectors have managed to achieve sustainable growth through careful adaptation and strategic implementation. A prominent example of this dynamic is observed in the technology sector. The rise of venture capital and private equity investments has played a crucial role in the rapid expansion of tech companies such as Uber, Airbnb, and Tesla. These firms have navigated the complexities of financialisation, using financial instruments such as convertible notes, preferred stock, and equity-based compensation to raise substantial capital while striving to balance innovation with sustainable growth. Venture capital funding, in particular, has provided the necessary financial resources for these companies to develop disruptive technologies and business models, allowing them to redefine industry standards and catalyse economic growth. For instance, Tesla's ability to innovate in electric vehicles and energy storage solutions has been largely driven by its access to capital markets, which allowed for large-scale research, development, and market expansion.

Another illustrative example is the real estate sector in major urban areas. Financialisation has transformed real estate into a financial asset class, attracting global capital flows and significantly boosting property development. The influx of investment through Real Estate Investment Trusts (REITs) and institutional investors has played a significant role in urban regeneration projects. For instance, cities like London, New York, and Hong Kong have experienced extensive property development driven by these financialised tools, which facilitated large-scale investments in both commercial and residential projects. In London, REITs have contributed to redeveloping derelict areas like King's Cross, transforming them into thriving business and residential hubs. However, despite the challenges associated with treating real estate as a purely financial asset, strategic investments from REITs and institutional investors have facilitated urban redevelopment, expanded the construction sector, and increased property values, all while striving to address the needs of local communities and mitigate negative social impacts through initiatives like affordable housing quotas and community engagement programmes.

Furthermore the impact of microfinance initiatives that have contributed to poverty alleviation and social equity. The Grameen Bank in Bangladesh, founded by Muhammad Yunus, is a notable case where financialised strategies such as microcredit have been used to provide small, collateral-free loans to impoverished individuals, particularly women. This has enabled them to start or expand small businesses, thereby improving their economic standing and fostering sustainable growth. The success of Grameen Bank has been replicated in over 40 countries, demonstrating how targeted financialised tools can promote social equity and economic empowerment. Similarly, the Kashf Foundation in Pakistan has facilitated sustainable growth by offering affordable financial services to low-income households, primarily focusing on women's economic empowerment. Despite the potential pitfalls of financialisation, microfinance strategies can help promote development and social equity when properly implemented.

4.4.2 Undermined Impact Objectives Due to Over-Financialisation

While financialisation has accelerated growth in some areas, it has also, at times, undermined impact objectives due to over-financialisation. The housing market crash of 2008 serves as a stark example of the dangers of excessive financialisation. In the lead-up to the crisis, housing was increasingly treated as a financial commodity rather than a social necessity. The proliferation of mortgage-backed securities, collateralised debt obligations (CDOs), and complex financial derivatives fuelled a housing bubble that ultimately collapsed, leaving millions of individuals without homes and causing widespread economic distress. The securitisation of mortgages led to a disconnect between homeowners and financial institutions, as lenders prioritised profit through the packaging and resale of mortgage debt over the well-being of borrowers. The over-financialisation of housing transformed what should have been a stable asset for individuals into a speculative vehicle for profit, undermining the social impact objective of providing secure housing for all.

Similarly, the agricultural sector in parts of the Global South has faced significant challenges due to over-financialisation. Financial instruments such as commodity futures and derivatives, originally intended to hedge risks for farmers, have increasingly attracted speculative investors seeking short-term profits. This influx of speculative capital has led to increased volatility in food prices, making it more difficult for small-scale farmers to maintain stable livelihoods and undermining food security in vulnerable regions. For example, during the 2007–2008 global food crisis, speculative trading in agricultural commodities like wheat, corn, and rice drove prices to unprecedented levels, exacerbating food insecurity for millions of people in the Global South. The financialisation of agricultural commodities has, in many instances, prioritised profit maximisation over the well-being of farming communities, compromising the broader impact objectives of sustainable agricultural development and poverty alleviation. Small-scale farmers, who often lack the financial literacy and resources to compete with large investors, have been disproportionately affected, struggling to manage the price volatility and sustain their livelihoods. These case studies illustrate the dual nature of financialisation. While it can be a powerful tool for economic growth, it also carries significant risks when applied without regard for social and environmental impact. The challenge for policymakers and investors is to strike a balance that leverages the benefits of financialisation while safeguarding against its potential to undermine critical impact objectives.

4.5 Reflections on Rethinking Value Creation for Sustainable Growth

As we navigate the complexities of modern financial systems, it becomes increasingly evident that traditional value creation models are no longer sufficient to address the multifaceted challenges of our time. The rethinking of value creation for sustainable growth requires a fundamental shift from narrow financial metrics to a broader paradigm that encompasses environmental restoration, social equity, and long-term resilience. This journey involves recalibrating financialisation, integrating stakeholder perspectives, leveraging technology, addressing structural inequities, and fostering a regenerative future. Together, these elements paint a picture of how finance can evolve to serve as a powerful engine of systemic transformation. At the heart of this transformation lies the need to redefine financialisation—a process often critiqued for prioritising short-term gains—into a force for sustainability. The potential of financial systems to drive positive change is immense, provided they incorporate adaptive and multidimensional metrics that go beyond traditional profit measures. For example, metrics that track biodiversity restoration or carbon sequestration enable investments to deliver lasting regenerative outcomes. Embedding such tools can enable the financial systems to ensure that capital flows are directed towards projects with enduring value. Technologies like blockchain and artificial intelligence further enhance this potential by providing transparency, traceability, and real-time adaptability, ensuring financial systems remain dynamic and responsive.

Equally significant is the role of inclusivity in financial systems. Stakeholder-centric approaches shift the focus from shareholders to the broader ecosystem of individuals and communities affected by financial decisions. Such inclusivity fosters trust, equity, and resilience. Participatory governance structures, like multi-stakeholder advisory boards and cooperatives, provide platforms for diverse voices to influence decision-making. This inclusive approach extends beyond governance; it informs the very design of financial products, ensuring they align with the needs and aspirations of marginalised populations. Community-driven renewable energy projects, for example, not only generate local employment but also foster ownership and long-term viability. The issue of structural inequities remains a critical barrier to financial inclusion. Addressing these inequities requires targeted and culturally sensitive interventions. Gender-responsive financial policies, concessional financing for underserved regions, and culturally tailored products such as Islamic finance are vital tools for bridging these gaps. However, addressing immediate access barriers is only one aspect; systemic change necessitates regulatory reforms and institutional innovations to ensure these efforts are scalable and sustainable. Technology emerges as both a tool and a catalyst in this reimagining of financial systems. While digital platforms and blockchain-powered solutions have already expanded financial access to millions, their true potential lies in their ability to address specific barriers faced by underserved communities. AI-driven credit-scoring models, for instance, can democratise access to credit by evaluating individuals based on alternative data, such as utility payments and mobile phone usage. At the same time, blockchain-based marketplaces for tokenised environmental assets offer a transparent and inclusive way to participate in sustainability markets. These innovations underscore the importance of continuous advancement and ethical deployment of technology to ensure equitable access and benefits.

Harmonising global financial standards emerges as another indispensable insight. Fragmented metrics and inconsistent reporting frameworks undermine the comparability and trust needed for global financial systems to function effectively. Initiatives like the Task Force on Climate-related Financial Disclosures (TCFD) highlight progress. Still, these frameworks need to be expanded to encompass broader sustainability metrics such as biodiversity and social equity. Universal standardisation, complemented by local adaptations, ensures transparency and accountability across jurisdictions, fostering a cohesive approach to sustainable finance. The vision for sustainable growth hinges on adopting a regenerative approach that prioritises the restoration of ecosystems, the empowerment of communities, and the resilience of economies. This vision requires a departure from extractive financial practices to models that generate positive externalities across environmental, social, and governance dimensions. Collaboration among governments, private-sector actors, civil society, and academic institutions is essential to achieve this transformation. Knowledge-sharing platforms, pilot projects, and interdisciplinary research can accelerate the diffusion of best practices and foster a culture of continuous improvement. The rethinking of value creation in sustainable finance is not merely a theoretical exercise but an urgent imperative for building a more equitable, resilient, and sustainable world. By embedding holistic methodologies, advancing adaptive technologies, and fostering inclusive frameworks, financial systems can transcend their traditional role and become catalysts for systemic change. This reimagined approach to finance holds the promise of a future where financial systems contribute to economic prosperity, societal well-being, and planetary health. The global community must seize this moment to reimagine finance as a driver of shared prosperity, ensuring that no one is left behind in the pursuit of sustainable growth.

4.6 Conclusion

This chapter has critically examined the evolving nature of sustainable finance, arguing for a fundamental redefinition of value creation that balances economic resilience, social equity, and environmental restoration. Consequently, the discussion has emphasised how financialisation, when structured within ethical and regulatory boundaries, can serve as a mechanism for positive transformation rather than an impediment. A key takeaway is the necessity of dynamic and adaptive methodologies. The introduction of regenerative metrics—which assess ecosystem restoration, climate resilience, and social well-being—represents a critical shift from conventional ESG indicators. These frameworks ensure that sustainability efforts are forward-looking, addressing both immediate concerns and long-term systemic needs. Moreover, the integration of real-time adaptability through digital innovations such as blockchain and artificial intelligence has been identified as a crucial step towards enhanced transparency, accountability, and financial responsiveness. These technologies enable dynamic decision-making, ensuring that capital is allocated towards impactful and verifiable sustainability initiatives rather than superficial compliance exercises. Equally important is the role of collaborative co-creation in shaping financial strategies. Sustainable finance cannot be driven solely by market mechanisms—it requires meaningful multi-stakeholder engagement, particularly with local communities, policymakers, and historically marginalised groups. Embedding inclusive governance structures fosters long-term legitimacy and enhances financial instruments' ability to address complex social and environmental challenges.

The discussion has also highlighted circular economic principles as a transformative approach to financial sustainability. Institutions can reduce environmental degradation by embedding resource efficiency, waste minimisation, and regenerative economic practices within financial models while fostering long-term economic stability. Finally, the chapter reinforces the urgency of universal standardisation and regulatory harmonisation. The current fragmentation in sustainability reporting and impact measurement impedes meaningful progress. Establishing enforceable, globally aligned frameworks is essential to ensuring credibility, preventing greenwashing, and fostering investor confidence in sustainable finance initiatives. In conclusion, redefining value creation in sustainable finance is not an abstract academic exercise but an operational necessity.

The transition to adaptive, transparent, and stakeholder-driven financial models is crucial in aligning economic activities with Sustainable Development Goals. Achieving this transformation will require ongoing technological innovation, policy coherence, and a firm commitment to integrating social and environmental considerations into the core of financial decision-making. Only through these concerted efforts can sustainable finance become a truly transformative force in addressing global challenges.

Note

1. See the framework in Jansen et al. (2024).

Reference

Jansen, A., Wang, R., Behrens, P., & Hoekstra, R. (2024, September 01). Beyond GDP: A review and conceptual framework for measuring sustainable and inclusive wellbeing. *The Lancet Planetary Health, 8*(9), e695–e705. https://doi.org/10.1016/S2542-5196(24)00147-5

PART III

Impact Investing and Inclusive Finance

CHAPTER 5

Current Trends in Impact Investing

5.1 Introduction

This chapter explores the contemporary landscape of impact investing, offering an analysis of its current state, emerging trends, and sectoral focus. Drawing from the latest GIIN market report, it evaluates the scale and dynamics of impact investing, highlighting its rapid growth and the diversity of investors participating in the market. The discussion provides a critical review of how impact investing continues to evolve as a tool for generating financial returns and positive social and environmental outcomes. The chapter details the significant expansion of Assets Under Management (AUM) in impact investing and the key players shaping the sector. It highlights the dominance of North America and Europe in capital allocation while acknowledging the increasing importance of emerging markets such as Sub-Saharan Africa, Southeast Asia, and Latin America. It also addresses challenges such as fragmented impact measurement frameworks, regulatory uncertainties, and macroeconomic factors like inflation and interest rate fluctuations that influence investment strategies. A major focus is given to Gender Lens Investing (GLI), a growing sub-sector of impact investing that integrates gender considerations into investment decisions. It explores the importance of directing capital towards women-led and gender-equitable businesses, assessing how GLI

N. Udohaya, *Impact Investing and Financial Inclusion*, Sustainable Development Goals Series,
https://doi.org/10.1007/978-3-031-87841-1_5

contributes to economic growth and social development. It further examines the motivations behind GLI, including alignment with organisational impact missions and the UN Sustainable Development Goals (SDGs), while also addressing existing barriers such as sectoral concentration and the need for more standardised impact measurement methodologies.

The chapter further delves into the sectoral focus of impact investing, providing an analytical breakdown of the key industries that attract impact capital. These include renewable energy, healthcare, education, agriculture, water and sanitation, affordable housing, financial services, and environmental conservation. Each sector is examined in terms of its role in sustainable development, the level of investment it attracts, and the financial mechanisms facilitating capital flows. Beyond these sectoral analyses, it delves deeper into the justifications for impact investments in financial services access, recognising the crucial role that financial inclusion plays in economic development. It examines the economic, social, and financial rationales for directing impact capital into financial services, highlighting how expanding access to credit, insurance, and digital financial tools fosters economic resilience and social mobility. The discussion explores how financial inclusion aligns with broader sustainable development objectives, contributing to poverty reduction, gender equality, and economic stability. Through this comprehensive analysis, the chapter provides insights into both the opportunities and challenges within impact investing, underscoring its transformative potential in advancing inclusive and sustainable global development.

5.2 The State of Impact Investing

The global impact investing market has achieved a remarkable scale, with $1.571 trillion in Assets Under Management (AUM) and a Compound Annual Growth Rate (CAGR) of 14 per cent over the past five years. Consequently, this growth signifies the increasing recognition of impact investing as a dual-purpose strategy that combines the pursuit of financial returns with the generation of positive social and environmental outcomes. Moreover, impact investors represent a diverse constituency, including private equity firms, institutional asset owners, and Development Finance Institutions (DFIs). A majority of these investors are headquartered in developed markets such as North America and Europe, which together account for 70 per cent of global AUM. Nonetheless, emerging markets, particularly Sub-Saharan Africa, Southeast Asia, and

Latin America, are gaining prominence as regions of increasing capital allocation. The range of AUM per organisation reflects the coexistence of small-scale, niche-focused players and large institutional investors such as pension funds and sovereign wealth funds that are reshaping their portfolios to align with impact-driven objectives. Additionally, investments are spread across various sectors, with a strong focus on climate resilience, renewable energy, affordable housing, and healthcare. This alignment with the Sustainable Development Goals (SDGs) has facilitated targeted interventions addressing systemic challenges such as biodiversity loss and food insecurity. However, persistent issues, including fragmented measurement frameworks, challenges in verifying impact data, and a lack of regulatory clarity, continue to hinder the industry's full potential. Nevertheless, progress in addressing these hurdles has been evident, with increased adoption of third-party verification and the integration of impact metrics into governance structures.

Furthermore, macroeconomic and political factors pose additional challenges to impact investment strategies. Inflation, rising interest rates, and the multifaceted effects of climate change are among the most significant influences on capital flows. These challenges disproportionately affect smaller investors, who often pivot towards domestic markets to mitigate risks. Conversely, larger institutional investors leverage their resources to expand international allocations, particularly in stable sectors like renewable energy. Simultaneously, regulatory changes and local political developments further amplify these challenges, necessitating adaptive strategies that include innovative financing mechanisms. In parallel, the future of impact investing is shaped by emerging trends that reflect the sector's adaptability and innovation. For instance, the increasing adoption of equity-like debt instruments illustrates a flexible approach to aligning financial performance with impact objectives. Similarly, blended finance models continue to play a critical role in mobilising private capital by de-risking investments in high-impact but traditionally underserved areas. Institutional investors, including pension funds and insurance companies, are central to this growth, leveraging their long-term investment horizons to drive market expansion and sustainability.

Despite its impressive growth, the market faces significant concentration risk. Specifically, a small number of large investors manage the majority of impact AUM, which could overshadow smaller, community-focused investments. Moreover, despite the overall capital growth, certain SDGs, such as clean water and sanitation, remain underfunded, revealing a disconnect between capital flows and global development needs. Consequently, to address these disparities, increased incentives and risk mitigation strategies are essential to redirect investments towards underserved sectors. Policy environments are also instrumental in shaping the trajectory of impact investing. Regulatory frameworks, such as the EU's Sustainable Finance Disclosure Regulation (SFDR) and the US Greenhouse Gas Reduction Fund, have established benchmarks for transparency and driven capital towards sustainability-focused initiatives. However, regulatory inconsistencies, particularly in emerging markets, present obstacles to further growth. Meanwhile, in developed markets, consumer advocacy has emerged as a powerful force, influencing corporate investment decisions and promoting sustainable practices.

Notably, investor satisfaction with impact investing remains high, with over 90% of respondents indicating that their financial and impact performance meets or exceeds expectations. This dual success not only validates the viability of impact investing but also strengthens its appeal to mainstream financial markets. Additionally, the willingness to experiment with innovative asset classes, such as equity-like debt, reflects a broader commitment to addressing financing gaps and supporting social enterprises and infrastructure projects. Nonetheless, standardising these instruments will be critical to fostering investor confidence and achieving scalability. The impact investing sector is marked by a dynamic interplay of growth, challenges, and opportunities. As the market continues to evolve, achieving scalability and standardisation will be vital for its long-term sustainability. Thus, strategic alignment among investors, policymakers, and frameworks is necessary to guide this transformative investment approach. As impact investing matures, its capacity to address global challenges and drive sustainable development will become increasingly significant. Moreover, one of the emerging and critical dimensions within this sector is the role of gender. Integrating a gender lens into impact investing holds immense potential to address systemic inequalities and foster inclusive growth, setting the stage for deeper exploration into this vital area of focus.

5.3 The Imperative for Improving the Lives of Women and Outcomes

The Gender Lens Investing (GLI) emerged as a tool that allows the integration of considerations for impact on women. The integration of gender into the discourse of impact investing underscores the need to address systemic inequalities while achieving measurable financial and social outcomes. Historically underrepresented, women and girls continue to face significant barriers in economic, political, and social domains (Iqbal et al., 2022). GLI seeks to bridge these gaps by embedding gender considerations into investment strategies, making it a critical component of the broader impact investing landscape. The World Economic Forum global gender gap report reveals persistent disparities between men and women in economic, social, and political domains, adding estimates that it will take approximately 134 years to achieve full gender parity, underscoring the enduring nature of these inequalities (Pal et al., 2024). Women disproportionately bear the burden of unpaid care work and face restricted access to economic opportunities. The disparities are evident in multiple domains, including economic participation, educational attainment, health, and political empowerment. Moreover, women disproportionately bear the burden of unpaid care work and face restricted access to economic opportunities, limiting their full participation in global economic systems. According to the UN's 2023 Gender Snapshot report, to achieve gender equality by 2030 an additional USD360 billion investment annually is required (UNWOMEN, 2023). Gender Lens Investing seeks to address this gap by integrating gender considerations into the investment process. It integrates gender-based analysis into investment decisions, recognising that empowering women and promoting gender equality can lead to substantial economic and social benefits. In 2023, the private gender lens investing market reached USD7.9 billion, a 30 per cent increase from the previous year and by Q1 2024, publicly traded gender lens equity funds stood at USD 4.6 billion as of Q1 2024. These reflect significant growth in GLI. However, substantial disparities persist, emphasising the need for strategic capital allocation and innovative solutions to advance gender equity. Investing in gender equality is a moral imperative as well as an economic one. As estimated by Mastercard (2022), increasing the participation of young women in Africa's workforce could contribute an estimated USD287 billion to the continent's economy by 2030.[1]

GLI is an intentional effort to redirect capital towards addressing structural inequities. It prioritises investments in women-led and women-owned enterprises. Beyond addressing gender disparities, GLI investments signify a recognition of women's economic and social contributions to the sectors they lead. In recent times, these allocations have focused primarily on sectors with high social and economic multipliers, such as financial services, renewable energy, and healthcare (Hand et al., 2024). These sectors provide critical opportunities to reduce inequalities while fostering sustainable development. However, the over-reliance on specific sectors risks reinforcing a limited narrative about women's contributions to the economy. Expanding GLI into traditionally male-dominated industries—such as technology, infrastructure, and manufacturing can redefine gender equity in broader economic terms. Geographically, GLI investments remain concentrated in developed markets, with the United States and Canada accounting for 63 per cent of allocations. These regions benefit from well-established financial systems and regulatory environments conducive to gender-focused investing. In contrast, emerging markets, including Sub-Saharan Africa, Latin America, and South Asia, received a smaller yet growing share of gender-focused capital. Investments in these regions are often characterised by smaller ticket sizes and higher perceived risks, which constrain their scalability. Addressing these barriers through innovative financial tools such as blended finance or risk-sharing mechanisms can unlock the latent potential of these underserved markets.

In terms of performance, 75 per cent of gender investors report achieving significant social impact through their strategies, 51 per cent also noting enhanced financial returns.[2] However, assessing the breadth and depth of these impacts remains challenging due to the lack of standardised metrics and methodologies for evaluating gender-specific outcomes. Fragmented frameworks and inconsistent reporting further complicate efforts to ensure transparency and accountability. Systemic barriers, such as the underrepresentation of women in leadership roles within financial institutions, present additional challenges to GLI. Women's participation in decision-making positions within the financial sector remains disproportionately low, limiting the diversity of perspectives that inform investment decisions. Addressing this gap necessitates systemic changes that prioritise inclusive leadership and capacity-building for women in finance. In reimagining the scope of GLI, a shift towards diversifying sectors, scaling investments in emerging markets, and creating robust impact assessment mechanisms is essential. These steps will ensure that GLI addresses the immediate need for gender equity and contributes to long-term structural changes that reshape the global financial ecosystem.

5.3.1 Motivations for Gender Lens Investing

Gender Lens Investing (GLI) is often driven by various motivations, each reflecting a distinct perspective on the integration of gender considerations into investment strategies. In the 2024 GIIN report, a significant proportion of investors cite alignment with organisational impact missions as the primary driver for adopting GLI strategies, with 85 per cent of investors emphasising this motivation. This mission-driven approach underscores a commitment to social responsibility and ethical investing. It is worth noting that the effectiveness of such alignment depends on the depth of integration of gender considerations into investment decisions. Superficial commitments can undermine the credibility and impact of GLI initiatives. Organisations must ensure that their gender-focused investment strategies are substantive and not merely symbolic. Another critical driver is alignment with global agendas such as the UN Sustainable Development Goals (SDGs), cited by approximately 41 per cent of gender investors as a motivated. While this alignment can drive capital towards pressing challenges, there is always the question of specificity and measurability of impact when it comes to aligning investment with the SDGs. The broad nature of the SDGs may lead to diluted efforts if investments are not strategically targeted. Consequently, investors should adopt rigorous frameworks to assess and ensure the tangible impact of their gender-focused investments. Financial and business incentives also play a role, with 30 per cent of investors citing these as motivations. Gender-diverse teams are often associated with enhanced innovation and financial performance. However, the correlation between gender diversity and financial returns is complex and may not always yield positive outcomes. Recent analyses indicate that some gender-focused funds have underperformed relative to benchmark,[3] prompting scrutiny of the financial rationale behind GLI. Investors should critically assess the financial implications of GLI and avoid overestimating the economic benefits without thorough analysis. Interestingly, only 22 per cent of investors cite client demand as a significant motivation for GLI. This could suggest that GLI is predominantly driven by internal organisational values rather than external market pressures. While this internal motivation can lead to authentic and impactful investment strategies, it may also result in a disconnect between investment offerings and client interests. To bridge this gap, investors can actively engage with clients to understand their values and preferences, ensuring that GLI strategies resonate with and are supported by the investor base. These motivations reflect the multifaceted nature of GLI, encompassing ethical, strategic, and economic considerations. However, to maximise its effectiveness, investors must critically evaluate their motivations and approaches, ensuring that gender lens strategies are robust, measurable, and aligned with both organisational and broader societal goals.

5.3.2 Benefits and Challenges

One of the most significant benefits of Gender Lens Investing (GLI) is its potential to enhance financial performance. Companies with diverse leadership teams often exhibit improved financial outcomes, as gender diversity fosters better decision-making and reduces risks. For instance, Lu et al. (2023) provides evidence that mixed-gender investment teams outperform their single-gender counterparts, highlighting the value of inclusivity in driving returns. In addition Young and DAI Bruett (2021) observed that increase in women's representation in senior management correlates with an increase in AUM allocated to women-led enterprises. Hence, GLI contributes directly to advancing gender equality by channelling capital into women-led businesses and initiatives that support women and girls. Furthermore, companies prioritising gender diversity tend to experience greater innovation and market expansion, as diverse teams are better equipped to serve various customers and stakeholders (Marime-Ball & Shaber, 2022). The approach aligns with global development goals, and enables investors to achieve both measurable social outcomes and broader societal benefits.

Despite its promise, GLI faces considerable challenges that complicate its broader adoption and effectiveness. Performance variability is a key concern, with some gender-focused funds underperforming relative to benchmarks (Thorarinsdottir et al., 2024). This has led to increased scrutiny over the financial rationale of GLI strategies and the need for more robust data to substantiate claims of superior performance. Data limitations and fragmented impact measurement frameworks hinder the ability of investors to evaluate the true scope of their gender-focused investments. Their absence creates challenges in assessing outcomes effectively, hence standardised metrics are essential for ensuring transparency and accountability. These challenges are exacerbated by systemic barriers, such as the underrepresentation of women in leadership and decision-making roles within the financial industry. These disparities limit the scalability of GLI and restrict its transformative potential. To address these challenges, the investment ecosystem must prioritise structural changes. This includes fostering gender diversity within financial institutions, creating enabling environments for women entrepreneurs, and advocating for improved data standards. By addressing these systemic issues, GLI can evolve into a more robust and impactful approach, ensuring its ability to drive both financial returns and meaningful social change.

5.3.3 Intersectionality and Broader Impacts

GLI has increasingly adopted intersectionality as a critical framework, recognising that gender does not exist in isolation but intersects with other social categories such as race, ethnicity, class, and sexual orientation. This perspective acknowledges the compounded disadvantages faced by individuals with intersecting identities, particularly women of colour and those in marginalised communities (Subramanian et al., 2022). By addressing these intersecting inequalities, GLI seeks to create investments that are not only inclusive but also transformative in addressing systemic barriers. Integrating intersectionality into GLI requires an intentional effort to understand and address the unique challenges faced by diverse groups of women. For example, women from marginalised racial or ethnic backgrounds may encounter systemic discrimination that limits their access to capital, leadership opportunities, and market participation. Addressing these challenges demands sophisticated frameworks, comprehensive data collection, and a commitment to equity-focused strategies. However, many investment firms struggle with the complexity of applying intersectional approaches due to a lack of expertise and resources. This can result in superficial applications that fail to generate substantive impact. Adopting an intersectional lens can lead to broader and more inclusive social outcomes. Investments that consider the interplay of gender, race, and class are better positioned to reduce inequalities and support systemic change. Framing GLI from the perspective of intersectionality aligns

GLI with broader social justice movements, enhancing its legitimacy and appeal to stakeholders seeking holistic impact.

The primary challenge of the broadening of focus inherent in intersectionality is the risk of mission drift, where expanded goals dilute the precision and efficacy of investment strategies. Similarly, impact washing, where intersectional language is adopted without substantive action, threatens the credibility of GLI. To address these risks, investors must commit to rigorous education on intersectionality, engage with affected communities to design relevant interventions, and develop robust metrics to measure outcomes. By doing so, GLI can evolve beyond a one-size-fits-all approach, ensuring that it effectively fosters equity across diverse identities. Such intentionality is crucial for GLI to achieve its vision of creating systemic change that uplifts marginalised groups and drives sustainable development.

5.3.4 *Future Directions for GLI*

GLI requires a concerted focus on scaling its impact, expanding its reach, and refining its methodologies. The future trajectory of GLI will be shaped by how effectively it addresses systemic challenges while adapting to emerging global trends. One of the critical areas for growth is increasing capital allocations to emerging markets. These regions, such as Sub-Saharan Africa and Southeast Asia, present significant opportunities for gender-focused investments, particularly in energy, infrastructure, and housing sectors. Expanding into these markets helps address regional inequalities and aligns with global development goals. Unlocking the potential of these regions will require innovative financial instruments, such as blended finance and equity-like debt, to de-risk investments and attract more capital to underserved areas. Leadership diversity will also play a pivotal role in shaping the future of GLI. Creating pathways for women to assume leadership roles within financial institutions is crucial for embedding gender equity in decision-making processes. Another promising direction lies in refining impact measurement frameworks. The current fragmentation in metrics and reporting standards undermines the ability to assess the true effectiveness of gender-focused investments. Developing standardised and robust methodologies will improve transparency and ensure accountability, enabling investors to measure the tangible outcomes of their initiatives. Moreover, integrating intersectional considerations into these metrics will broaden the scope of GLI, ensuring that investments address overlapping dimensions of inequality. Intersectionality itself is an area that warrants further exploration within GLI. Recognising and addressing the compounded disadvantages faced by individuals at the intersection of gender, race, ethnicity, and other social categories can drive deeper and more inclusive impact. This approach demands intentionality and collaboration with affected communities to design interventions that are both relevant and transformative. Technology and data innovation will be central to the future of GLI. Leveraging advanced analytics, AI-driven insights, and digital tools can enhance the precision of investment strategies and streamline impact assessment processes. These innovations can help bridge data gaps and provide investors with actionable insights to refine their approaches.

Finally, fostering a culture of collaboration among investors, policymakers, and community stakeholders will be key to sustaining and amplifying the impact of GLI. By building partnerships and sharing best practices, the ecosystem can address systemic barriers more effectively and ensure that gender lens investing remains a dynamic and adaptive strategy for driving equity. The future of GLI is both promising and complex. With deliberate efforts to address current limitations and embrace innovation, it has the potential to become a cornerstone of equitable and inclusive economic development worldwide.

5.4 Sectoral Focus of Impact Investing

As the field of impact investing has grown, so too has the diversity of sectors it targets, ranging from renewable energy and healthcare to education and financial inclusion. Each sector presents unique opportunities and challenges for impact investors, who must navigate the complexities of delivering both financial performance and meaningful impact. This section further reviews the report for a discussion on the sectoral focus of impact investing. It examines the key sectors that attract impact capital and the rationale.

5.4.1 Energy

The energy sector lies at the heart of sustainable development, serving as a critical enabler of economic growth, environmental protection, and social equity. According to the GIIN report, impact investors are increasingly prioritising climate-resilient infrastructure, which not only enhances energy efficiency and modernises grids but also contributes to broader climate adaptation goals (Hand et al., 2024). This dual focus underscores the strategic importance of energy investments in tackling global challenges while fostering local economic growth. At 21 per cent of total impact AUM the energy sector commands the largest share of impact AUM, driven by the global push for renewable energy and energy efficiency. Investments target solar, wind, and other clean energy technologies that align with decarbonisation goals. This dominance reflects the sector's scalability, clear financial returns, and its pivotal role in achieving climate-related goals. Emerging markets represent a critical opportunity in the energy sector. Though underrepresented compared to developed markets, these regions present substantial potential due to unmet energy needs and the opportunity to scale renewable solutions. Impact investors' growing interest in these markets reflects a concerted effort to address the global financing gap for sustainable development, which remains a pressing challenge. Financial mechanisms such as blended finance have been deployed to advance energy investments, particularly in underfunded regions. Policy frameworks and consumer demand are further shaping the trajectory of energy investments. Supportive policies, including incentives and regulations, are instrumental in driving capital allocation towards renewable energy. Additionally, the increasing awareness and advocacy for sustainability among consumers influence investment strategies, emphasising the alignment between market forces and environmental objectives. Despite challenges, such as fragmented regulations and varying market conditions, the overall momentum for energy investments remains robust. The potential for growth in energy-focused impact investing is considerable. Commitments to climate solutions and sustainable development are expected to drive further capital allocation (Hand et al., 2024). By leveraging innovative financial mechanisms and targeting underrepresented regions, impact investors can play a transformative role in advancing the global energy transition.

Key Investment Areas

- **Renewable Energy Projects**: Investment in solar, wind, geothermal, and hydroelectric projects is at the forefront of the clean energy transition. These initiatives offer scalable, cost-effective alternatives to fossil fuels and have wide-ranging benefits for urban and rural areas alike.

- **Energy Efficiency Innovations**: Supporting technologies that enhance energy efficiency in buildings, industrial processes, and transportation reduces energy consumption and lowers greenhouse gas emissions, aligning financial returns with environmental impact.
- **Decentralised Energy Systems**: Microgrids, solar home systems, and other decentralised solutions expand energy access in remote and underserved areas. These systems enable community resilience and reduce dependence on centralised energy grids.
- **Energy Storage Solutions**: Addressing the intermittency of renewable energy sources like solar and wind, advanced storage technologies ensure a reliable and continuous power supply, bolstering the feasibility of clean energy projects.
- **Emerging Clean Energy Technologies**: Investments in innovative technologies, such as hydrogen fuel cells and carbon capture systems, pave the way for a sustainable energy future while creating significant market opportunities.

5.4.2 Healthcare

Healthcare is a cornerstone of human development and societal well-being, with access to quality medical services directly influencing life expectancy, economic productivity, and social equity. Healthcare investments are a vital component of the impact investing landscape, addressing critical global challenges related to health equity, access, and innovation. At 9 per cent of the total impact AUM Healthcare has the third highest allocation. The focus on healthcare aligns closely with international goals, such as achieving universal health coverage and improving health outcomes in underserved communities—*SDG 3.8*. Impact investors in this sector prioritise scalable solutions that deliver botanical returns and measurable health impacts. A significant trend in healthcare impact investing is the emphasis on innovation. Investors are increasingly supporting advancements in medical technology, digital health solutions, and scalable healthcare delivery models. These investments aim to address systemic inefficiencies, enhance patient outcomes, and expand access to quality care across diverse populations. Emerging markets present unique opportunities for healthcare investments due to the significant gaps in access and infrastructure. These regions often face dual burdens of infectious diseases and chronic illnesses, making them critical targets for impactful interventions. The growing interest of impact investors in these markets reflects a strategic focus on addressing unmet health needs while fostering local capacity-building and economic development. The role of blended finance mechanisms in healthcare is pivotal. By leveraging public and private capital, these approaches mitigate investment risks and enable the scaling of impactful healthcare projects. Such strategies are particularly effective in funding health infrastructure, expanding access to essential medicines, and supporting community-based health initiatives.

Policy and regulatory environments significantly influence healthcare investments. Supportive policies, such as subsidies, tax incentives, and government partnerships, are crucial in enabling impactful healthcare solutions. Additionally, evolving consumer expectations around health and wellness drive demand for investments that align with social and environmental values. The momentum for impact investing in healthcare remains strong in spite of challenges, such as fragmented healthcare systems and varying regulatory standards. The sector's growth is driven by the increasing recognition of health as a fundamental pillar of sustainable development and social progress. Impact investors are well-positioned to catalyse transformative changes in healthcare delivery and innovation. As the field matures, strategic collaboration and targeted resource allocation will be essential to maximising the impact of healthcare investments (Hand et al., 2024).

Key Investment Areas

- **Access to Healthcare**: Expanding healthcare infrastructure, such as clinics, hospitals, and mobile health units, is critical for improving access. Investments in telemedicine technologies further ensure that remote and underserved populations can receive timely and high-quality care.
- **Medical Technology and Diagnostics**: Supporting advancements in diagnostic tools, treatment devices, and wearable health monitoring systems enhances the quality and efficiency of care delivery. For example, portable diagnostic devices enable early detection of diseases in low-resource settings.
- **Pharmaceuticals and Vaccines**: Investments in research, development, and distribution of affordable medicines and vaccines address

global health challenges, including infectious diseases, chronic conditions, and pandemics. These efforts ensure equitable access to life-saving treatments.

- **Health Workforce Development**: Training and capacity-building initiatives are essential for addressing shortages of skilled healthcare professionals, particularly in low-income and rural areas. Programmes that focus on upskilling and retaining workers improve the reach and quality of health services.
- **Preventative and Community Health Programmes**: Community-focused health initiatives, such as sanitation drives, health education campaigns, and disease prevention programmes, address the root causes of poor health outcomes. These programmes reduce long-term healthcare costs and enhance community resilience.

5.4.3 Education

Education is the cornerstone of societal progress and individual empowerment, providing the foundation for economic growth, innovation, and social equity. It equips individuals with the skills and knowledge required to thrive in an increasingly complex and interconnected world while enabling communities to overcome systemic challenges such as poverty and inequality. This is why education investments occupy a critical role in impact investing. It addresses systemic inequities and fostering long-term economic and social development. These investments align with global priorities, such as ensuring inclusive and equitable quality education and promoting lifelong learning opportunities for all—SDG4. Impact investors in this sector focus on scalable, sustainable solutions that yield measurable improvements in educational outcomes. A key focus in education impact investing is innovation. Investors are increasingly supporting advancements in edtech, vocational training, and alternative learning models. These initiatives aim to bridge gaps in access, improve the quality of education, and equip learners with skills tailored to the demands of the modern workforce. Education has a current allocation of 6 per cent of total impact AUM. The rise of edtech and its potential to bridge education gaps makes this sector increasingly attractive to impact investors. Emerging markets represent significant opportunities for education-focused investments due to persistent disparities in access and quality. These regions often face challenges such as inadequate infrastructure and teacher shortages, making them prime targets for impactful educational interventions. By addressing these gaps impact investors will enhance learning outcomes and contribute to broader social and economic progress. Innovative financing, mechanism such as blended finance, is particularly effective in funding school infrastructure, providing teacher training, and enhancing digital learning initiatives. Policy and regulatory frameworks significantly shape the landscape of education investments. Supportive policies, including funding for Public–Private Partnerships and incentives for innovation, are critical to enabling impactful solutions. Additionally, the growing recognition of education as a cornerstone of sustainable development drives increased collaboration among stakeholders to advance shared goals (Hand et al., 2024). The sector is buoyed by a collective understanding of education's transformative potential in addressing poverty, inequality, and social mobility. Impact investors are uniquely positioned to drive meaningful change by aligning capital with innovative and sustainable education solutions.

Key Investment Areas

- **Early Childhood Education**: Foundational learning during early childhood is critical for cognitive, social, and emotional development. Investments in early education address developmental disparities and provide children with the skills needed to succeed academically and socially.
- **Workforce Development and Vocational Training**: Programmes that focus on skills development and vocational training align educational outcomes with evolving labour market needs. These initiatives equip individuals with the competencies required to succeed in high-demand industries such as technology, healthcare, and renewable energy.
- **Educational Technology** (EdTech): Digital platforms and tools are revolutionising education delivery by expanding access, personalising learning, and improving engagement. Innovations such as AI-driven tutoring, virtual classrooms, and gamified learning experiences address gaps in traditional education systems.
- **Teacher Training and Capacity-Building**: Investments in teacher training programmes enhance the quality of education by equipping educators with modern teaching methodologies and resources.

These initiatives are particularly impactful in underserved regions where teacher shortages are common.

- **Infrastructure Development**: Building and upgrading schools, libraries, and other learning facilities ensure that students have access to safe and conducive learning environments. Infrastructure investments also include providing digital connectivity and resources to bridge the digital divide.

5.4.4 *Agriculture*

Agriculture is a cornerstone of human survival and economic stability, playing a critical role in ensuring food security, sustaining rural livelihoods, and promoting environmental sustainability. It directly supports the livelihood of nearly 2.5 billion people worldwide, many of whom are smallholder farmers in developing countries. As the global population is projected to reach 10 billion by 2050, ensuring sustainable agricultural practices has become paramount for meeting the world's food demands while addressing climate change and resource degradation. Investment in agriculture represents an opportunity to tackle systemic challenges while fostering inclusive growth and sustainable development. Food and Agriculture has a current allocation of 10 per cent of the total impact AUM investments—the second highest. The sector's significance lies in its dual impact on economic resilience and environmental sustainability, particularly in emerging markets with high agricultural dependency. Agricultural impact investments are increasingly focused on transformative approaches that address both productivity and sustainability. Emphasis is placed on improving supply chain efficiencies, introducing climate-resilient farming practices, and promoting agroecological methods that balance productivity with environmental conservation. These interventions aim to safeguard natural resources while meeting the rising global demand for food. Emerging markets are especially significant within this sector, given their high reliance on agriculture for both economic and social stability. Investments in these regions often target infrastructural deficiencies, enhance market access for smallholder farmers, and provide tools and technologies to increase resilience against climate risks. These initiatives drive local economic activity and contribute to food security at a global scale. Blended finance is also leveraged to mobilise private capital into high-risk agricultural markets. Policy alignment is integral to the success of agricultural investments. Supportive regulatory frameworks, incentives for sustainable farming, and partnerships between public and private actors create an enabling environment for impactful investments. Governments and international organisations play a crucial role in addressing structural barriers and coordinating efforts across value chains to maximise outcomes. Challenges persist, including fragmented value chains, unpredictable weather patterns, and regulatory inconsistencies across regions. Nonetheless, the momentum for agricultural impact investing continues to grow, driven by a heightened awareness of agriculture's potential to address global challenges. Investors are increasingly recognising the interconnectedness of agricultural sustainability with broader development goals, such as poverty reduction, climate resilience, and biodiversity conservation (Hand et al., 2024). The journey towards resilient and equitable agricultural systems will require sustained collaboration and innovative approaches, underscoring the transformative power of impact capital in shaping the future of food and farming.

Key Investment Areas

- **Sustainable Farming Practices**: Promoting regenerative agriculture, organic farming, and climate-resilient techniques reduces environmental impact, enhances soil health, and ensures long-term productivity. Techniques such as crop rotation, agroforestry, and water-efficient irrigation systems are foundational to sustainable food systems.
- **Agricultural Technology (Agritech)**: Innovations in precision farming, drone-based monitoring, and data analytics optimise resource use, improve yields, and minimise waste. Agritech solutions also enable smallholder farmers to adapt to climate variability, market fluctuations, and pest outbreaks with greater efficiency.
- **Market Access and Infrastructure**: Building robust supply chains, storage facilities, and transportation networks ensure farmers can access fair markets and reduce post-harvest losses. Investments in cooperatives, digital marketplaces, and fair-trade initiatives further enhance equitable value distribution and market transparency.
- **Financial Inclusion for Farmers**: Providing tailored financial products, such as microloans, crop insurance, and digital payment

systems, enables farmers to invest in modern technologies, diversify income streams, and manage risks effectively. Financial inclusion fosters resilience and drives innovation among smallholder farmers.

- **Sustainable Livestock and Aquaculture**: Supporting eco-friendly livestock farming and aquaculture practices addresses the growing global demand for protein-rich foods while reducing greenhouse gas emissions and promoting ethical treatment of animals. Innovations in feed, waste management, and breeding practices enhance efficiency and sustainability.

5.4.5 Water and Sanitation

Water and sanitation are foundational to human well-being, public health, and sustainable development. Investments in this sector address critical challenges, including access to clean water, sanitation infrastructure, and water resource management. For impact investors, water and sanitation represent a strategic avenue to achieve measurable social and environmental impact while contributing to global priorities such as SDG6. Impact investments in water and sanitation frequently target innovative solutions to close gaps in access and affordability. These include technologies for efficient water purification, scalable sanitation systems, and integrated water resource management strategies. Such interventions aim to improve the quality and reliability of services, particularly in underserved and water-stressed regions, where the need for sustainable solutions is most acute. Despite its critical importance, WASH receives a disproportionately small share of investments 1 per cent of total impact AUM. The sector's low allocation highlights a gap in addressing basic needs, presenting a significant opportunity for targeted investment. Emerging markets stand at the centre of many water and sanitation investment initiatives. Here, the lack of infrastructure and institutional capacity amplifies the challenges of delivering essential services. Impact investors are stepping in to address these gaps by funding projects that expand access to clean water and adequate sanitation facilities while fostering local capacity for sustainable resource management. Blended finance structures are also being deployed to de-risk investments in water and sanitation. The policy environment significantly shapes the potential for impact in water and sanitation. Supportive regulations, such as subsidies for infrastructure development and incentives for adopting water-efficient technologies, enhance the viability of investments. International cooperation and government-led frameworks also play a key role in scaling impactful interventions and aligning stakeholder efforts. Despite notable progress, the sector faces persistent challenges, including climate-induced water scarcity, inequities in service provision, and fragmented governance structures (Hand et al., 2024). Overcoming these obstacles requires concerted action and collaboration across sectors. Investors can drive transformative change in this critical area by focusing on systemic solutions and leveraging cross-sectoral partnerships.

Key Investment Areas

- **Water Infrastructure Development**: Building and upgrading water supply systems, including pipelines, reservoirs, and treatment plants, ensures reliable access to safe and clean water. Innovations in water recycling and desalination are also gaining traction as sustainable solutions to address water scarcity.
- **Sanitation Solutions**: Investing in scalable and affordable sanitation technologies, such as eco-friendly toilets and waste management systems, improves hygiene and reduces health risks in underserved regions. These solutions are particularly impactful in urban slums and rural areas with inadequate sanitation facilities.
- **Decentralised Systems**: Community-based water and sanitation systems provide tailored solutions for remote and underserved areas. These systems are often more cost-effective and sustainable than centralised infrastructure.
- **Water Resource Management**: Supporting initiatives that promote sustainable water use, such as rainwater harvesting, watershed management, and irrigation efficiency, enhances resilience to climate variability and secures long-term water availability.
- **Technology and Innovation**: Smart water management systems, IoT-enabled monitoring tools, and mobile apps for water delivery and sanitation tracking improve service delivery and accountability while reducing operational costs.

5.4.6 Affordable Housing and Infrastructure

Affordable housing and infrastructure underpin economic growth, social equity, and environmental resilience. For impact investors, these sectors represent opportunities to address pressing global challenges, such as urbanisation, housing shortages, and inadequate infrastructure in underserved communities. Investments in affordable housing focus on creating equitable access to safe, quality, and cost-effective living spaces. These initiatives are designed to tackle systemic challenges such as housing affordability crises, urban overcrowding, and homelessness. Impact investors in this space increasingly target scalable housing solutions that integrate energy-efficient designs and community-oriented services to enhance residents' quality of life. Infrastructure investments, on the other hand, concentrate on improving foundational systems that drive social and economic well-being. These include transportation networks, energy grids, healthcare facilities, and educational institutions. Impact investors prioritise projects that deliver long-term benefits by fostering connectivity, resilience, and inclusivity in both urban and rural settings. Affordable Housing has a current allocation of 7 per cent of the total impact AUM. Emerging markets are particularly prominent in the affordable housing and infrastructure landscape. Rapid urbanisation and population growth in these regions create significant demand for sustainable development solutions. Impact capital is being channelled towards addressing deficits in infrastructure and expanding access to affordable housing, often through innovative Public–Private Partnerships and financial structures. Blended finance mechanisms play a transformative role in driving investments in housing and infrastructure. Policy frameworks and regulatory environments are central to scaling impact in these sectors. Supportive policies, such as zoning reforms, tax incentives for affordable housing developers, and subsidies for infrastructure projects, create an enabling environment for impactful investments. International organisations and governments often collaborate with private investors to align development goals with market opportunities. Land acquisition complexities, high upfront costs, and regulatory barriers remain some of the primary challenges. Addressing these obstacles requires innovative thinking, strategic partnerships, and a long-term commitment to inclusive development. By leveraging multi-stakeholder approaches and focusing on sustainable design principles, impact investors can mitigate risks and maximise outcomes.

Key Investment Areas

- **Affordable Housing Development**: Supporting the construction of housing units tailored to low- and middle-income families addresses the growing demand for secure and affordable living spaces. Advances in modular construction, 3D printing, and alternative materials have made it possible to reduce costs and accelerate timelines, improving project feasibility.
- **Urban Renewal and Slum Upgrading**: Transformative projects that revitalise underserved neighbourhoods through mixed-use developments integrate residential, commercial, and public spaces. These initiatives often engage local communities to ensure that developments align with their needs and foster long-term inclusivity.
- **Green and Sustainable Infrastructure**: Promoting environmentally conscious construction methods, including energy-efficient designs and renewable energy integration, reduces operational costs and mitigates environmental impacts. Green infrastructure projects contribute to climate resilience by addressing urban heat islands and improving resource efficiency.
- **Public Transportation Systems**: Investments in accessible and efficient public transportation networks connect low-income populations to education, healthcare, and employment opportunities. By reducing traffic congestion and greenhouse gas emissions, these systems enhance urban sustainability.
- **Smart Cities and Digital Connectivity**: Investing in technology-driven urban solutions, such as smart grids, IoT-enabled utilities, and high-speed digital connectivity, improves service delivery, fosters innovation, and enhances urban living standards.

5.4.7 Environment and Conservation

Environment and conservation investments are central to addressing the planet's most pressing ecological challenges, hence investments in environment and conservation represent a cornerstone of sustainable development and ecological stewardship. These investments address critical issues such as biodiversity loss, ecosystem degradation, and climate change mitigation. For impact investors, this sector offers the dual benefit of preserving natural systems and delivering measurable social and financial returns. Impact capital in the environment and conservation sector often supports projects aimed at protecting and restoring ecosystems. This includes reforestation, marine conservation, and land preservation initiatives. By targeting vulnerable habitats and biodiversity hotspots, these investments contribute to global efforts to halt environmental degradation while fostering sustainable resource use. A growing emphasis is placed on innovative approaches that integrate technology and conservation. Examples include data-driven monitoring systems, climate-resilient agricultural practices, and renewable energy projects designed to reduce environmental footprints. These innovations enhance the efficiency of conservation efforts and create opportunities for scalability and replication across regions. Forestry and Conservation have a current allocation of 1 per cent of the total impact AUM. The sector's modest share reflects underinvestment relative to its environmental importance, though rising interest in biodiversity conservation is expected to boost allocations. Emerging markets are at the forefront of environment and conservation investments due to their rich biodiversity and acute vulnerability to climate change impacts. In these regions, impact investors play a pivotal role in funding initiatives that balance economic development with ecological preservation. Efforts such as community-based conservation programmes and sustainable livelihoods projects demonstrate the potential for aligning environmental goals with local economic needs. Blended finance mechanisms are also encouraging participation from diverse stakeholders, including governments, NGOs, and private enterprises, to align objectives and share responsibilities in protecting natural resources. Supportive policies, such as carbon pricing, incentives for renewable energy, and regulations to combat illegal wildlife trade, create favourable conditions for impactful projects. Additionally, international agreements and frameworks, such as the Paris Agreement, guide investment priorities and encourage cross-border collaboration. Nonetheless challenges such as funding gaps, governance complexities, and the slow pace of policy implementation persist. Addressing these barriers requires coordinated efforts from global and local actors to prioritise conservation, enforce regulations, and foster a culture of sustainability. Impact investors can catalyse change by aligning their strategies with long-term environmental objectives and leveraging collaborative partnerships.

Key Investment Areas

- **Reforestation and Afforestation**: Supporting large-scale tree-planting initiatives restores degraded landscapes, enhances carbon sequestration, and mitigates the impacts of climate change. These projects also provide income opportunities for local communities through sustainable forestry practices.
- **Sustainable Land Management**: Investments in practices such as agroforestry, rotational grazing, and organic farming balance agricultural productivity with ecological preservation, promoting long-term sustainability.
- **Marine Conservation and Sustainable Fisheries**: Protecting marine ecosystems through coral reef restoration, mangrove replanting, and sustainable fishing practices ensures the health of oceans and the resilience of coastal communities.
- **Biodiversity Conservation**: Supporting the protection of endangered species and their habitats through wildlife corridors, anti-poaching initiatives, and community-based conservation programmes helps to maintain ecological balance and promote tourism.
- **Clean Energy Integration in Conservation**: Leveraging renewable energy solutions, such as solar and wind power, for conservation projects reduces reliance on fossil fuels and enhances the sustainability of operations.

5.4.8 Financial Services

The financial services sector holds a central role in impact investing, serving as a catalyst for economic development and social transformation. The sector prioritises expanding access to essential financial tools—such as credit, savings, insurance, and payment systems—for underserved individuals and businesses, particularly in emerging markets. By fostering financial resilience, investments in this domain align with broader goals of economic equity and sustainable development. A defining characteristic of financial services impact investing is its dual focus on innovation and inclusivity. Traditional approaches like microfinance remain a cornerstone of financial inclusion, providing small-scale loans, savings opportunities, and micro-insurance to low-income populations. Simultaneously, fintech solutions, such as mobile banking platforms, peer-to-peer lending, and digital payment systems, have revolutionised service delivery. These technologies enable cost-effective access to financial tools for previously excluded populations, enhancing economic participation and creating pathways out of poverty. Financial services impact investment holds 8 per cent of the current total impact investment AUM. The sector's impact on poverty alleviation and economic empowerment drives continued investor interest, especially in regions with low financial access. Emerging markets are at the forefront of financial services investments. Regions with significant unbanked populations, such as Sub-Saharan Africa, Southeast Asia, and Latin America, present vast opportunities for impactful interventions. Investments often address systemic barriers, including limited infrastructure and regulatory hurdles, to bring financial services to underserved communities while supporting local economic growth. Blended finance mechanisms are particularly being used in financial services impact investing. Policy and regulatory environments play a critical role in shaping the success of financial services investments. Supportive policies such as incentives for digital financial services, regulatory frameworks for fintech, and measures to promote gender equality in financial access create an enabling environment for impactful solutions. Moreover, partnerships between governments, development organisations, and private investors further strengthen the sector's capacity to deliver results (Hand et al., 2024). Many regions continue to face high levels of financial exclusion, exacerbated by systemic inequalities and limited infrastructure. However, the sector's ongoing innovation and the growing commitment of impact investors provide a strong foundation for addressing these gaps.

Key Investment Areas

- **Microfinance**: Impact investors have significantly supported Microfinance Institutions (MFIs) that provide small-scale loans to individuals and micro-enterprises, particularly in developing countries. These loans empower entrepreneurs to start or grow businesses, fostering economic independence.
- **Digital Financial Inclusion**: Fintech innovations like mobile banking and digital payment systems have revolutionised access to financial services. Platforms like M-Pesa in Africa have showcased how mobile technology can bridge the gap for unbanked populations.

- **Small and Medium Enterprise (SME) Financing**: SMEs are critical drivers of job creation and economic development. Impact investments in SME-focused funds help address capital shortages and support business growth in emerging markets.
- **Insurance Solutions**: Affordable insurance products tailored for low-income populations, such as crop or health insurance, help mitigate risks and build resilience against economic shocks.

Economic Justifications for Impact Investing in Financial Inclusion

Economic Growth and Development: Financial inclusion is a critical driver of economic growth, particularly in developing and emerging markets (World Bank, 2024). Access to financial services enables individuals and businesses to save, invest, and manage risks more effectively, which in turn stimulates economic activity and creates a virtuous cycle of growth. When Small and Medium-sized Enterprises (SMEs) have access to credit, they can expand their operations, hire more employees, and contribute to local and national economies. Studies have shown that financial inclusion correlates with higher GDP growth rates.[4] According to the International Monetary Fund (IMF), increasing financial inclusion can lead to an additional 1 per cent annual GDP growth in developing economies (Khera et al., 2021). This is because financial inclusion enhances resource allocation efficiency, increases labour and capital productivity, and promotes entrepreneurship. Moreover, financial inclusion plays a crucial role in formalising the informal economy, which is prevalent in many developing countries. By providing access to formal financial services, such as bank accounts and insurance, impact investing helps integrate informal businesses into the formal economy, increasing their visibility, access to credit, and opportunities for growth.

Financial Resilience and Stability: Financial inclusion is also essential for enhancing financial resilience among individuals and communities, particularly in low-income and vulnerable populations.[5] Access to savings accounts, insurance, and credit allows individuals to manage financial risks, such as medical emergencies, job loss, or natural disasters, without falling into poverty. This financial resilience is critical for reducing vulnerability to economic shocks and promoting long-term economic stability (Pomeroy et al., 2020). Furthermore, financial inclusion contributes to the stability of financial systems by broadening the base of financial participation. When more people use formal financial services, it enhances the depth and liquidity of financial markets, making them more resilient to external shocks. This, in turn, can reduce the likelihood of financial crises and promote more stable economic growth.

Social Justifications for Impact Investing in Financial Inclusion

Reducing Poverty and Inequality: Financial inclusion is a powerful tool for poverty reduction and reducing economic inequality. By providing access to financial services, impact investing empowers low-income individuals to build assets, invest in education and healthcare, and improve their overall quality of life.[6] Access to microfinance will allow a poor entrepreneur to start or expand small businesses, generating income that can lift them out of poverty. In addition, financial inclusion helps address economic inequalities by providing marginalised groups, such as women, rural populations, and ethnic minorities, with the tools they need to participate fully in the economy. Women, in particular, benefit significantly from financial inclusion, as they are often excluded from formal financial systems due to legal, cultural, and economic barriers. By empowering women with access to credit, savings, and insurance, financial inclusion promotes gender equality and enhances the economic security of families and communities. The social impact of financial inclusion extends beyond individual empowerment. By reducing poverty and inequality, financial inclusion contributes to social stability and cohesion. When people have access to financial resources and opportunities, they are less likely to experience economic hardship and social exclusion, which can reduce social tensions and promote peace and stability.

Empowering Marginalised and Vulnerable Populations: Impact investing in financial inclusion is particularly justified by its potential to empower marginalised and vulnerable populations who are often excluded from traditional financial systems. This includes rural populations, who may live far from banking infrastructure; refugees and displaced persons, who often lack the documentation required to access financial services; and informal workers, who do not have a steady income or collateral to secure loans. Financial inclusion initiatives that leverage technology, such as mobile banking and digital financial services, are particularly effective in reaching these populations. Mobile money platforms like M-Pesa in Kenya have enabled millions of previously unbanked individuals to access financial services through their mobile phones, transforming their ability to save, borrow, and transact (Mulili, 2022). By providing these populations with access to financial services, impact investing helps to break down barriers to economic participation, promote social inclusion, and enhance the economic resilience of some of the world's most vulnerable people.

Financial Justifications for Impact Investing in Financial Inclusion

Financial Returns with Social Impact: One of the key justifications for impact investing in financial inclusion is the potential to achieve competitive financial returns alongside measurable social impact. The financial inclusion sector offers a diverse range of investment opportunities, from Microfinance Institutions and fintech startups to SME financing and micro-insurance. Many of these investments have demonstrated strong financial performance, driven by the high demand for financial services in underserved markets and the scalability of technology-driven solutions. For example, Microfinance Institutions (MFIs) have shown that achieving financial sustainability while serving low-income clients is possible. Some MFIs have grown into profitable, large-scale operations, attracting significant investment from both impact and traditional investors. Similarly, fintech companies that focus on financial inclusion have attracted substantial venture capital investment, as they offer scalable solutions with the potential for high returns. Also, the financial inclusion sector is characterised by a growing customer base, particularly in emerging markets where millions of people are gaining access to financial services for the first time. This expanding market presents significant growth opportuni ties for investors, particularly those who can leverage technology to reach underserved populations.

Diversification and Risk Mitigation: Investing in financial inclusion also offers benefits in terms of portfolio diversification and risk mitigation. Financial inclusion investments often involve exposure to emerging markets, which can provide diversification benefits for investors with portfolios concentrated in developed markets. Additionally, the financial inclusion sector includes a range of asset classes, including debt, equity, and alternative investments, allowing investors to diversify their portfolios across different risk-return profiles. Moreover, financial inclusion can contribute to risk mitigation by enhancing the financial resilience of individuals and communities. When people have access to financial services, they are better able to manage risks such as health emergencies, crop failures, or economic downturns, reducing the likelihood of default on loans or financial distress. This, in turn, can reduce credit risk and improve the overall stability of financial systems.

Impact investing in financial inclusion is justified by a compelling combination of economic, social, and financial benefits. Impact investors can promote economic growth, reduce poverty and inequality, and empower marginalised populations by expanding access to financial services. At the same time, financial inclusion investments offer competitive financial returns, diversification benefits, and alignment with ESG and sustainable investment goals while enhancing global economic stability. As the field of impact investing continues to evolve, financial inclusion will remain a cornerstone of efforts to build a more inclusive, resilient, and sustainable global economy.

5.5 Conclusion

This chapter has explored the significant advancements and ongoing challenges within impact investing, which continues to integrate financial performance with measurable social and environmental benefits. With global assets under management surpassing $1.5 trillion, the sector's increasing relevance is undeniable, offering opportunities to mobilise capital towards pressing global challenges. Impact investing faces key hurdles, including the lack of uniform impact measurement standards, regulatory fragmentation, and disparities in capital distribution across regions and sectors. While North America and Europe remain dominant players, capital flows towards Sub-Saharan Africa, Southeast Asia, and Latin America are rising. However, closing persistent financing gaps—especially in underfunded sectors like clean water, sanitation, and gender equity- remains a priority. Financial inclusion emerged as a critical pillar of impact investing, underscoring the transformative role of expanding access to credit, insurance, and digital financial tools. Beyond economic gains, financial inclusion fosters resilience, particularly in underserved communities, reinforcing the importance of targeted investments in sustainable financial services. Additionally, GLI has gained traction, addressed systemic inequalities and proved that integrating gender considerations into investment decisions yields financial and social returns. Innovative financing mechanisms, such as blended finance and equity-like debt, continue to bridge risk gaps, allowing private capital to flow into traditionally underserved areas. To sustain momentum and maximise impact, the sector must prioritise:

- **Developing consistent impact measurement frameworks** to enhance transparency and accountability.
- **Strengthening policy and regulatory clarity** to encourage broader investor participation.
- **Scaling financial inclusion efforts** to ensure equitable access to essential financial services.
- **Leveraging technological innovation** to optimise impact assessment and improve efficiency in capital allocation.

Impact investing is poised to play a defining role in the global transition towards sustainable finance and sustainability. Its success will depend on continued collaboration between investors, policymakers, and financial institutions to drive systemic change. The sector's resilience and capacity for innovation signal a promising trajectory towards a more equitable and sustainable global economy.

Notes

1. Such projections highlight the untapped potential of integrating gender equity into global investment frameworks.
2. See Hand et al. (2024).
3. See Thorarinsdottir et al. (2024).
4. See recent studies like (Odame et al., 2024) and Pal et al. (2025).
5. See Klapper and Lusardi (2020) and Lyons et al. (2020).
6. This is extensively documented in the development finance literature—see Bugg-Levine and Emerson (2011), Mehra et al. (2012), Fanconi and Scheurle (2017), and Agrawal (2018).

References

Agrawal, A. (2018). Effectiveness of impact-investing at the base of the pyramid: An empirical study from India. In *Social entrepreneurship and sustainable business models: The case of India* (pp. 207–246). Palgrave Macmillan.

Bugg-Levine, A., & Emerson, J. (2011). *Impact investing: Transforming how we make money while making a difference*. Wiley. https://books.google.co.uk/books?id=LhxIwPYYhfEC

Fanconi, P. A., & Scheurle, P. (2017). *Small money big impact: Fighting poverty with microfinance*. Wiley.

Hand, D., Sunderji, S., Ulanow, M., Remsberg, R., & Xiao, K. (2024). *State of the market 2024: Trends, performance and allocations*. https://thegiin.org/publication/research/state-of-the-market-2024-trends-performance-and-allocations/

Iqbal, A., Hassan, S., Mahmood, H., & Tanveer, M. (2022, November 01). Gender equality, education, economic growth and religious tensions nexus in developing countries: A spatial analysis approach. *Heliyon, 8*(11), e11394. https://doi.org/10.1016/j.heliyon.2022.e11394

Khera, P., Ogawa, M. S., & Sahay, M. R. (2021). *Is digital financial inclusion unlocking growth?* International Monetary Fund.

Klapper, L., & Lusardi, A. (2020). Financial literacy and financial resilience: Evidence from around the world. *Financial Management, 49*(3), 589–614. https://doi.org/10.1111/fima.12283

Lu, Y., Naik, N. Y., & Teo, M. (2023). Diverse hedge funds. *The Review of Financial Studies, 37*(2), 639–683. https://doi.org/10.1093/rfs/hhad064

Lyons, A. C., Kass-Hanna, J., Liu, F., Greenlee, A. J., & Zeng, L. (2020). *Building financial resilience through financial and digital literacy in South Asia and Sub-Saharan Africa*.

Marime-Ball, P., & Shaber, R. (2022). *The XX Edge: Unlocking higher returns and lower risk*. Simon and Schuster.

Mastercard. (2022). *Young women in africa: Agents of economic growth and transformation by 2030*.

Mehra, R., Patel, P., Shetty, A., Golla, A., Kanesathasan, A., Jacobs, K., & Young, M. (2012). *Financial services for low-income women: Opportunities for economic empowerment*. International Center for Research on Women.

Mulili, B. M. (2022). Digital financial inclusion: M-PESA in Kenya. In *Digital business in Africa: Social media and related technologies* (pp. 171–191). Springer.

Odame, C., Appiah, K. O., & Gyimah, P. (2024). The financial inclusion–economic growth nexus: What is new now? *Journal of Money and Business, 4*(2), 121–137.

Pal, K. K., Piaget, K., Zahidi, S., & Baller, S. (2024). *Global gender gap report 2024*. World Economic Forum.

Pal, S., Vankila, S., & Fernandes, M. N. (2025, January 16). Interplay of financial inclusion and economic growth in emerging economies. *World Development Sustainability*, 100201. https://doi.org/10.1016/j.wds.2025.100201

Pomeroy, R., Arango, C., Lomboy, C. G., & Box, S. (2020). Financial inclusion to build economic resilience in small-scale fisheries. *Marine Policy, 118*, Article 103982.

Subramanian, T., Muirow, A., & Anderson, J. (2022). Evolving the gender analysis in gender lens investing: Moving from counting women to valuing gendered experience. *Journal of Sustainable Finance & Investment, 12*(3), 684–703.

Thorarinsdottir, F. V., Óladóttir, Á. D., Darmstadt, G. L., & Benediktsdóttir, S. (2024). Gender lens investing: A scoping review from theoretical foundations to public equity applications. *Sustainability, 16*(20), 8916.

UNWOMEN. (2023). *Global gender equality in 2023: Urgent efforts needed to reach 2030 goals.* https://www.unwomen.org/en/news-stories/feature-story/2023/09/global-gender-equality-in-2023-urgent-efforts-needed-to-reach-2030-goals

World Bank. (2024). *Financial Inclusion: A key enabler to reducing poverty and boosting prosperity.* The World Bank. https://www.worldbank.org/en/topic/financialinclusion/overview

Young, R., & DAI Bruett, T. (2021). 3 How gender-lens investing measures up. *The catalytic effects of DFI investment–gender equality, climate action and the harmonisation of impact standards*, 30.

CHAPTER 6

Financial Inclusion

6.1 Introduction

Financial inclusion has become a fundamental aspect of global development efforts, offering a pathway to poverty reduction, economic empowerment, and sustainable growth. This chapter examines the intersection of financial inclusion and impact investing, exploring how targeted capital allocation can bridge systemic financial service access gaps. The discussion highlights the transformative potential of impact investing in fostering inclusive finance, empowering marginalised communities, and driving progress towards the Sustainable Development Goals (SDGs). The significance of financial inclusion lies in its capacity to integrate underserved populations into the formal financial system, enabling individuals and businesses to access essential services such as savings, credit, insurance, and payments. These services are not merely economic tools but fundamental socio-economic stability and resilience enablers. However, traditional financial markets often overlook these populations due to perceptions of high risk and low profitability, leaving millions unbanked or underbanked. Impact investing is critical in addressing these market failures by deploying patient, risk-tolerant capital to innovative financial models that prioritise social and financial returns. This chapter is structured to comprehensively analyse how impact investing can drive financial inclusion. It begins with a foundational discussion on financial inclusion,

N. Udohaya, *Impact Investing and Financial Inclusion*, Sustainable Development Goals Series,
https://doi.org/10.1007/978-3-031-87841-1_6

defining its key components and examining its role in fostering economic empowerment. It explores the alignment between financial inclusion and the SDGs, illustrating how inclusive financial systems contribute to the SDGs. It underscores the importance of impact investing in scaling financial inclusion solutions, particularly in regions where traditional banking infrastructure is absent or insufficient. Furthermore, the chapter explores how technological advancements and business initiatives are reshaping financial inclusion and the opportunities to leverage impact investing to drive financial inclusion. The discussion in this chapter positions financial inclusion as a critical pillar of sustainable development and economic empowerment. Through a nuanced analysis of impact investing's role in bridging financial gaps, the chapter underscores the need for strategic, scalable solutions that prioritise social impact and financial sustainability.

6.2 Financial Inclusion

The World Bank defines financial inclusion as the process of ensuring that all individuals, households, and businesses—regardless of their socio-economic status, location, or other potential barriers—have access to a broad range of affordable and responsible financial products and services. These services include savings, credit, insurance, and payments, which are essential for supporting economic participation and resilience (World Bank, 2014). Financial inclusion is not a static concept; it is constantly progressing in response to an evolving financial service landscape shaped by innovations such as digital finance, fintech, mobile money, and new policy initiatives. Financial inclusion aims to extend the reach of financial services to those historically excluded or underserved, such as low-income individuals, women, rural communities, and small businesses and carries every individual along as demands and services evolve. The core principle of inclusive finance is to leave no one behind, recognising the importance of financial access to foster social and economic empowerment, reduce poverty, and promote shared prosperity. By removing barriers to accessing financial services and adapting those services to the diverse needs of the population, inclusive finance helps build resilience, supports livelihoods, and creates opportunities for more equitable growth. This section examines how financial inclusion has become a valuable tool for sustainable growth.

6.2.1 *Economic Empowerment Through Financial Inclusion*

Access to financial services is a critical enabler of economic empowerment, particularly for marginalised groups. Financial services, such as microloans, savings, and insurance, allow individuals to build resilience, invest in their futures, and take advantage of economic opportunities that would otherwise be out of reach (Lal, 2021). Microfinance Institutions (MFIs) and fintech innovations have emerged as key impact-driven actors in bridging the gap for those excluded from traditional banking systems. MFIs provide small-scale loans that enable individuals to start or expand micro-enterprises and a broader range of financial services tailored to underserved populations, contributing to household income and improving economic stability. Fintech solutions, such as mobile money and digital banking platforms, have revolutionised the delivery of financial services, making them more accessible, efficient, and affordable. These innovations facilitate access and foster financial literacy, equipping underserved populations with the knowledge needed to make informed financial decisions. The benefits of inclusive finance extend beyond individual economic resilience. Financial inclusion has a multiplier effect on local economies, stimulating entrepreneurship, job creation, and increased demand for goods and services. It fosters economic dynamism at the grassroots level, driving local economic development by supporting small businesses and entrepreneurs. Small and Medium-sized Enterprises (SMEs), which are often the backbone of local economies, benefit from increased access to credit, allowing them to expand, innovate, and contribute to broader economic growth. Women entrepreneurs, in particular, stand to gain significantly from access to finance. Financial exclusion disproportionately affects women, limiting their ability to engage in entrepreneurial activities and improve their economic standing (Fareed et al., 2017; Henry et al., 2022; Zogning, 2023). However, when women are provided with access to credit, savings, and other financial services, they are empowered to invest in their businesses, enhance their productivity, and achieve economic independence. Borges (2007) showed that women tend to reinvest up to 90 per cent of their income back into their households, contributing to improved health, education, and overall well-being of their families. This supports Thomas (1990), who found evidence that maternal income increased family nutrition four to seven times more than paternal income, which indicates that women's empowerment in the household leads to an improvement in children's health and more human capital accumulation than men's. Thus, access to finance enhances individual economic resilience and strengthens entire communities' socio-economic fabric. Additionally, financial inclusion reduces vulnerability to economic shocks (Koomson et al., 2020; Liu et al., 2024). Access to savings accounts and insurance products enables individuals to build safety nets, reducing their susceptibility to unforeseen events such as illness, natural disasters, or economic downturns. Financial inclusion helps individuals and families maintain stability and avoid falling into poverty by mitigating these risks. Resilience is particularly crucial in developing economies, where social safety nets may be limited or non-existent.

Impact investing in financial inclusion encourages the development of products and services tailored to underserved populations' specific needs. Products designed for low-income individuals or those with irregular incomes help ensure that financial services are accessible and useful to all segments of society. Impact investing helps create a more inclusive financial ecosystem that supports sustainable economic empowerment for everyone by addressing the unique financial needs of marginalised groups.

6.2.2 *Financial Inclusion and the Promotion of the SDGs*

Financial inclusion lies at the heart of sustainable development, playing a critical role in advancing the interconnected goals of poverty reduction, economic growth, and societal equity. Its ability to bridge systemic gaps in access to financial services positions it as a catalyst for achieving multiple Sustainable Development Goals (SDGs). Central to this impact is the way financial inclusion unlocks opportunities for underserved populations, enabling them to participate meaningfully in local and global economies. The transformative potential of financial inclusion is most evident in its ability to alleviate poverty and enhance livelihoods, directly contributing to SDG 1 (No Poverty).[1] Access to affordable credit allows smallholder farmers and micro-entrepreneurs to invest in tools, technologies, and practices that improve productivity and income stability. These investments, in turn, bolster food security (SDG 2: Zero Hunger), create jobs (SDG 8: Decent Work and Economic Growth), and foster economic resilience. For rural populations, where traditional financial infrastructure is often absent, mobile banking and digital wallets provide vital access to financial services, reducing geographical barriers and transaction costs. Financial inclusion also intersects significantly with gender equality (SDG 5: Gender Equality). Systemic barriers have long excluded women from formal financial systems, limiting their economic participation. Initiatives that prioritise gender equity, such as microfinance programmes tailored for women or mobile payment platforms designed with accessibility in mind, empower women to manage resources, invest in their families, and participate in entrepreneurial ventures. These efforts not only enhance individual agency but also strengthen societal progress by increasing women's contributions to economic development. Another critical dimension is the role of financial inclusion in fostering economic growth and reducing inequalities. Inclusive financial systems enable marginalised groups, such as informal workers and persons with disabilities, to access credit, insurance, and savings mechanisms. These services provide a safety net during economic shocks and create pathways for upward mobility, aligning with SDG 10 (Reduced Inequalities). Moreover, financial products linked to sustainability—such as green bonds or micro-insurance for climate adaptation—highlight how financial inclusion supports environmental resilience alongside economic inclusion, advancing SDG 13 (Climate Action).

The integration of digital technologies has amplified the impact of financial inclusion, revolutionising how underserved populations engage with financial systems. Mobile banking, e-wallets, and fintech platforms have expanded financial access in remote areas, fostering economic participation where traditional banking services have been unable to reach. These innovations have enabled more efficient financial flows, reduced costs, and improved security, empowering communities to save, transact, and invest with confidence while addressing SDG 9 (Industry, Innovation, and Infrastructure). Education plays a pivotal role in the success of financial inclusion. Financial literacy initiatives equip individuals with the knowledge to use savings, credit, and insurance effectively, enhancing their ability to make informed decisions. These programmes are particularly impactful when tailored to specific demographic needs, such as youth or women, ensuring that financial products are understood and utilised to their fullest potential. By improving financial literacy, these initiatives strengthen the foundation for sustainable economic participation and directly contribute to SDG 4 (Quality Education).

The intersection of financial inclusion with climate action presents another critical opportunity. Climate-resilient financial products, such as agricultural insurance or green loans, enable vulnerable communities to adopt sustainable practices and adapt to environmental challenges. These investments mitigate climate risks and ensure that economic development is sustainable and equitable by aligning financial inclusion with SDG 13 (Climate Action). The impact of financial inclusion on achieving the SDGs is far-reaching and multifaceted, but its success depends on strategic collaboration and innovation. Public–Private Partnerships have emerged as a key mechanism for scaling financial inclusion efforts, leveraging the resources and expertise of diverse stakeholders to address systemic challenges. Policymakers also play a crucial role by fostering regulatory environments that encourage innovation while ensuring consumer protection and equity. As the financial inclusion landscape continues to evolve, its potential to drive systemic change and advance the SDGs remains unparalleled. Financial inclusion offers a pathway to a more equitable, resilient, and sustainable global economy by addressing barriers, leveraging technology, and fostering inclusive practices.

6.2.3 Investor Demand for Financial Inclusion

Investor interest in financial inclusion has grown significantly over the past decade, driven by a combination of social, economic, and technological factors. With billions of people unable to access essential services such as credit, savings, and insurance, this lack of financial services is particularly acute in developing economies, where the lack of formal financial systems disproportionately affects rural areas and low-income populations. For these communities, managing risks, seizing economic opportunities, and building financial security often hinges on access to finance. Recognising this gap as both a challenge and an opportunity, investors have increasingly channelled capital into solutions that foster economic resilience and reduce systemic inequalities. The advent of digital technologies has transformed the financial inclusion landscape, creating innovative pathways to address systemic barriers. Mobile banking, digital payment platforms, and blockchain-enabled solutions have revolutionised the delivery of financial services by making them more affordable, accessible, and efficient. These advancements have gained particular traction in regions where traditional banking infrastructure is limited, enabling millions of previously unbanked individuals to participate in formal financial systems. The scalability of these digital solutions has made them highly attractive to investors seeking both financial returns and measurable social impact.

Impact capital in financial services aims to close the financial inclusion gap. Private equity, venture capital, and debt financing are increasingly directed towards scalable solutions that target underserved populations. Microfinance Institutions (MFIs) continue to play a central role in providing small loans and other financial services tailored to the needs of low-income individuals and entrepreneurs. However, the rise of fintech companies has diversified the sector, bringing a wave of digital innovation that complements traditional approaches. With their ability to reduce transaction costs and reach remote areas, Fintech platforms have become a focal point for investors eager to deploy capital in impactful and technologically advanced ways. One of the key factors sustaining investor confidence in financial inclusion is the sector's resilience during economic downturns. During crises such as the COVID-19 pandemic, Microfinance Institutions and digital financial platforms demonstrated their ability to adapt and continue serving communities effectively. They underscored their role as stabilisers in times of uncertainty by providing emergency credit, facilitating digital payments, and supporting small businesses. The resilience reinforced the perception of financial inclusion investments as viable and impactful assets capable of delivering consistent returns and significant social benefits. Blended finance models have further catalysed investments in financial inclusion by mitigating risks and attracting a broader range of capital. These models combine public, private, and philanthropic funding, often with public or philanthropic entities assuming the first-loss position. This structure de-risks investments for private investors while amplifying the social impact of initiatives. Such approaches have been instrumental in scaling financial inclusion projects and fostering confidence among stakeholders.

Investors are drawn to a variety of vehicles and strategies within the financial inclusion space. Fintech innovations have garnered significant attention, leveraging technology to deliver digital payment systems, mobile banking services, and blockchain-based solutions that reduce costs and extend access. Meanwhile, Microfinance Institutions remain a cornerstone of the sector, addressing community-specific needs through equity and debt financing. Impact-linked finance products, such as social impact bonds and sustainability-linked loans, align financial returns with measurable social outcomes, creating opportunities for both profitability and impact. Collaborative partnerships, particularly those involving Development Finance Institutions (DFIs), have also played a vital role in de-risking investments and attracting private co-investment. The future of financial inclusion is poised for growth as the sector continues to innovate and adapt. The proliferation of digital technologies will further enhance the reach and scalability of financial services, enabling even the most marginalised communities to benefit. As blended finance models mature and regulatory frameworks evolve, the sector is likely to attract a more diverse range of investors. Impact investing in financial inclusion offers transformative opportunities to address systemic gaps, drive inclusive growth, and foster resilience in communities worldwide.

6.3 Why Impact Investing Matters for Financial Inclusion

Impact investing in financial inclusion allow investment to address systemic gaps in traditional financial systems and catalysing the development of innovative solutions that empower underserved populations. These systemic gaps, often stemming from factors such as limited infrastructure, lack of tailored financial products, and socio-economic inequalities, have left millions excluded from formal financial systems. Such investment creates a pathway to sustainable and equitable access to financial services by deliberately aligning financial objectives with social impact.

Financial Exclusion and the Scope of the Problem

Globally, financial exclusion remains a significant barrier to socio-economic development. According to the World Bank, over 1.4 billion adults lack access to formal financial services, such as bank accounts, savings, credit, and insurance (World Bank, 2023). This exclusion curtails their economic opportunities and prevents them from fully participating in the formal economy. However, financial exclusion is not limited to those who are entirely unbanked; it also includes those who are underserved by the financial services sector and those at risk of being left behind due to rapidly evolving financial markets. The underserved population includes individuals who may have limited access to financial products that truly meet their needs. For instance, they may have a bank account but lack access to affordable credit, insurance, or other financial tools that would enable them to improve their economic standing. These individuals often face high fees, limited product options, or barriers to accessing more sophisticated financial services, leaving them unable to fully benefit from financial inclusion. The lack of tailored financial products that address the unique circumstances of low-income individuals or those with irregular incomes further exacerbates their vulnerability, preventing them from achieving economic stability.

Moreover, the rapidly evolving financial market, driven by technological advancements, has introduced new challenges. Digital finance and fintech innovations are transforming how financial services are delivered, making them more efficient and accessible for many. However, there is a risk that certain segments of the population—particularly those without digital literacy or access to technology—may be left behind. The digital divide is a significant barrier for rural communities, older adults, and those with limited education, who may struggle to access or use digital financial services. As financial markets continue to evolve, it is essential to ensure that these groups are not excluded from the benefits of financial innovation. The challenges are even more pronounced for vulnerable populations, such as women, rural communities, and informal workers. Women face systemic barriers to accessing finance, including legal, cultural, and educational factors that often limit their economic autonomy (Demirgüç-Kunt et al., 2018). Rural communities face challenges due to a lack of infrastructure, geographic isolation, and limited financial institutions. Informal workers, who make up a large proportion of the labour force in many developing countries, often struggle to access formal credit and insurance, making them particularly vulnerable to economic shocks. Financial exclusion for these groups perpetuates cycles of poverty, reinforcing social inequality and economic stagnation.

The traditional financial ecosystem is frequently constrained by risk aversion and a profit-driven focus, which sidelines populations in regions deemed unprofitable or high-risk. Rural communities, informal workers, and low-income groups often fall outside the purview of conventional financial institutions due to inadequate credit histories, geographic isolation, or systemic biases. This exclusion exacerbates existing socio-economic disparities and limits opportunities for growth and resilience. Impact investing disrupts this cycle by introducing patient, risk-tolerant capital that prioritises long-term social returns alongside financial viability. It does so by filling funding gaps and fostering partnerships, innovation, and inclusivity at every level of financial service delivery. By supporting initiatives focusing on underrepresented populations and regions, impact investing ensures that financial inclusion efforts are far-reaching and transformative. Moreover, the catalytic nature of impact investing encourages collaboration between stakeholders, including governments, non-profits, and private-sector entities. This multi-stakeholder approach ensures that solutions are holistic and sustainable, addressing the root causes of exclusion rather than merely treating the symptoms. The following section explores the key mechanisms through which impact investing drives meaningful progress towards universal access to financial services, highlighting its indispensable role in fostering equity, resilience, and empowerment within underserved communities.

6.3.1 Unlocking Capital for Underserved Markets

Impact investing mobilises capital for populations and regions often overlooked by traditional financial systems, effectively addressing systemic funding gaps. The challenge of providing financial services to underserved markets stems from perceptions of high risk, low profitability, and inadequate infrastructure. These systemic barriers often deter traditional financial institutions, creating a persistent cycle of exclusion. Impact investors address these challenges through several critical mechanisms:

Risk-Tolerant Capital: Traditional financial institutions often prioritise immediate returns and low-risk profiles, which inherently disadvantages regions or groups lacking formal financial infrastructure or credit histories. Impact investors adopt a broader perspective, recognising the untapped potential in these markets. They provide the necessary foundation for other financial players to enter these markets by absorbing initial risks. For example, by using alternative data sources, such as community-level indicators or behavioural data, impact investors can create more nuanced risk assessments, ensuring that capital flows to where it is most needed. This approach has catalysed credit access for individuals and small enterprises that were previously excluded from formal financial systems.

Blended Finance Structures: Blended finance models are a cornerstone of impact investing's ability to address funding gaps. These models combine concessional capital from development agencies with private-sector funds, reducing risks for commercial investors while maintaining a focus on social outcomes. Blended finance has been instrumental in sectors such as housing finance for low-income groups and renewable energy initiatives tied to financial inclusion. To amplify their impact, these models can incorporate performance-linked funding, where financial returns are tied directly to measurable improvements in access and equity.

Sector-Specific Investments: Impact investors often focus on sectors that demonstrate significant potential for social and financial returns. For example, digital financial platforms have enabled low-cost remittance services, critical to supporting families in developing economies. Agricultural finance initiatives tailored to smallholder farmers' seasonal needs have provided not only credit but also stability, allowing for sustainable growth. Impact capital has also enabled the development of niche financial products, such as insurance for climate-related risks. These investments highlight the importance of tailoring financial solutions to meet the unique needs of underserved populations.

Strengthening Financial Intermediaries: Financial intermediaries, such as Microfinance Institutions (MFIs), cooperatives, and community-based savings groups, are vital in connecting underserved populations with financial services. However, these entities often face challenges related to funding, operational capacity, and scalability. Impact investors actively strengthen these intermediaries by providing both capital and technical support. For instance, investments in digital transformation for MFIs have enabled them to reach more clients efficiently, while partnerships with cooperatives have enhanced their ability to offer diversified financial products. Strengthening these intermediaries ensures that financial services reach even the most remote and excluded communities.

6.3.2 *Bridging Gaps in Infrastructure and Accessibility*

In many regions, inadequate financial infrastructure and geographic isolation prevent communities from accessing essential services. These challenges create significant barriers to financial inclusion, particularly in rural and underserved areas where traditional financial institutions have limited presence. Impact investing plays a crucial role in addressing these structural issues by financing innovative solutions and projects that expand accessibility and reduce the cost of financial services. This section explores how impact investing bridges infrastructure gaps and why these efforts are vital to advancing financial inclusion.

Building Last-Mile Distribution Networks: Geographic isolation and the lack of physical infrastructure often mean that financial services are inaccessible to millions. Traditional financial institutions typically find establishing physical branches in remote areas cost-prohibitive, leaving vast populations unserved. Impact investing addresses this by funding projects that create last-mile distribution networks. For instance, investments in agent-based banking systems enable local representatives to provide essential banking services like deposits, withdrawals, and remittances in areas without formal branches. Mobile money agents, particularly in Sub-Saharan Africa, have proven how transformative such networks can be, connecting rural communities to the formal financial system. Expanding these networks with targeted investments in training and technology could enhance their efficiency and coverage, ensuring that no community remains isolated.

Supporting Fintech Innovations: The advent of fintech has revolutionised the delivery of financial services by eliminating the reliance on traditional banking infrastructure. Impact investors are pivotal in scaling these innovations, which are critical to bridging accessibility gaps. For instance, investments in digital onboarding technologies have enabled remote account creation and identity verification, reducing the need for physical documentation and in-person visits. Similarly, funding for mobile payment platforms has created low-cost, accessible alternatives to cash transactions, empowering underserved populations to participate in the digital economy. Technologies like blockchain enable impact investing to enhance trust and security in financial systems, addressing long-standing barriers to adoption. The potential for fintech to innovate beyond current capabilities underscores the need for continued investment to keep pace with evolving accessibility challenges.

Expanding Digital Payment Ecosystems: Digital payment systems have emerged as a cornerstone of financial inclusion, reducing transaction costs and broadening access to financial services. Impact investing has been instrumental in building the infrastructure needed to support these systems. For example, investments in cloud-based platforms and interoperability frameworks have made it possible for users to seamlessly transact across different service providers, fostering a more inclusive ecosystem. Impact investors also support partnerships between fintech companies, telecom operators, and governments to scale digital payment solutions. These collaborations have facilitated the delivery of integrated services such as microloans and savings plans. Addressing gaps in financial literacy—often through complementary investments in education programmes—is essential to ensuring these ecosystems achieve their full potential and are accessible to all.

Bridging infrastructure and accessibility gaps is foundational to achieving financial inclusion. Underserved populations remain disconnected from the tools and services that enable economic empowerment without adequate infrastructure. Impact investing matters in this context because it provides the necessary financial resources, drives innovation, builds partnerships, and ensures that solutions are sustainable and scalable. By tackling these barriers, impact investors are creating pathways for millions of people to participate in the formal economy, laying the groundwork for more inclusive and equitable growth.

6.3.3 *Empowering Social Enterprises as Catalysts*

Impact investing plays a pivotal role in supporting social enterprises that design and deliver financial products tailored to the unique needs of underserved groups. Social enterprises often operate in environments where traditional financial models fail to address the complexities of poverty, inequality, and exclusion. They enable access to services that improve financial inclusion and foster long-term socio-economic resilience by targeting these gaps. Providing affordable credit, creating savings schemes, offering insurance products, and enabling experimentation and innovation are ways impact investing empowers social enterprises.

Providing Affordable Credit: Access to affordable credit is one of the most significant barriers low-income borrowers face. Traditional financial institutions often exclude these individuals due to the perceived risks and high service costs. Social enterprises bridge this gap by designing microloans and flexible credit products tailored to the cash flow patterns of underserved communities. For instance, loans with small principal amounts and repayment schedules aligned with seasonal income fluctuations enable borrowers to repay without excessive strain. Impact capital is critical in supporting these enterprises, enabling them to take on higher risks and experiment with innovative lending models. Peer-to-peer lending platforms backed by impact investors have demonstrated how alternative credit assessment methods, such as analysing mobile payment histories or community reputation, can broaden access to credit. Scaling such initiatives requires sustained impact investments to address ongoing challenges like repayment defaults and operational costs, ensuring these models remain viable and scalable.

Creating Savings Schemes: Savings play a foundational role in financial stability, yet many low-income individuals lack access to secure and incentivised savings products. Social enterprises supported by impact investors have developed tailored savings schemes that align with the specific needs of their target populations. Examples include group savings accounts for women entrepreneurs or savings products linked to specific goals like education or healthcare. These enterprises often incorporate behavioural incentives, such as matched savings contributions or interest rate bonuses for consistent deposits, to encourage participation and instil a culture of saving. Impact investing can further scale these efforts by funding the development of mobile applications that facilitate savings, providing easy-to-use interfaces and financial literacy resources to encourage adoption. Moreover, investments in partnerships between enterprises and local financial institutions can extend the reach of savings products, ensuring accessibility even in remote areas.

Offering Insurance Products: Vulnerable populations are disproportionately affected by economic shocks such as natural disasters, health emergencies, and crop failures. Due to high costs and limited market understanding, traditional insurance models often exclude these groups. Social enterprises have pioneered micro-insurance products designed to address these vulnerabilities, providing low-cost coverage for health, agriculture, and other critical risks. For example, crop insurance products with premiums tied to weather patterns or health insurance schemes bundled with affordable preventive care services have proven effective. Impact investors play a key role in supporting these initiatives by funding pilot programmes, underwriting initial losses, and helping enterprises build the technical capacity to analyse risk and price premiums effectively. Additionally, investments in data-driven tools like satellite monitoring for agriculture or AI-based health diagnostics can further refine insurance offerings, ensuring they are responsive to the needs of underserved populations.

Enabling Experimentation and Innovation: One of the defining characteristics of social enterprises is their ability to innovate and adapt to the unique challenges of their target communities. However, this requires significant upfront investment and a tolerance for risk that traditional investors are often unwilling to provide. Impact capital fills this gap by enabling social enterprises to test and refine their solutions. For instance, funding from impact investors has allowed enterprises to explore digital platforms that deliver financial education alongside products, enhancing user engagement and understanding. Similarly, grants or concessional loans enable enterprises to trial new models, such as pay-as-you-go financing for solar home systems, which combine credit with energy access. By supporting such experimentation, impact investing fosters the continuous evolution of inclusive financial solutions, ensuring they remain relevant and effective in addressing emerging challenges.

Empowering social enterprises is critical to advancing financial inclusion because these organisations catalyse systemic change. They address gaps left by traditional financial institutions, offering products and services that are accessible, affordable, and tailored to the needs of underserved populations. Impact investing matters in this context because it provides the necessary resources and risk tolerance to enable social enterprises to thrive. By supporting these enterprises, impact investors create pathways for broader economic participation, resilience, and empowerment, laying the foundation for a more inclusive financial ecosystem.

6.3.4 Fostering Inclusion Through Innovation

Technological innovation is at the heart of many financial inclusion initiatives, and impact investing accelerates its development and adoption. Impact investors nurture innovations that would otherwise struggle to attract commercial capital by providing patient and flexible funding. These investments enhance the reach and efficiency of financial services and contribute to building trust among communities historically marginalised by traditional financial systems. This section explores the critical role impact investing plays in fostering inclusion through innovation.

Alternative Credit Scoring: Traditional credit-scoring systems often exclude individuals without formal credit histories, leaving millions unable to access loans and other financial products. Impact investors support innovations that use alternative data—such as mobile phone usage, utility payment records, or social media activity—to assess creditworthiness. These systems expand lending opportunities by providing a more holistic view of an individual's financial behaviour. For instance, fintech startups in emerging markets have used mobile payment histories to evaluate loan eligibility for small-scale entrepreneurs who lack conventional financial records. This approach enables access to credit and incentivises borrowers to engage with formal financial systems. Impact investors can further drive progress by funding advanced analytics tools and AI-based platforms that refine credit risk assessments, ensuring these systems remain inclusive, efficient, and scalable.

Blockchain for Transparency: Trust is a significant barrier to financial inclusion, particularly in regions where fraud and corruption undermine confidence in formal financial systems. Blockchain technology offers a transformative solution by creating secure, transparent, and tamper-proof transaction records. Impact investors have played a key role in funding blockchain-based platforms that enhance transparency in financial transactions, such as peer-to-peer lending or remittance services. For example, blockchain has been used to facilitate cross-border remittances, reduce transaction costs, and ensure funds reach intended recipients without intermediaries. Future investments could focus on integrating blockchain with local financial institutions to build decentralised identity verification systems, enabling underserved populations to establish secure digital identities. Such advancements can reduce barriers to accessing credit, insurance, and other financial products while fostering trust in formal systems.

Digital Wallets and Payments: Digital wallets and payment platforms have revolutionised financial inclusion by offering low-cost, accessible alternatives to cash-based systems. Impact investors have been instrumental in scaling these innovations, which cater specifically to low-income populations. Mobile money services, such as those in Sub-Saharan Africa, have enabled individuals without bank accounts to save, transfer, and receive funds securely and conveniently. In addition to facilitating basic transactions, impact capital has supported the integration of additional features like micro-savings tools, credit access, and micro-insurance products into digital wallets. These enhancements empower users to build financial resilience and plan for long-term goals. Further investments in this area could address gaps in financial literacy and digital infrastructure, ensuring even the most remote communities benefit from digital payment systems. Subsidising mobile data costs and developing user-centric educational programmes can significantly boost adoption and sustained usage.

Fostering inclusion through innovation is essential to addressing the systemic barriers that exclude millions from financial systems. Impact investing accelerates the adoption of transformative technologies that extend the reach of financial services and create new pathways for trust and participation. By supporting alternative credit-scoring, blockchain-based solutions, and digital payment ecosystems, impact investors empower underserved populations to engage meaningfully with formal financial systems. These efforts promote financial inclusion and contribute to broader socio-economic development, ensuring that innovation becomes a driver of equity and empowerment.

6.3.5 *Mitigating Inequality and Driving Systemic Change*

Financial exclusion often intersects with broader systemic inequalities, including those related to gender, ethnicity, and geography. These overlapping barriers perpetuate cycles of poverty and limit access to economic opportunities for marginalised groups. Impact investing explicitly targets these inequities by channelling resources towards initiatives that prioritise inclusivity. Impact investing expands access to financial services and drives broader social and economic transformation by addressing structural inequalities.

Funding Women-Led Enterprises and Gender-Inclusive Products: Gender inequality remains one of the most pervasive barriers to financial inclusion, with women disproportionately excluded from formal financial systems. Women often face structural challenges such as lack of collateral, cultural restrictions, and limited financial literacy, which hinder their ability to access credit, savings, and insurance products. Impact investors are crucial in closing this gap by funding women-led enterprises and supporting products designed to meet women's specific financial needs. For instance, microfinance programmes tailored for women have successfully empowered female entrepreneurs and improved household welfare. These programmes often combine credit with capacity-building initiatives, such as financial literacy training and mentorship, ensuring that women can maximise the impact of their investments. Additionally, gender-focused fintech solutions, such as mobile savings accounts with privacy features or digital platforms offering flexible repayment schedules, provide women with tools to overcome systemic barriers. Scaling these efforts requires sustained impact capital to address gender-specific challenges and create an enabling environment for financial empowerment.

Supporting Underserved Communities Through Holistic Approaches: Financial exclusion in remote and underserved regions often stems from a combination of geographic isolation, infrastructure deficits, and socio-economic disparities. Addressing these challenges requires more than just extending financial infrastructure; it involves creating comprehensive and context-sensitive solutions. Impact investing targets these areas by funding community-focused projects, such as cooperatives and Microfinance Institutions, that are specifically designed to serve marginalised groups. For example, investments in agricultural finance programmes can provide smallholder farmers access to credit and insurance tailored to seasonal income cycles, enhancing both productivity and resilience. Additionally, impact investors support initiatives that integrate financial services with other development objectives, such as education and healthcare access. This holistic approach improves financial inclusion and addresses the root causes of exclusion, fostering long-term community development. Partnerships with local governments and NGOs are critical to scaling these efforts and ensuring their sustainability.

Partnering with Governments and NGOs to Address Structural Barriers: Financial exclusion is often rooted in systemic issues such as discriminatory practices, lack of financial literacy, and inadequate legal frameworks. These structural barriers disproportionately affect marginalised groups, including ethnic minorities, refugees, and individuals without formal identification. Impact investors collaborate with governments and NGOs to develop holistic solutions that address these underlying causes. For example, partnerships have facilitated the rollout of national identity programmes that enable individuals without formal documentation to access financial services. These programmes are particularly impactful in regions where legal identity is a prerequisite for accessing credit, opening bank accounts, or receiving social benefits. Impact capital also funds educational initiatives to improve financial literacy among marginalised groups, empowering them to make informed decisions about their financial futures. Additionally, advocacy efforts funded by impact investors can lead to reforms in discriminatory policies and promote regulatory environments that encourage inclusive finance. Addressing these systemic barriers requires a multi-stakeholder approach, combining financial resources with technical expertise and policy advocacy to create lasting change.

Mitigating inequality and driving systemic change are essential to creating a more equitable financial ecosystem. Financial exclusion not only limits individual opportunities but also undermines broader socio-economic development. Impact investing matters because it prioritises the inclusion of marginalised populations, ensuring that financial services are accessible, affordable, and tailored to their needs. By funding women-led enterprises, supporting underserved regions, and addressing systemic barriers, impact investors catalyse transformative change, paving the way for a more just and inclusive economy.

6.4 Innovations Driving Financial Inclusion

Financial innovation has been pivotal in enhancing financial inclusion and providing access to financial services for the underbanked and unbanked populations, particularly in emerging and developing countries. This section looks at some of the innovative solutions and technologies that are helping to promote access to financial services. It also examines the success of the innovations, exploring their funding atmosphere to understand how their potential impact is being hindered by funding gap.

6.4.1 Technology and Products

Mobile Money Services

Mobile Money Services (MMS) are financial services accessed and conducted using mobile devices like mobile phones or tablets. They allow users to store, send, and receive money, as well as make payments for goods and services, often without needing a traditional bank account. MMS has emerged as a key innovation in promoting financial access and inclusion, especially in regions with limited formal banking infrastructure. Defined as financial services that allow users to conduct transactions using mobile devices, MMS aim to overcome barriers to financial services—which could be in the form of remote location, cost of travel to bank branch, cost of services, documentation requirements, gender and social norms or financial literacy and trust, for the unbanked and under-served populations.[2] Empirical studies show the critical role MMS plays in reducing financial exclusion and contributing to poverty reduction, economic growth, and improved well-being.[3] The success of MMS lies mainly in its ability to capitalise on the widespread availability of mobile phones. Unlike traditional banking services that require physical branches, MMS leverages mobile technology and agent networks to enable easy access to financial services, including savings, transfers, and credit. As Suri and Jack (2016) demonstrated, the spread of MMS has resulted in significant improvements in household welfare, with reduced transaction costs and increased user financial resilience. MMS has also played a vital role in empowering women and small businesses. By facilitating micro-transactions and providing access to savings and credit, MMS has enabled women, particularly in rural areas, to gain greater financial independence and participate more actively in the economy (Dorfleitner & Nguyen, 2024). Women can often manage their finances, save for emergencies, and invest in small income-generating activities, improving their overall economic standing.

For small businesses, MMS has provided a convenient and cost-effective way to conduct transactions, reducing the reliance on cash and enabling more efficient management of business finances. Small and Medium-sized Enterprises (SMEs) have been able to leverage MMS to accept payments, pay suppliers, and access credit facilities, which has enhanced their growth and resilience (Konte & Tetteh, 2023; Tiwasing et al., 2024). The ability to make seamless transactions allows SMEs to expand their customer base, including those who do not have access to traditional banking. Kikulwe et al. (2014) found that remittances received from relatives and friends through MMS contribute to income directly and help reduce risk and liquidity constraints, promoting the commercialisation of farm produce among farmers. Mobile money users apply more purchased farm inputs, market a larger proportion of their output, and have higher profits than non-users of this technology. Domestic remittances are typically crucial source of income for many low-income households. The ease of transferring money using mobile phones has improved financial flows between urban and rural areas, enhancing the ability of rural families to access cash when needed and contributing to improved food security and educational opportunities for children (Babatunde, 2018; Fourmy, 2020; Moniruzzaman, 2022).

a. **Industry Growth and Adoption Trends**

GSMA (2024) reports that the mobile money industry has witnessed remarkable growth over the last decade, with the number of registered accounts reaching 1.75 billion globally, facilitating transactions worth $1.4 trillion annually. This growth is driven mainly by increasing demand for accessible financial services, especially in regions where access to traditional banking remains challenging. The Sub-Saharan Africa region remains a major driver of this growth, accounting for nearly three-quarters of all global mobile money accounts. Although adoption rates remain strong, the growth trajectory is beginning to mature, evidenced by a deceleration in new account registrations. In regions like West Africa, however, growth remains robust, driven by countries like Nigeria, Ghana, and Senegal, which are seeing a surge in new registrations and greater adoption of mobile money services. The rapid expansion of agent networks in these regions has significantly contributed to the uptake of services, improving the reach and access to mobile money facilities (GSMA, 2024). The increased competition in the mobile money market, particularly from non-mobile network operators (non-MNOs) such as banks and giant tech companies, has also reshaped the ecosystem in certain regions. In West Africa, non-MNOs are emerging as viable challengers to traditional MNO-led mobile money services, driving greater innovation and offering enhanced services that cater to diverse customer needs. Regulatory adjustments and financial inclusion policies have further bolstered growth in this region, promoting an environment that supports the proliferation of agent networks and new services.

The growth of MMS in the past decade can be seen it its economic impact, contributing $600 billion to the Gross Domestic Product (GDP) of countries with active MMS from 2013 to 2022, representing around 1.5 per cent GDP growth across those nations. This growth is a reflection of increased transaction volumes and an indication of enhanced financial inclusion and deeper integration of previously excluded populations into the formal economy. MMS continues to transform the economic landscape by enabling millions of individuals secure and convenient digital transactions. For many users, mobile money accounts have become their first formal financial instrument, allowing them to save, borrow, and invest in previously inaccessible ways. This economic empowerment is particularly evident in Small and Medium-sized Enterprises (SMEs), which rely on mobile money to facilitate payments, manage supply chains, and access microloans to expand their operations.[4] The scope of mobile money has extended far beyond person-to-person (P2P) transfers and bill payments. New use cases, including international remittances, merchant payments, and digital savings products, are gaining momentum and redefining the capabilities of mobile money platforms.

- **International Remittances**: In 2023, international remittance flows through mobile money channels grew by 33 per cent year-on-year, reaching $29 billion. This trend highlights the growing preference for mobile money over traditional remittance methods, which are often costly and inaccessible for many low-income households. Mobile money has made cross-border transfers more affordable and accessible, contributing significantly to the financial resilience of millions of households.

- **Merchant Payments**: Merchant payments through mobile money also saw substantial growth, with transaction values reaching $74 billion in 2023. The adoption of mobile money by merchants has transformed payment ecosystems, allowing for greater digitisation of transactions. This shift is particularly beneficial for small businesses that face challenges in accessing traditional banking services and point-of-sale systems.
- **Digital Savings and Credit**: Mobile money platforms increasingly offer digital savings and microcredit products, allowing users to save and borrow conveniently. These services are vital for underserved populations who lack access to formal financial institutions. By integrating savings and lending features, mobile money is helping individuals build financial resilience and meet emergency needs.

Technological advancements are playing a key role in shaping the current state of MMS. The integration of mobile money platforms with other financial technologies, such as digital credit scoring, blockchain, and artificial intelligence, is enhancing service offerings and improving user experiences. Digital credit scoring, for instance, allows MMS providers to offer microloans to users based on their transaction histories, making credit more accessible to those without formal credit records (CGAP, 2023). Blockchain technology is also being explored as a means to enhance the security and transparency of transactions, addressing concerns related to fraud and data breaches. The COVID-19 pandemic also had a significant impact on the growth of MMS. During the pandemic, the use of contactless payments surged, with mobile money becoming a preferred method for conducting transactions in many regions. Governments and humanitarian organisations leveraged MMS to distribute relief funds, highlighting the role of mobile money as a tool for financial resilience during crises (GSMA, 2022). The pandemic accelerated digital adoption, prompting many users who were previously reliant on cash to transition to mobile money platforms for everyday transactions.

Despite these successes, the adoption of MMS is uneven across regions. Latin America and Southeast Asia, for example, have lagged behind Sub-Saharan Africa and South Asia in terms of mobile money adoption. In Latin America, factors such as market fragmentation, regulatory constraints, and competition from traditional banking services have limited the uptake of MMS (GSMA, 2023). Countries like Mexico and Brazil have seen slower growth due to stringent regulatory requirements and a well-established banking sector that competes with mobile money offerings. In Southeast Asia, while countries like the Philippines and Indonesia have seen some growth in MMS, regulatory barriers and limited mobile infrastructure challenges have hindered widespread adoption. Regulatory challenges remain a critical factor influencing the growth of MMS. In some countries, regulations are either too restrictive or lack the clarity needed to support MMS growth. For instance, in India, the Reserve Bank of India's stringent requirements for mobile money operators have limited the expansion of services like M-Pesa, which eventually exited the market (Ghosh, 2018). Conversely, countries that have adopted enabling regulatory frameworks, such as Kenya's approach to fostering innovation while ensuring consumer protection, have seen substantial success in expanding mobile money adoption (AFI, 2023). Infrastructure limitations, particularly in rural areas with poor network coverage, remain a significant barrier to the growth of MMS, as well. The success of MMS depends on the presence of a robust agent network that can manage cash liquidity effectively. However, maintaining such a network in remote areas is costly and logistically challenging, affecting the scalability of MMS (Suri & Jack, 2016). Additionally, customer trust remains a critical issue, especially in regions with low financial literacy. Fraud, misinformation, and lack of awareness about the benefits of MMS hinder adoption, particularly among older populations and those in rural areas. Addressing these issues requires sustained efforts in consumer education and awareness-building to enhance trust and drive adoption.

b. **Successes and Failures of Mobile Money Services**

The mobile money landscape is marked by a blend of significant successes and notable failures that have shaped the evolution of financial inclusion globally. While services like M-Pesa in Kenya and bKash in Bangladesh serve as exemplary models of how MMS can transform economies, other cases, like M-Pesa in South Africa and mobile money services in India, reveal the challenges and limitations of these initiatives. This discussion reviews two successful and two failed case studies of mobile money services, focusing on their strategies, impacts, and the key reasons behind their outcomes.

i. **M-Pesa in Kenya—Success**

Launched in 2007 by Safaricom and Vodafone, M-Pesa has become one of the most prominent success stories in mobile money. The history of M-Pesa dates back to early 2003, when the concept of a mobile-based financial service was conceived to help microfinance borrowers in Kenya manage loan repayments more effectively. Initially, the idea was to create a tool for Microfinance Institutions to track loan disbursements and collections, but it soon evolved into a broader platform for peer-to-peer money transfers that could address the financial needs of Kenya's unbanked population. The service was rolled out as a pilot project in 2005, with initial funding from Vodafone, Safaricom, and support from the United Kingdom's Department for International Development (DFID). The pilot was successful, and M-Pesa was officially launched in 2007, providing a secure, affordable, and convenient way for users to send money to friends and family without needing a traditional bank account. M-Pesa addressed a critical gap in the financial services market in Kenya, where over 80 per cent of the population lacked access to formal banking. This was particularly important in a cash-driven economy, where carrying money posed significant risks and sending money over long distances could be costly and insecure. As M-Pesa gained popularity, it rapidly expanded its offerings to include additional financial services such as savings, loans, and insurance. In collaboration with the Commercial Bank of Africa (CBA), M-Pesa launched M-Shwari, a service that allowed users to save money and access microloans directly through their mobile phones. This expansion helped create a financial safety net for millions, promoting savings behaviour and providing small-scale credit that was previously out of reach. The success of M-Pesa can be attributed to several key factors,[5] including the following.

- **Market Fit and Need**: Studies showed that a significant portion of the Kenyan population lacked access to formal banking services. This made M-Pesa an attractive solution. M-Pesa catered to the unmet needs of the unbanked population, providing a means to store and transfer money securely. The innovation was able to tap into the large unbanked population by providing a secure platform for saving money, transferring funds, and conducting transactions without the need for traditional banking services.[6] This particularly impacted rural and underserved communities, enabling them to gain financial security and independence.

- **Regulatory Support**: Regulatory flexibility played a major role in M-Pesa's success.[7] The Kenyan Central Bank took a supportive approach, allowing Safaricom to roll out the service without excessive regulatory burdens. M-Pesa was permitted to operate without the same constraints as traditional banks. For example, the Central Bank of Kenya allowed Safaricom to operate M-Pesa under a "special payment services" category, rather than necessitating a full banking licence. This enabled M-Pesa to launch and scale rapidly without being subject to the more stringent requirements typically imposed on banks, such as minimum capital requirements or restrictions on transaction types. This regulatory flexibility provided M-Pesa with a significant advantage, allowing it to expand its offerings (such as microloans through M-Shwari) without the delays and administrative burdens often associated with banking licences. It encouraged rapid scalability and innovation in the types of services offered.
- **Extensive Agent Network**: M-Pesa's agent network was another major driver of its success. The role of agents was critical in ensuring the widespread adoption of M-Pesa. Safaricom's strategy to develop an extensive network of agents in both urban and rural areas enabled users to access cash-in and cash-out services. They built a network of over 250,000 agents to support service accessibility to users even in remote areas. Agents provided the bridge between digital and physical currency, making mobile money practical for the masses.
- **Strong Brand Trust**: Safaricom, a leading telecommunications provider in Kenya, was already well-established and trusted. According to Morawczynski and Pickens (2012), the fact that M-Pesa was backed by a well-established telecommunications company helped build consumer trust in a new financial product. This played a significant role in driving the adoption of M-Pesa, as users were more willing to try out a financial service backed by a reliable brand.

M-Pesa has had a profound impact on Kenya's economy and society. According to Suri and Jack (2016), M-Pesa, users are more financially resilient and able to withstand financial shocks better than non-users. The service has played a significant role in women's economic empowerment by giving them greater control over their finances and boosting their participation in business activities. Today, over 50 million users across Africa rely on M-Pesa, making it one of the most successful mobile money services globally. Its success has inspired similar initiatives in other countries and has demonstrated the transformative potential of mobile money in achieving financial inclusion.

ii. **bKash in Bangladesh—Success**

Soon after its launch in 2011, bKash became the leading mobile money service in Bangladesh, with over 70 million users (World Bank, 2023). The idea for bKash was born out of a recognition of the large unbanked population in Bangladesh and the need for accessible financial services. At the time, traditional banking was out of reach for many Bangladeshis, especially those living in rural areas. The founders saw the potential for mobile technology to bridge this gap, particularly as mobile phone usage was increasing rapidly in the country. The service was developed through a partnership between BRAC Bank and Money in Motion, with significant backing from international stakeholders such as the International Finance Corporation (IFC) and the Bill & Melinda Gates Foundation. BRAC Bank, already well-known for its work in microfinance, provided the credibility and local expertise necessary to navigate the market, while Money in Motion brought in technical and strategic expertise to develop the mobile money infrastructure. The involvement of IFC and the Bill & Melinda Gates Foundation not only provided essential funding but also brought insights on scaling financial inclusion based on global best practices. Initially, bKash launched as a basic person-to-person (P2P) money transfer service, which allowed users to send and receive money through their mobile phones without needing a traditional bank account. This was particularly valuable for people in rural areas who relied on domestic remittances from family members working in cities. Over time, bKash expanded its services to include bill payments, mobile top-ups, merchant payments, and savings, transforming itself into a comprehensive digital financial platform. bKash's success can be attributed to several key factors, including the following.

- **Strong Partnerships**: As mentioned above, bKash's partnership with BRAC Bank and support from organisations like IFC and the Bill & Melinda Gates Foundation were instrumental in its growth. Strong Partnerships are crucial for mobile money services like bKash for several reasons. First, *the issue of* credibility and trust. In a market like Bangladesh, where many individuals were sceptical of digital financial services, partnerships with established institutions such as BRAC Bank and NGOs helped to establish credibility. BRAC Bank was already well-known for its focus on microfinance and had significant trust within the community. This association assured safety and reliability, which was vital for encouraging adoption. Gas (2017) emphasises the importance of partnerships with reputable institutions for building trust in financial services, especially among the unbanked population. Second, *access to funding and resources*. The support from organisations like the International Finance Corporation (IFC) and the Bill & Melinda Gates Foundation was instrumental in securing the capital necessary for bKash's initial growth. These partnerships also brought technical assistance and insights into best practices for scaling financial inclusion. IFC and the Gates Foundation are both known for their focus on scalable solutions for poverty alleviation, which gave bKash a strategic advantage. Financial backing from such reputable foundations can significantly boost the scalability and credibility of mobile money ventures in emerging markets (Baptista & Heitmann, 2010). Third, Technical Expertise and Knowledge Transfer. Money in Motion brought essential technical and strategic expertise that helped in setting up a robust digital platform. This partnership ensured that the technology behind bKash was reliable and efficient, helping avoid many of the pitfalls that other mobile money services faced. Croxson et al. (2022) note that having a partner with a deep understanding of both technology and the local market dynamics can provide the know-how required to create and maintain a successful digital financial platform. Finally, Community Engagement. Partnerships with NGOs enabled bKash to reach underserved rural communities. NGOs have deep connections with local populations, and their endorsement helped in gaining the trust of rural users who were unfamiliar with digital financial products. According to GSMA (2020), collaboration with community-based organisations is a key driver of success for mobile money services in regions with limited access to formal financial institutions. Strong partnerships provided bKash with the necessary trust, funding, technical expertise, and local knowledge required for its successful rollout and expansion.

- **Regulatory Support**: Similar to M-Pesa discussed earlier, the Bangladesh Bank, the country's central bank, played a proactive role in promoting mobile financial services. It provided an enabling regulatory framework that allowed non-bank entities like bKash to offer financial services, while also maintaining customer security. This supportive regulatory environment ensured that bKash could operate without the full constraints typically imposed on traditional banks. In particular, unlike traditional banks, bKash was allowed to operate under a non-banking financial institution (NBFI) licence, which meant that it did not have to meet the stringent capital and compliance requirements. This regulatory flexibility enabled bKash to launch and scale rapidly, by allowing a mobile financial service provider to operate independently without holding a full banking licence, Bangladesh Bank opened the door for innovative financial services to reach underserved populations, ultimately accelerating financial inclusion. The central bank also played a crucial role in setting interoperability standards for mobile financial services. The Bangladesh Bank's early push for a level playing field and openness allowed different mobile money services to interconnect with banks and financial institutions. This initiative was intended to ensure that customers could access broader financial services beyond the platform of a single provider. The policy played a vital role in allowing users to integrate their MMS accounts with formal bank accounts, enhancing the utility of mobile money. In addition, the Know Your Customer (KYC) requirements for mobile money in Bangladesh were simplified and relatively straightforward compared to those for traditional banking. Bangladesh Bank adopted a risk-based approach to KYC, which allowed the use of simplified documentation for onboarding low-risk users. This further made it easier for individuals in rural areas to open accounts. The simplified KYC process not only lowered the barrier to entry but also reduced onboarding costs for bKash, making it feasible to serve millions of customers. Finally, the

Bangladesh Bank also focused on ensuring consumer protection by instituting specific guidelines for handling customer grievances and dispute resolution. This was critical in building trust among users, many of whom were first-time participants in the formal financial system. By establishing clear processes for dispute resolution and requiring mobile financial service providers like bKash to maintain transparency in their dealings, the central bank helped to create a secure environment that encouraged adoption. The combination of regulatory flexibility, simplified procedures for onboarding, and transparent dispute resolution created an enabling environment that was central to bKash's success in Bangladesh.

- **Extensive Agent Network**: Another key factor behind the success of bKash in Bangladesh is the establishment of a widespread agent network, which played a crucial role in making mobile money services accessible across the country. Just like in the M-Pesa, these agents serve as intermediaries who facilitate cash-in and cash-out transactions, which are vital for users to convert physical cash into digital money and vice versa. bKash's agent network now includes over 300,000 agents, who are present in both urban and rural areas. This agent network was built strategically to ensure financial inclusion in regions where formal banking infrastructure was either underdeveloped or non-existent. They are often small business owners such as grocery shop owners, pharmacy owners, or other local retailers who can offer mobile money services in addition to their regular business operations. The agents sometimes educate the customers about how to use bKash's services. Having a trusted local agent who could explain how to send and receive money, top up their balance, or pay bills was invaluable in helping build trust and increase adoption.
- **High Mobile Penetration**: The rapid increase in mobile phone penetration in Bangladesh provided fertile ground for the growth of mobile money services like bKash. By leveraging the high mobile usage rate, bKash reached millions of unbanked individuals and provided them with an easy-to-use platform for financial transactions. The simplicity of the service, which required only basic mobile phones, meant that it was accessible even to those without smartphones or internet access.
- **User-Centric Services**: bKash continuously expanded its service offerings based on the needs of its users. Beyond P2P transfers,

bKash offered bill payments, merchant payments, and digital savings products, which enabled users to integrate financial services into their daily lives. This evolution helped make bKash an indispensable tool for individuals and businesses alike, fostering customer loyalty and driving widespread adoption.

bKash has had a transformative impact on financial inclusion in Bangladesh. Enabling easy domestic remittances has improved the flow of money from urban to rural areas, significantly contributing to rural households' economic stability. The ability to receive remittances quickly and securely has improved the financial resilience of many families. Furthermore, the integration of bKash into everyday payments—such as utility bills and merchant purchases—has helped integrate more individuals into the formal economy, promoting economic growth and reducing poverty levels. Today, bKash is an integral part of financial services in Bangladesh and continues to innovate to meet the evolving needs of its users.

iii. **Failure Case: M-Pesa in South Africa**

Despite the remarkable success of M-Pesa in Kenya, its expansion into South Africa in 2010 proved to be a significant failure. Unlike Kenya, where M-Pesa became a widely adopted solution for financial inclusion, the service struggled to gain traction in South Africa and was ultimately shut down in 2016. The failure of M-Pesa in South Africa can be attributed to several interrelated factors that highlight the importance of understanding local market dynamics and regulatory environments when launching a product in a new region. Here is a look at some of the factors in detail.

- **High Banking Penetration:** One of the main challenges that M-Pesa faced in South Africa was the high level of banking penetration compared to Kenya. By the time M-Pesa launched in South Africa, over 70 per cent of the population had already had access to traditional banking services. The South African financial sector was well-developed, with a robust network of bank branches, ATMs, and established relationships between consumers and banks. Unlike Kenya, where the majority of the population lacked access to formal financial services, South Africans had a wide range of banking options that already met their needs. This meant that the market demand for an alternative like M-Pesa was far less pronounced. In Kenya, M-Pesa addressed a critical gap by providing an easy way for people to send money, make payments, and access financial services without needing a bank account. In South Africa, however, the perceived value of M-Pesa was limited, as many consumers already had established relationships with traditional banks and were comfortable using them. Additionally, the existing financial infrastructure, including widespread ATM networks and accessible bank branches, reduced the need for a mobile money solution.

- **Regulatory constraints** were another major factor that hindered M-Pesa's success in South Africa. Unlike in Kenya, where regulators adopted a **flexible and supportive approach**, South Africa's regulatory environment was more stringent and presented significant hurdles. The **South African Reserve Bank (SARB)** required M-Pesa to comply with regulations that were similar to those faced by traditional banks, including **Know Your Customer (KYC)** requirements and Anti-Money Laundering (AML) compliance. The SARB was concerned about the potential risks associated with mobile money services, including fraud and money laundering. However, the regulations meant that customers had to provide extensive documentation to open and maintain M-Pesa accounts, which made it difficult for M-Pesa to onboard new users quickly and posed a significant barrier to entry. This regulatory inflexibility in South Africa also limited M-Pesa's ability to innovate and adapt its services to meet the needs of the local population and that negatively impacted adoption rates. It also, increased operational costs for M-Pesa and limited the service's ability to differentiate itself from traditional banking products. The cumbersome regulatory requirements ultimately hindered M-Pesa's growth and prevented it from achieving the same level of success that it had in Kenya.
- **Pricing and Market Competition**: The **pricing structure** of M-Pesa in South Africa also contributed to its lack of success. The **costs associated with using M-Pesa** were higher compared to the fees charged by traditional banks and other digital payment solutions. In a market where most people already had access to affordable banking services, M-Pesa's pricing was perceived as uncompetitive. Consumers were unwilling to switch to a more expensive service when they already had access to low-cost or free banking services,

significantly hindering M-Pesa's ability to attract users. Additionally, M-Pesa faced **strong competition** from well-established financial institutions and emerging digital banking services. Banks in South Africa had already introduced **low-cost digital banking products** that offered similar capabilities to M-Pesa, such as mobile banking apps, internet banking, and card-based transactions. These products were often offered at lower fees, and many consumers were already familiar with using them. Unlike in Kenya, where M-Pesa was the first widely accessible mobile money solution, the South African market was already saturated with digital financial services, making it difficult for M-Pesa to differentiate itself. Also, major banks in South Africa, such as **Standard Bank, First National Bank (FNB), and Absa,** had established customer bases and were perceived as more trustworthy than a new entrant like M-Pesa. These banks were also quick to respond to M-Pesa's entry by promoting their own mobile and digital payment solutions, thereby reducing M-Pesa's market share and making it challenging to gain a foothold in the market.

- **Cultural and Consumer Behaviour Differences**: Another significant factor that could be observed from the case of M-Pesa in South Africa when compared to the context of Kenya is the difference in **cultural and consumer behaviour.** In Kenya, the informal economy was a major driver of M-Pesa's adoption, as many people relied on cash transactions and informal financial networks. M-Pesa provided a convenient way to formalise these transactions and offered security and ease of use that was previously unavailable. In South Africa, however, a larger proportion of the population was already integrated into the formal economy, and consumer behaviour was more aligned with traditional banking practices. South African consumers were accustomed to using bank accounts, debit cards, and ATMs for their financial needs, and they did not see the added value of switching to a mobile money solution like M-Pesa. Additionally, trust in traditional banks was high, and consumers were less inclined to adopt a new financial The failure of M-Pesa in South Africa underscores the importance of understanding local market dynamics, regulatory environments, and consumer behaviour when expanding a product to new regions. Unlike in Kenya, where M-Pesa filled a critical gap in financial services, South Africa's well-developed financial infrastructure and regulatory constraints made it difficult for M-Pesa to succeed.

Impact and Lessons Learned from M-Pesa in South Africa

One of the key lessons from M-Pesa's failure in South Africa is the need for **market adaptation**. A successful product in one market cannot simply be replicated in another without considering the unique characteristics of the new market. For M-Pesa, this meant that the factors that drove success in Kenya—such as low banking penetration and supportive regulations—were not present in South Africa, and the service could not adapt effectively to these differences. Another lesson is the importance of **competitive pricing**. Pricing becomes a critical differentiator in a market with high banking penetration and established financial services. M-Pesa's relatively high fees made it unattractive compared to the low-cost options offered by traditional banks and digital financial services in South Africa. Lastly, the experience highlights the significance of **regulatory engagement**. To succeed, mobile money providers need to work closely with regulators to create an enabling environment that balances security and compliance with the need for innovation and accessibility. The stringent regulatory requirements in South Africa, while aimed at protecting consumers, limited M-Pesa's ability to offer a compelling alternative to traditional banking services. The lesson from South Africa serves as an essential reminder that successful expansion into new markets requires a deep understanding of local conditions, including consumer needs, competitive landscapes, and regulatory requirements. Without this understanding, even the most successful products can struggle to gain traction and ultimately fail to achieve their goals.

iv. **Mobile Money Services in India—Failure Case**

Mobile money services have emerged as a transformative force in developing economies, offering an alternative to traditional banking for unbanked populations. By enabling users to conduct financial transactions through mobile phones, these services have successfully bridged financial gaps in many regions, fostering economic growth and empowering underserved communities. However, the Indian experience with mobile money services presents a complex narrative of underachievement. This case of failure is rooted in a confluence of strategic missteps and external pressures that collectively hindered the potential of these services to thrive. India, with its vast population and significant rural demographic, seemed an ideal candidate for the adoption of mobile money services. A substantial proportion of the population lacked access to formal banking services, creating a clear need for alternative financial solutions. With over a billion mobile phone connections and a significant push towards digital transformation, the conditions appeared ripe for mobile money services to flourish. However, this optimism quickly turned into disillusionment as systemic issues and strategic misalignments derailed the initiative. From the outset, mobile money services in India faced regulatory restrictions that constrained their growth and adoption. The Reserve Bank of India (RBI), concerned about potential risks, imposed stringent conditions on mobile money operators, such as limiting cash-out capabilities and enforcing complex Know Your Customer (KYC) norms. These regulatory barriers significantly reduced the accessibility and utility of mobile money platforms, especially for rural populations that relied heavily on cash transactions. Simultaneously, the introduction of the Unified Payments Interface (UPI) in 2016 drastically altered the competitive landscape. UPI's seamless integration with bank accounts and its cost-effective transaction model overshadowed the utility of standalone mobile wallets. While UPI enabled digital payments in India, it inadvertently marginalised the mobile money sector, which lacked the interoperability and simplicity offered by UPI.

The strategy for promoting mobile money services also underestimated the cultural and infrastructural realities of the Indian market. Many potential users, particularly in rural areas, lacked the digital literacy required to adopt mobile money platforms confidently. Moreover, India's entrenched reliance on cash as a medium of transaction further impeded the transition to digital solutions. Despite the promising potential of mobile money, these socio-cultural barriers were not adequately addressed through targeted education and awareness campaigns.

The competitive environment further compounded these challenges. The emergence of fintech giants like Paytm, Google Pay, and PhonePe, which offered aggressive marketing campaigns and attractive cashback incentives, left traditional mobile money services struggling to compete. These platforms quickly captured the urban and semi-urban markets, leaving limited room for mobile money operators to establish a foothold. Infrastructure gaps also played a crucial role in the failure of mobile money services. Patchy network connectivity and inconsistent power supply in rural areas hindered the functionality of mobile money platforms. During the initial rollout, high data costs and limited smartphone penetration further restricted adoption among low-income users who needed the services the most. Finally, the Indian government's direct financial inclusion initiatives, such as the Pradhan Mantri Jan Dhan Yojana (PMJDY) and the Aadhaar-enabled Payment System (AePS), offered a more direct and reliable alternative to mobile money. These programmes allowed users to access banking services without intermediaries, diminishing the relevance of private mobile money platforms.

In hindsight, the failure of mobile money services in India can be attributed to a lack of strategic alignment with the country's unique socio-economic and regulatory conditions. Mobile money services could not sustain their initial promise by failing to account for cultural preferences, infrastructural limitations, and the rapidly evolving digital payment ecosystem. This experience underscores the importance of context-specific strategies and a deep understanding of local market dynamics in designing successful financial inclusion initiatives. Several critical factors contributed to the failure of mobile money services in India, and they are discussed below.

- **Regulatory Rigidities:** The Reserve Bank of India (RBI) imposed stringent regulations on mobile money services, significantly limiting their operational scope. For example, mobile money operators were restricted from offering cash-out services, a feature that rural users, who depend heavily on physical cash for daily transactions, found indispensable. Without the ability to access liquid cash, the practicality of mobile money platforms was severely diminished. Additionally, the mandatory Know Your Customer (KYC) norms required detailed documentation and identity verification, which increased the onboarding costs for operators and added complexity for users, many of whom lacked formal identification. This regulatory environment, though aimed at maintaining financial security, inadvertently stifled innovation and constrained service providers from effectively addressing the needs of unbanked populations.
- **Lack of Interoperability:** Mobile money platforms in India largely operated as closed-loop systems, meaning transactions could only occur within a single operator's network. This siloed approach reduced the utility of these platforms, as users could not send or

receive money across different networks. In contrast, the Unified Payments Interface (UPI), launched by the National Payments Corporation of India (NPCI), provided an open ecosystem, allowing seamless transactions between bank accounts and across multiple service providers. This feature made UPI significantly more appealing to users and businesses alike. The inability of mobile money platforms to integrate or compete with such interoperable systems rendered them obsolete in the eyes of a rapidly digitising user base.

- **Cultural Resistance to Change:** India's economy has long been dominated by cash transactions, especially in rural and semi-urban areas where cash is perceived as tangible and secure. This cultural reliance on cash created a psychological barrier to adopting digital payment systems. Many potential users viewed mobile money platforms with scepticism, fearing fraud, misuse, or loss of funds. Additionally, there was limited trust in technology due to the lack of familiarity with digital interfaces. Efforts to build trust and demonstrate the reliability of mobile money services were insufficient, leaving large sections of the population unconvinced about their benefits. This resistance was further exacerbated by a lack of visible role models or community champions advocating for the use of mobile money.
- **Infrastructural Shortcomings:** Infrastructure challenges in India's rural and semi-urban areas significantly impeded the adoption of mobile money services. Network connectivity in these regions was often unreliable, leading to failed transactions and a lack of trust in the system. Frequent power outages further compounded the issue as users struggled to keep their mobile devices charged. Additionally, during the early stages of mobile money adoption, high data costs placed a financial burden on low-income users, making it difficult for them to access these services consistently. The absence of robust and reliable infrastructure meant that mobile money platforms were often seen as an unreliable solution, unable to meet the demands of users in critical moments.
- **Market Competition:** The entry of fintech giants like Paytm, Google Pay, and PhonePe intensified competition in the digital

payment space. These platforms introduced features such as user-friendly interfaces, seamless integration with bank accounts, and attractive incentives like cashback and discounts. They also invested heavily in advertising and promotions, capturing the attention of urban and tech-savvy users. Traditional mobile money platforms, which lacked the resources to compete on marketing or technological innovation, struggled to retain users. The aggressive strategies of these competitors left mobile money services marginalised, particularly as users gravitated towards platforms that offered a more comprehensive and integrated suite of financial services.

- **Government-Led Alternatives:** The Indian government's financial inclusion initiatives, such as the Pradhan Mantri Jan Dhan Yojana (PMJDY) and Aadhaar-enabled Payment System (AePS), provided a direct pathway to formal banking for millions of previously unbanked citizens. These initiatives allowed users to open zero-balance bank accounts and perform transactions using biometric authentication, bypassing the need for intermediaries like mobile wallets. These programmes reduced the demand for private mobile money platforms by offering secure, government-backed solutions. Additionally, government subsidies and benefits were increasingly disbursed through these accounts, making them more relevant and reliable for users seeking financial inclusion.
- **Limited User Education:** A critical barrier to the adoption of mobile money services was the lack of awareness and education among potential users. Many individuals, particularly in rural areas, were unfamiliar with how mobile money platforms worked or how to use them effectively. Few targeted campaigns addressed these knowledge gaps, leaving users confused or apprehensive about adopting the technology. Furthermore, the absence of hands-on training or local support systems meant that users who faced initial difficulties often abandoned the platforms altogether. Without adequate education on the safety, benefits, and usability of mobile money services, these platforms failed to gain the trust and confidence of their intended audience.

Impact and Lessons Learned from Mobile Money Services in India

The story of mobile money services in India offers several crucial lessons that are both practical and relevant for stakeholders seeking to advance financial inclusion in similar contexts. First, regulations must balance fostering innovation and ensuring user safety. In India, overly restrictive policies such as limits on cash-out services and stringent Know Your Customer (KYC) norms stifled growth. A more adaptive regulatory framework, designed to accommodate the realities of underserved populations, could have supported a broader adoption of mobile money services. Second, interoperability is essential for scalability. Closed-loop systems isolated users and providers, whereas the success of the Unified Payments Interface (UPI) underscored the benefits of an open ecosystem. Future initiatives should prioritise creating platforms that allow seamless integration across networks, enhancing usability and expanding the reach of digital payment systems. Infrastructure investments are another critical takeaway. Reliable network connectivity, consistent power supply, and affordable data access are foundational to ensuring the smooth operation of mobile money services, especially in rural and semi-urban areas. Without these, even the most innovative solutions risk being deemed unreliable by users. User education is equally vital. The Indian experience highlighted the importance of building trust and familiarity with digital payment systems, particularly in cash-reliant economies. Outreach campaigns tailored to local contexts, involving community influencers, and providing hands-on training could help address fears around fraud and misuse while showcasing the tangible benefits of mobile money. Lastly, collaboration between government initiatives and private operators is key. Government-led programmes like the Pradhan Mantri Jan Dhan Yojana (PMJDY) and the Aadhaar-enabled Payment System (AePS) significantly advanced financial inclusion. However, private players could have played a complementary role by offering innovative and user-friendly services that extend the reach of these public schemes. Synergies between the two can lead to more sustainable and inclusive financial ecosystems. These lessons emphasise the importance of context-sensitive strategies, collaborative approaches, and a user-first mindset in the design and implementation of financial inclusion initiatives. By addressing these aspects, future efforts can avoid the missteps observed in the Indian case and build systems that truly empower underserved populations.

c. How Impact Investing Can Support Mobile Money Services (MMS) in Promoting Financial Inclusion

Drawing from the success of MMS in countries like Kenya and the lessons from failures in India and South Africa, impact investing can facilitate the conditions necessary for MMS to thrive and reach the unbanked population.

Flexible Regulatory Environments: One of the significant factors contributing to the success of MMS, such as **M-Pesa in Kenya**, was the presence of a **supportive and flexible regulatory environment**. The Kenyan government adopted a **proportional regulatory framework** that allowed M-Pesa to scale quickly while ensuring consumer protection. This regulatory flexibility enabled M-Pesa to reach millions of unbanked individuals and become a household name. In contrast, the failure of MMS in countries like **India** and **South Africa** can be attributed, in part, to **restrictive regulatory frameworks.** In these countries, MMS providers were required to partner with banks, adding complexity and limiting their ability to operate independently. Impact investors can promote **regulatory advocacy** to create an enabling environment for MMS. Impact investors can work with regulators to develop frameworks that support innovation while managing risks by funding policy advocacy initiatives. Additionally, impact investors can support the establishment of **regulatory sandboxes,** which provide a controlled environment for testing new MMS solutions, fostering growth while ensuring compliance.

Infrastructure Development and Agent Network Strengthening: The success of MMS is also closely linked to the availability of **reliable infrastructure** and a robust **agent network**. In Kenya, M-Pesa's success was largely due to its extensive network of agents who provided cash-in and cash-out services, even in remote areas. In contrast, the lack of infrastructure, such as unreliable electricity and limited internet connectivity, was a major barrier to MMS scalability in regions like rural India. Impact investing can address these infrastructure challenges by funding **internet connectivity, electricity,** and **agent network expansion** in underserved areas. Impact investors can ensure that MMS can operate effectively in remote regions by investing in the development of essential infrastructure, thereby increasing financial inclusion. Furthermore, impact investors can support **agent training programmes** to enhance service quality and ensure agents are equipped to manage liquidity effectively, particularly in rural settings where cash flow can be inconsistent.

Addressing Funding Gaps and Supporting Startups: A key challenge for smaller MMS startups is the **lack of access to capital,** which limits their ability to compete with established players and scale their operations. In successful cases like M-Pesa, early funding and support from organisations like **Vodafone** and the **Bill & Melinda Gates Foundation** enabled rapid expansion and innovation. However, in many regions, smaller MMS startups face significant funding constraints, preventing them from expanding their services and reaching more users. Impact investors can help **bridge the funding gap** by providing capital to early-stage MMS providers. This type of investment can cover high operational costs, such as maintaining agent networks and investing in technology, and meet regulatory compliance requirements. Impact investors can enable more competition in the MMS sector by supporting these startups, driving innovation and ensuring that MMS providers can reach underserved populations.

Building Customer Trust and Promoting Financial Literacy: The success of MMS depends not only on infrastructure and regulatory support but also on the users' trust. In regions with low financial literacy, misinformation and fears of fraud can deter individuals from using mobile money services. In Kenya, the success of M-Pesa was partly due to extensive **consumer education** initiatives that built trust and helped users understand the benefits of MMS. In contrast, a lack of trust and financial literacy were significant barriers to adoption in countries like South Africa and India. Impact investing can support **financial literacy initiatives** to educate potential users about the benefits of MMS, how to use digital financial services, and the importance of secure financial practices. Impact investors can help build trust in MMS by funding **consumer protection measures** and **awareness campaigns**, particularly in low-income and rural communities. This will encourage more people to adopt MMS, thereby expanding financial inclusion.

Supporting Resilience and Cross-Sectoral Impact: MMS has demonstrated resilience during **economic shocks,** such as the **COVID-19 pandemic,** by enabling contactless transactions and facilitating the rapid disbursement of government aid. This resilience is critical in times of crisis, especially for vulnerable populations that rely on financial services to meet their basic needs. Impact investors can further strengthen this resilience by funding the development of MMS infrastructure that can be scaled quickly during emergencies, ensuring that financial services remain accessible to those who need them most. Impact investing can also enhance the **cross-sectoral impact** of MMS by supporting their integration into sectors such as **agriculture, healthcare, and education**. For example, MMS can facilitate payments for agricultural inputs, healthcare services, and school fees, thereby contributing to broader socio-economic development. Impact investors can create a multiplier effect that benefits multiple aspects of society by investing in projects that leverage MMS for cross-sectoral applications.

Driving Innovation and Expanding Financial Services: Successful MMS, such as M-Pesa, have gone beyond basic money transfers to offer a range of financial services, including **microloans, insurance, and savings products**. These services have had a profound impact on improving the economic resilience of underserved populations. In contrast, the failure of MMS in India was partly due to the inability to differentiate their offerings in a competitive market dominated by digital wallets and traditional banks. Impact investors can support the **expansion of financial services** available through MMS by providing funding for the development of new products and services. This could include **micro-insurance, digital credit scoring**, and **savings products,** all of which can enhance the value proposition of MMS and attract more users. By fostering **partnerships** between mobile network operators, financial institutions, and governments, impact investors can build a more inclusive financial ecosystem that offers diverse financial services to those currently excluded from traditional banking systems.

Digital Banking Platforms and Neo-Banks

Digital banking platforms and neo-banks represent a rapidly evolving sector in the financial industry characterised by innovation, convenience, and a customer-centric approach. Both aim to provide seamless financial services to individuals and businesses, often without the need for physical branches and leverage technology to offer a superior banking experience. These institutions operate primarily or exclusively online, offering a suite of financial services through digital platforms. This shift caters to the evolving preferences of consumers who increasingly demand convenience, speed, and personalised banking experiences. These platforms are changing the way people interact with their finances, providing more accessible, efficient, and personalised services. Digital banking platforms and neo-banks represent two distinct but complementary approaches to delivering financial services in the digital age.

Digital Banking Platforms: Digital banking platforms are online and mobile interfaces provided by traditional banks or financial institutions to allow their customers to access banking services electronically. These traditional banking institutions integrate digital technologies to enhance their service delivery. They offer online and mobile banking services alongside their conventional branch-based operations, providing customers with a hybrid model that combines the reliability of established banks with the convenience of digital access. These platforms offer features such as account management, money transfers, bill payments, loan applications, and investment services. Customers can use digital banking platforms to perform most banking activities without visiting a branch, making banking more accessible and convenient. While these banks may still maintain physical branches, their digital platforms provide an alternative to in-person banking. Traditional banks, such as JPMorgan Chase, Wells Fargo, and HSBC, have all invested significantly in building robust digital banking platforms that enable customers to access services 24/7 from anywhere.

Neo-Banks: In contrast, neo-banks are entirely digital entities that operate without any physical presence. They are often fintech startups that provide a range of banking services through mobile applications or web platforms. Neo-banks typically focus on specific customer segments, such as tech-savvy individuals, small businesses, or freelancers, offering tailored services that address the unique needs of these groups. Unlike traditional banks, neo-banks are not constrained by legacy infrastructure, which allows them to be more agile, responsive, and customer-centric. Neo-banks often partner with licensed banks or obtain their own limited banking licences to operate in various jurisdictions. They focus on providing transparency, simplicity, and cost-effective services. These banks typically focus on providing a streamlined user experience, offering services like instant account opening, fee-free international payments, digital savings tools, budgeting tools, low or no fees and cryptocurrency trading, all accessible through a single app. Neo-banks provide personalised services and a superior banking experience by focusing on customer needs and leveraging data analytics.

Traditional banks are responding to the rise of neo-banks by enhancing their digital offerings. Many established banks launched their own digital banking platforms, incorporating features like mobile check deposits, budgeting tools, and instant transfers to compete with neo-banks. The competition has led to a convergence in the quality of digital services provided by traditional banks and neo-banks, benefiting customers with more choices and better services. In emerging markets, digital banks are crucial to advancing financial inclusion. In regions such as Southeast Asia and Africa, neo-banks and digital banking platforms are providing essential financial services to the unbanked and underbanked. They have gained popularity by offering low-cost banking services and simplified account-opening processes, making banking more accessible to previously excluded people. While traditional banks continue to adapt to the digital age, neo-banks are pushing the boundaries of what is possible in banking, introducing new products and features at a pace that challenges the established norms. Consumers around the world are increasingly accepting digital banking platforms and neo-banks, which indicates that the future of banking will likely be dominated by digital-first and customer-centric financial solutions.

a. **Global Growth and Market Dynamics**

The neo-banking sector has experienced exponential growth over the past decade. The global neo-banking market has witnessed substantial growth in recent years. The market was valued at approximately USD 148.7 billion in 2023 (Market.US, 2023). This significant valuation underscores the rapid adoption of digital-only banking solutions worldwide. Projections indicate that this growth trajectory will continue, with the market anticipated to reach approximately USD 5.3 trillion by 2033, exhibiting a Compound Annual Growth Rate (CAGR) of 49 per cent from 2024 to 2033. This remarkable expansion is driven by factors such as the increasing demand for convenient banking solutions, the proliferation of smartphones, and the widespread adoption of internet services. Regionally, Europe has been a dominant player in the neo-banking market, accounting for a significant share of 38 per cent in 2023. This region benefits from a robust regulatory framework, including the Revised Payment Services Directive (PSD2), which has promoted open banking and increased competition. Leading neo-banks such as Revolut,

Monzo, and N26 have capitalised on these favourable conditions to attract millions of users with innovative products and services. North America also holds a significant position in the neo-banking sector, driven by a tech-savvy population and high financial inclusion rates. Neo-banks like Chime and Varo have gained traction in the US market by offering low-cost banking solutions, real-time payments, and financial management tools tailored to underserved segments, including gig workers and small businesses. In the Asia-Pacific region, neo-banking is experiencing rapid expansion. Markets such as India, China, and Southeast Asia are seeing increased adoption of digital banking services, driven by rising smartphone penetration and government initiatives supporting digitalisation. Notable players include Tencent-backed WeBank in China and RazorpayX in India, which have tailored their offerings to meet the diverse needs of their customer bases. Emerging markets, particularly in Africa and Latin America, are also showcasing significant potential for neo-banking growth. In Africa, platforms like TymeBank in South Africa and Kuda Bank in Nigeria are focusing on financial inclusion by offering simplified account setups and services catering to the unbanked population. Similarly, Latin America has witnessed the rise of Nubank, a Brazilian neo-bank that has transformed the regional banking landscape with its customer-centric approach and transparent fee structures. The global neo-banking market is becoming increasingly diversified, with growth trajectories shaped by regional characteristics, regulatory landscapes, and the evolving demands of consumers.

b. **Successes and Failures of Neo-Banks**

i. **Revolut—Success Case**

Revolut's journey from a small-scale currency exchange platform to a global neo-banking leader illustrates the transformative potential of technology in financial services. Founded in 2015 by Nikolay Storonsky and Vlad Yatsenko, the company entered a financial landscape riddled with inefficiencies. Traditional banks were characterised by high fees, cumbersome processes, and limited accessibility, especially for international transactions. Recognising these pain points, Revolut launched to revolutionise the way people interact with money, beginning with fee-free currency exchange and interbank rates for international transfers. In the mid-2010s, mobile adoption and app-based services were growing rapidly, creating opportunities for innovative digital solutions. Revolut leveraged this trend by offering a mobile-first platform that resonated with frequent travellers, expatriates, and tech-savvy millennials. Its transparent pricing and intuitive design distinguished it from traditional banks, fostering rapid user adoption and loyalty. Initially focused on addressing the high foreign exchange costs, Revolut expanded into a comprehensive financial ecosystem. The app now integrates multi-currency accounts, cryptocurrency trading, stock investments, budgeting tools, and more. This diverse range of services allows millions of users worldwide to manage their finances seamlessly through a single platform. By simplifying onboarding processes and eliminating hidden fees, Revolut has made banking services accessible to individuals excluded from traditional systems. The app provides essential financial tools through smartphones for users in regions with limited banking infrastructure. Features like real-time spending notifications and savings vaults help users take control of their finances while promoting financial literacy through effective money management.

Operating in over 200 countries, Revolut has transformed cross-border transactions by reducing friction. Its multi-currency functionality allows users to hold and exchange funds in multiple currencies without incurring exorbitant fees. This capability has been particularly beneficial for small businesses and individuals engaged in international trade, filling a critical gap in the financial ecosystem. Revolut's journey has not been without challenges. Navigating complex regulatory landscapes has required the company to secure banking licences and comply with Anti-Money Laundering (AML) and data protection laws in diverse jurisdictions. Profitability has been another concern, as the company has prioritised growth and user acquisition over immediate financial returns. Additionally, reports of high-pressure workplace environments have highlighted the need for cultural and organisational improvements. Despite these hurdles, Revolut has reshaped the financial landscape. By leveraging technology to remove barriers and enhance accessibility, it has set new benchmarks for modern banking. Its contributions to financial inclusion, from empowering the underbanked to simplifying global transactions, have established it as a transformative force in the industry. Revolut's success can largely be attributed to a few critical factors that have shaped its growth and impact on the financial industry. Below are the key factors enabling its success, discussed in detail.

- **Customer-Centric Innovation**: Revolut's success is deeply rooted in its ability to anticipate and meet user needs. Revolut empowers its users to take control of their finances by offering features such as real-time spending notifications, savings vaults, and personalised budgeting tools. These innovations cater to various financial behaviours and have helped the platform attract and retain millions of customers. Furthermore, the integration of cryptocurrency trading and investment options highlights its commitment to staying ahead of market trends and appealing to diverse user demographics.
- **Global Accessibility and Scalability**: One of Revolut's defining strengths is its ability to provide financial services globally. Operating in over 200 countries, the platform facilitates cross-border transactions with minimal friction, making it invaluable for expatriates, international workers, and small businesses. Its multi-currency functionality allows users to hold, manage, and exchange funds seamlessly, reducing the costs and complexities associated with international banking. This accessibility has been pivotal in extending financial services to underserved populations and creating a meaningful global financial ecosystem.
- **Transparent Pricing Model**: Revolut's transparent pricing model has been a key differentiator in the financial services market. Unlike traditional banks that often rely on hidden fees, Revolut provides clear and upfront costs for its services, particularly in areas like currency exchange and international money transfers. This approach has built a strong foundation of trust among users, encouraging adoption and fostering long-term loyalty. Revolut has redefined the standard for fair and equitable banking practices by eliminating financial opacity.
- **Technological Sophistication**: The platform's reliance on cutting-edge technology has enabled it to deliver a seamless and secure user experience. Leveraging cloud computing and advanced analytics, Revolut processes transactions in real time while maintaining robust data protection measures. This technological infrastructure supports scalability and allows Revolut to introduce new features rapidly, ensuring that it remains competitive in the dynamic fintech landscape.
- **Strategic Funding and Branding**: Revolut's ability to secure substantial funding has been instrumental in its growth. Notable

funding rounds, such as the $1 billion raised in 2021, have provided the resources necessary to expand into new markets, innovate its product offerings, and enhance its infrastructure. Additionally, Revolut's modern branding and sleek user interface have established it as a leading choice for digitally savvy users, distinguishing it from traditional banks and other fintech competitors.

Through these factors, Revolut redefined modern banking and significantly contributed to financial inclusion. Its ability to empower the underbanked, simplify global transactions, and foster financial literacy underscores its transformative role in the industry. As Revolut continues to innovate and navigate its challenges, it exemplifies the possibilities of digital finance in an evolving world.

ii. **Nubank—Success Case**

Nubank, founded in 2013 by David Vélez, Cristina Junqueira, and Edward Wible, exemplifies how innovation and a customer-focused strategy can revolutionise financial services in emerging markets. Based in Brazil, the company set out to simplify banking for the unbanked and underbanked populations of Latin America. Leveraging technology and addressing local challenges, Nubank has become one of the largest independent digital banks globally, with a customer base exceeding 75 million as of 2024. At the time of Nubank's inception, the Brazilian financial landscape was riddled with inefficiencies. Traditional banking services were characterised by high fees, opaque processes, and limited accessibility, leaving a significant portion of the population unbanked or underserved. Nubank identified these systemic shortcomings as an opportunity to redefine financial services. Its first offering, a no-fee credit card, served as an entry point for customers to access credit and participate in the formal economy. This approach appealed to tech-savvy millennials and addressed the broader issue of financial exclusion. The Brazilian market was particularly well-suited for Nubank's digital-first model due to the widespread adoption of smartphones and increasing internet penetration. The absence of physical branches allowed Nubank to operate at lower costs, enabling the delivery of affordable and accessible financial products. Its mobile app provided a seamless experience, combining intuitive design with features that simplified financial management. These efforts resonated deeply with users frustrated by the complex and often exclusionary practices of traditional banks. Nubank's commitment to financial inclusion has been transformative. The streamlined account-opening process removed barriers such as excessive documentation and credit history requirements, enabling millions to access financial services for the first time. The no-fee structure eliminated a significant pain point for customers, many of whom were previously deterred by the high costs associated with conventional banking. For small businesses, Nubank offered much-needed financial tools to support growth in a region where traditional credit options were often inaccessible or prohibitively expensive. The company's regional expansion into Mexico and Colombia further underscored its ability to adapt to diverse market conditions while retaining its core principles. Tailoring its offerings to local financial ecosystems, Nubank replicated its success by addressing similar challenges of financial exclusion and high banking costs. This strategic growth not only increased its customer base but also diversified its revenue streams, enhancing its resilience in a competitive fintech landscape.

Nubank redefined digital finance in Latin America by setting new standards for efficiency and accessibility. Its app integrates AI-driven customer support and data-driven credit assessments, offering a personalised and efficient user experience. Transparency remains a cornerstone of its approach, fostering trust and loyalty among users. The company has also prioritised financial literacy, equipping customers with the knowledge to make informed financial decisions and take control of their economic futures. Operating in emerging markets comes with its share of challenges. Nubank has had to navigate economic instability, complex regulatory environments, and intensifying competition from other fintech players. Maintaining its customer-focused ethos while scaling operations will be critical to sustaining its growth. Despite these hurdles, Nubank's innovative approach and adaptability position it as a leader in financial inclusion and a transformative force in the industry. Nubank's impact extends beyond banking, fostering economic empowerment and social mobility throughout Latin America. It has demonstrated the power of digital innovation to create meaningful change by addressing systemic barriers and prioritising customer needs. Nubank's journey is a compelling example of how neo-banks can revolutionise financial services in emerging markets, paving the way for a more inclusive financial future. Below are the key factors enabling its success, discussed in detail.

- **Financial Inclusion Focus**: Nubank's commitment to addressing financial exclusion in Latin America is central to its success. By simplifying account-opening processes and eliminating traditional banking barriers such as high fees and stringent credit requirements, Nubank has brought financial services to millions of individuals who were previously excluded from the formal financial system. The introduction of its no-fee credit card offered an entry point for customers to build credit and participate in the formal economy, fostering financial empowerment and inclusion. This focus on removing barriers has allowed Nubank to target a broad demographic, including younger users and the unbanked population.
- **Technological Innovation**: At the heart of Nubank's operations is its cutting-edge technology, which enables rapid scalability and efficient service delivery. The use of advanced data analytics allows Nubank to assess credit risk even for customers with limited financial histories, enabling it to extend credit responsibly. AI-powered customer support ensures that users receive timely, personalised assistance, enhancing customer satisfaction and retention. The app's intuitive design provides a seamless user experience, making financial management straightforward and accessible, particularly for those unfamiliar with traditional banking processes.
- **Customer-Centric Approach**: Nubank's relentless focus on customer needs differentiates it from traditional financial institutions. The bank prioritises simplicity, transparency, and user empowerment. Features such as instant spending notifications, real-time transaction tracking, and flexible credit limits allow users to take control of their finances. Nubank also invests heavily in financial literacy, educating users on managing their money effectively. This commitment to customer empowerment builds trust and loyalty, ensuring long-term engagement with the platform.
- **Strategic Market Expansion**: Nubank's ability to expand beyond Brazil while maintaining its core values has been instrumental in its growth. Its entry into Mexico and Colombia demonstrates its capacity to adapt to local financial ecosystems while addressing similar challenges of financial exclusion and high banking costs. By tailoring its offerings to meet regional needs and replicating

its customer-centric model, Nubank has strengthened its presence across Latin America. This expansion has not only increased its customer base but also diversified its revenue streams, reducing dependence on a single market.

- **Trust and Brand Loyalty**: Trust is a cornerstone of Nubank's success. By offering transparent terms, no hidden fees, and responsive customer support, the bank has built a reputation for reliability and fairness. Its branding strategy, which emphasises empowerment, innovation, and accessibility, resonates deeply with its target audience, particularly younger demographics. This strong brand identity and customer trust translate into high user retention and positive word-of-mouth, further solidifying its market position.
- **Strategic Funding**: Nubank's ability to attract significant investment has been critical in driving its growth. With backing from global investors and a successful IPO in 2021, Nubank has secured the resources needed to enhance its technological capabilities, expand into new markets, and develop innovative products. These financial resources have allowed Nubank to sustain its growth trajectory while continually investing in customer-focused innovations and operational efficiencies.

Nubank's journey from a nascent fintech startup to a regional financial powerhouse exemplifies the transformative potential of customer-centric innovation and digital technology. By addressing the systemic barriers of traditional banking and prioritising financial inclusion, Nubank has reshaped the financial landscape of Latin America. Its success is a testament to the impact of accessibility, transparency, and adaptability in fostering economic empowerment and social mobility. As Nubank continues to expand and innovate, it stands as a model for how neo-banks can drive meaningful change in emerging markets, paving the way for a more equitable financial future.

iii. **Volt Bank (Australia)—Failure Case**

Volt Bank, established in 2017, was heralded as one of Australia's most promising neo-banks. It gained prominence as the first Australian neo-bank to secure a full banking licence in 2019, symbolising the potential of digital-only financial institutions in transforming traditional banking. The rise of digital banking globally created a favourable environment for such initiatives, with Australia being no exception. High smartphone penetration and a tech-savvy population provided a fertile ground for Volt's vision of redefining banking through customer-centric innovation and streamlined digital services. Volt positioned itself as a modern alternative to traditional banks, focusing on transparency, accessibility, and innovation. The bank aimed to eliminate traditional banking fees, offer competitive interest rates on savings, and provide a seamless digital experience through its mobile-first platform. Its strategy targeted tech-savvy customers and early adopters of digital banking who were dissatisfied with the high fees and complex processes of traditional financial institutions. Securing a full banking licence in 2019 added to Volt's credibility, making it one of the first digital-only banks in Australia to achieve this milestone. This allowed Volt to expand its offerings and gain the trust of investors and customers alike. The bank raised significant initial funding to support its operations and laid ambitious plans to grow its customer base and develop innovative products.

Despite its early promise, Volt faced several challenges as it attempted to scale its operations. The Australian banking sector, dominated by a few well-established traditional banks, posed significant competitive barriers. These incumbents, equipped with extensive resources, rapidly adapted to digital trends, offering seamless mobile banking solutions that rivalled those of emerging neo-banks like Volt. Additionally, the entry of other neo-banks intensified competition in the digital banking space, making differentiation increasingly difficult. Volt's business model, heavily reliant on achieving scale to generate sustainable revenue, encountered hurdles. Its relatively small customer base and limited product suite restricted its ability to grow revenue streams. Competitors offering a broader range of services, such as loans, credit cards, and investment products, were better positioned to attract and retain customers. Furthermore, Volt navigated stringent regulatory and compliance requirements as a fully licensed bank. The associated costs placed additional strain on its financial resources, diverting attention and funds from product development and customer acquisition efforts. The onset of the COVID-19 pandemic exacerbated Volt's challenges. Economic uncertainty led to greater caution among investors, making it difficult for the bank to secure additional funding. This lack of financial backing severely limited Volt's ability to execute its growth plans, enhance its offerings, and compete effectively in a saturated market. In June 2022, Volt announced its closure, returning its banking licence and refunding all customer deposits. The decision marked a significant setback for Australia's neo-banking sector and underscored the challenges of sustaining a digital-only banking model in a competitive and highly regulated market. Volt's ambitious vision and eventual failure highlight the complexities of operating in the financial sector, offering valuable lessons for future neo-banks.

Key Factors of Failure

- **Funding Shortfalls**: One of the most significant challenges Volt Bank faced was its inability to secure sustained funding, a critical factor for any neo-bank operating in a competitive and resource-intensive industry. Initially, Volt's funding rounds were successful, providing the bank with enough capital to launch its platform and begin operations. The enthusiasm from investors during these early stages reflected optimism about the potential of digital-only banking and Volt's promise to disrupt the Australian financial sector. However, as the bank attempted to scale, its financial needs grew substantially, and the optimism that initially attracted investors began to wane. The inability to demonstrate rapid growth or achieve profitability became a stumbling block in Volt's efforts to secure additional investment. Investors were increasingly cautious about committing further capital without clear evidence of the bank's ability to sustain itself. This hesitancy was further amplified during the COVID-19 pandemic, which created widespread economic uncertainty and tightened global investment flows. Many investors shifted their focus towards safer, more established ventures, leaving startups like Volt struggling to raise funds. The lack of financial resources created a cascading effect on Volt's operations. Without sufficient funding, the bank was unable to invest in essential areas such as product development, marketing, and customer acquisition. These areas are vital for a neo-bank's growth, as they enable the institution to attract and retain customers, introduce competitive offerings, and differentiate itself in a crowded market. Volt's inability to secure these investments meant that it could not keep pace with its competitors, many of whom had access to larger funding pools or more diverse revenue streams. This financial shortfall also limited Volt's ability to innovate, a critical factor in maintaining relevance and competitiveness in the fintech space. The bank's constrained resources forced it to scale back on ambitious projects, further reducing its appeal to both customers and potential investors. The lack of sustained funding created an insurmountable barrier for Volt, leaving it unable to execute its vision or adapt to the rapidly evolving demands of the market.

- **Market Competition**: The competitive landscape in Australia presented one of the most significant challenges for Volt Bank, underscoring the difficulty of carving out a niche in a highly concentrated banking sector. Australia's financial industry is dominated by a handful of traditional banks, often called the "big four", which possess extensive customer bases, robust infrastructures, and substantial resources. These incumbent institutions were quick to recognise the digital shift and responded by enhancing their own mobile banking services. By leveraging their established market presence and economies of scale, they were able to offer advanced digital experiences that rivalled, and in some cases surpassed, those of emerging neo-banks like Volt. In addition to the dominance of traditional banks, Volt faced mounting pressure from other neo-banks entering the Australian market around the same time. Competitors such as Xinja and 86 400 brought similar propositions to the table, further saturating the digital banking space. This intensified competition made it increasingly difficult for Volt to differentiate itself. While Volt initially positioned itself as a transparent and customer-centric alternative to traditional banking, these qualities alone were not enough to set it apart in a crowded market where many players were vying for the same digitally savvy customer base. Volt's product suite also contributed to its competitive disadvantage. Unlike some of its rivals that offered a diverse range of services, including loans, credit cards, and wealth management options, Volt's offerings were more limited in scope. This lack of diversification restricted its ability to attract a broader demographic and cross-sell products, which are crucial strategies for driving customer retention and generating sustainable revenue streams. The relatively narrow focus of its services reduced its appeal to customers seeking comprehensive financial solutions, allowing competitors with more robust offerings to capture greater market share. The competitive pressures on Volt were further exacerbated by its inability to scale rapidly. In the neo-banking sector, achieving scale is essential for profitability and creating a network effect that reinforces customer loyalty and brand strength. Volt's slower growth trajectory meant it struggled to build the critical mass needed to compete effectively. As other banks continued to innovate and expand their offerings, Volt found itself lagging behind, unable to keep pace with the evolving expectations of consumers in a highly dynamic industry. The combination of entrenched traditional banks, aggressive new entrants, and Volt's own operational limitations created an environment where sustaining a competitive edge proved insurmountable. The inability to differentiate its offerings and achieve the necessary scale left Volt vulnerable in a market that demanded constant innovation and adaptability, contributing significantly to its eventual closure.

- **Operational Sustainability**: Volt Bank's operational sustainability emerged as a critical challenge that ultimately undermined its long-term viability. Unlike traditional banks and some of its neo-bank competitors, Volt struggled to achieve the scale required to sustain its operations and generate consistent revenue streams. The bank's relatively small customer base limited its ability to cover operational costs, let alone invest in developing new products or expansion initiatives. Volt's inability to grow rapidly placed it at a significant disadvantage in an industry where scale often equates to survival. A core issue in Volt's operational sustainability was its narrow product suite, which primarily focused on savings accounts and payment solutions. This limited scope restricted its ability to attract a diverse customer base and generate multiple revenue streams. In contrast, competitors offering a broader range of financial products—including loans, credit cards, and investment options—were better positioned to meet the varied needs of their customers. This lack of diversification left Volt overly reliant on a small market segment, making it vulnerable to shifts in customer preferences and competitive pressures. Operational inefficiencies further compounded Volt's challenges. The bank faced difficulties streamlining its processes and quickly adapting to changing market demands. This lack of agility hindered its ability to roll out new features or promptly address customer feedback, eroding trust and satisfaction among its user base. As competitors continued to innovate and improve their offerings, Volt's slower pace of adaptation became a glaring disadvantage. Moreover, the costs associated with operating as a fully licensed bank placed additional strain on Volt's financial resources. Compliance with stringent regulatory requirements, such as maintaining capital reserves and adhering to cybersecurity standards, demanded significant investment. While these measures were essential for ensuring the safety and stability of its operations, they consumed resources that could have been directed towards growth initiatives. For a nascent institution like Volt, these operational costs created a delicate balancing act between compliance and innovation. Volt's struggles with operational sustainability underscored the importance of achieving efficiency and scale in the neo-banking sector. Its limited product offerings, inability to grow its customer base, and challenges in adapting to market demands left it ill-equipped to compete in a rapidly evolving industry. This lack of sustainability, coupled with external pressures such as funding shortfalls and intense competition, contributed significantly to its eventual closure.

- **Regulatory and Compliance Costs**: As a fully licensed bank, Volt Bank was required to comply with stringent regulatory standards, which presented both operational challenges and financial strain. The Australian banking sector is governed by rigorous compliance requirements, including maintaining capital reserves, ensuring robust cybersecurity measures, and adhering to Anti-Money Laundering (AML) and Counter-Terrorism Financing (CTF) regulations. While these regulations are essential for maintaining trust and stability in the financial system, they also impose significant costs on banks, particularly newer entrants like Volt. For Volt, meeting these compliance requirements diverted valuable resources away from other critical areas such as product development and customer acquisition. The costs associated with regulatory compliance were substantial, encompassing not only financial outlays but also the allocation of managerial and operational capacity. Ensuring compliance required Volt to invest in specialised personnel, advanced technology, and ongoing audits, all of which added to its financial burdens. The complexity of navigating Australia's regulatory landscape further compounded Volt's challenges. Unlike established banks with extensive compliance departments and experience in managing regulatory obligations, Volt had to build these capabilities from the ground up. This required significant time and effort, which detracted from the bank's ability to focus on innovation and growth. The dual pressure of meeting regulatory standards and competing in a dynamic market created a resource imbalance that was difficult to overcome. Moreover, the high cost of compliance disproportionately affected

Volt due to its limited scale. Larger banks could spread these costs across a much larger customer base, reducing their per-customer impact. In contrast, Volt's smaller customer base meant that these costs represented a much larger proportion of its overall expenditures, further constraining its ability to invest in areas critical to its growth. The cumulative effect of these regulatory and compliance costs was a significant barrier to Volt's success, contributing to its eventual decision to cease operations.

Lessons Learned from the Failure of Volt Bank

The failure of Volt Bank provides a clear lens through which to understand the critical challenges and missteps that neo-banks must avoid to thrive in a highly competitive financial landscape. At its core, Volt's collapse underscores the importance of balancing ambition with practicality, ensuring that innovation is supported by robust financial planning, operational resilience, and an adaptive approach to market dynamics. One of the most striking lessons from Volt's downfall is the indispensability of sustainable funding. Neo-banks operate in a capital-intensive environment where technological development, customer acquisition, and compliance are significant cost drivers. While Volt succeeded in securing initial funding, it failed to establish a continuous pipeline of financial support to sustain its growth. The inability to demonstrate profitability or significant progress discouraged further investment, creating a funding gap that hindered its ability to scale. This situation highlights the critical need for neo-banks to not only secure diverse funding sources but also maintain investor confidence by demonstrating clear pathways to revenue generation and long-term viability. Volt's challenges also reveal the importance of competitive differentiation in a crowded market. Transparency and fee-free banking, while attractive, were insufficient to distinguish Volt from well-established traditional banks and other agile neo-banks. Customers are drawn to institutions that offer comprehensive, user-friendly solutions tailored to their needs. For neo-banks, this means going beyond basic features to deliver personalised services, innovative tools, and value propositions that address specific customer pain points. Differentiation is not just about standing out; it's about meeting unmet needs in ways that foster customer loyalty and trust. Operational inefficiency was another Achilles' heel for Volt. A narrow product suite and slow response to market demands underscored the bank's inability to adapt to the dynamic expectations of modern consumers. In the fast-paced fintech industry, operational agility is not optional—it's a requirement for survival. Neo-banks must build scalable and flexible infrastructures that allow them to pivot quickly in response to new opportunities and challenges. Investing in robust technology platforms and fostering a culture of innovation are essential steps towards achieving this agility. The regulatory environment in which Volt operated posed significant challenges, yet it also offered lessons on the importance of compliance readiness. Adhering to stringent regulatory requirements can be resource-intensive, particularly for newer entrants. However, viewing compliance as a burden rather than an opportunity was a missed strategic angle for Volt. Proactively embracing regulatory standards and integrating compliance into operational strategies can build credibility with customers and stakeholders, transforming a potential liability into a competitive advantage. Finally, Volt's reliance on achieving scale as a primary business strategy exposed its vulnerability when growth stalled. A resilient business model is one that diversifies revenue streams and focuses on creating long-term customer value. Neo-banks must balance customer acquisition efforts with retention strategies, leveraging cross-selling opportunities and building loyalty through exceptional service. Dependence on aggressive growth without addressing underlying structural weaknesses is a precarious path, as demonstrated by Volt's trajectory. The lessons from Volt Bank's failure are a reminder that success in the neo-banking sector requires more than just innovation and vision. It demands meticulous planning, the ability to adapt to challenges, and a customer-centric approach that prioritises trust, value, and sustainability. By understanding and addressing these critical factors, future neo-banks can avoid the pitfalls that led to Volt's demise and position themselves for enduring success in an evolving industry.

iv. **GloriFi—Failure Case**

GloriFi, launched in 2022, aimed to disrupt the neo-banking space by targeting a niche market of conservative consumers in the United States. Branding itself as a "pro-America" financial institution, GloriFi sought to align its banking services with traditional American values, positioning itself as an alternative to mainstream financial institutions. Despite its ambitious vision and initial media attention, the company failed to establish itself as a viable neo-bank, ceasing operations within months of its launch. The story of GloriFi highlights the challenges of overly niche branding, operational missteps, and financial instability in the highly competitive neo-banking sector. The founders of GloriFi identified a gap in the market for a financial institution that catered specifically to conservative Americans. The bank's mission was rooted in values such as patriotism, free enterprise, and self-reliance, which it believed were underrepresented in the offerings of traditional banks. Its marketing campaigns prominently featured themes of American pride, aiming to attract customers who felt alienated by what they perceived as the increasingly progressive stances of larger financial institutions. GloriFi's initial plans offered a range of financial products, such as checking accounts, savings accounts, credit cards, and insurance services. It promised a seamless digital banking experience while emphasising its commitment to traditional values. The bank's founders envisioned a strong, loyal customer base drawn from conservative demographics, particularly in rural and suburban areas of the United States. However, this highly targeted strategy would ultimately prove to be a double-edged sword.

The company faced significant challenges from the outset. While its message resonated with some conservative customers, the hyper-focused branding alienated a significant portion of the potential market. This emphasis on political and cultural identity created unnecessary polarisation, limiting GloriFi's appeal and making it difficult to attract a broader demographic essential for achieving the scale required for profitability. The decision to lean heavily into niche branding constrained its market potential and undermined its ability to build a sustainable customer base. Operational inefficiencies compounded GloriFi's struggles—delays in launching key products, such as its much-anticipated credit card, eroded trust among early adopters. Additionally, technical issues undermined the bank's promise of a seamless digital experience, leaving users frustrated and dissatisfied. These challenges created significant reputational damage before the bank could establish a strong foothold in the market. The inability to deliver on its operational promises further weakened its competitive position in a highly dynamic industry. Financial instability sealed GloriFi's fate. Despite raising over $50 million in initial funding, the company quickly encountered cash flow problems due to high operational costs and poor financial management. The inability to secure additional funding left GloriFi unable to sustain its operations.

Cost-cutting measures, including layoffs, were insufficient to address the underlying financial challenges. The company's financial instability was exacerbated by its overestimation of the size and loyalty of its target market. While there was initial interest from conservative consumers, this enthusiasm did not translate into a sustainable customer base. Traditional banks and other neo-banks continued to attract customers with broader, more inclusive value propositions and superior service offerings.

In November 2022, GloriFi announced its closure, citing financial difficulties and an inability to achieve the necessary scale to remain operational. The company's assets were liquidated, and its customers were advised to transition to other financial institutions. The rapid rise and fall of GloriFi serves as a cautionary tale about the risks of overly niche branding, operational inefficiencies, and misaligned market strategies in the neo-banking sector. GloriFi's failure underscores the complexities of operating in the neo-banking space. While its vision was bold and unique, its execution fell short, providing valuable lessons for aspiring neo-banks. The case of GloriFi highlights the importance of balancing innovation with operational excellence, financial stability, and broad market appeal to thrive in an increasingly competitive landscape.

Key Factors of Failure

- **Overly Niche Branding**: GloriFi's hyper-focused branding strategy was a double-edged sword. While it resonated strongly with a specific demographic, it simultaneously alienated a vast segment of the broader market. The emphasis on conservative values and political messaging limited its appeal to customers outside its niche target group. In the competitive neo-banking sector, achieving scale is critical for sustainability, and GloriFi's exclusionary branding made it difficult to attract a diverse customer base. By prioritising cultural alignment over broad market appeal, the company significantly constrained its growth potential and created polarisation that worked against its long-term goals.
- **Operational Inefficiencies**: From the outset, GloriFi's operations were plagued by inefficiencies that undermined its ability to deliver on its promises. Delays in product launches, particularly its much-anticipated credit card, created frustration among early adopters and diminished trust in the brand. These delays were compounded by technical issues that disrupted the user experience, directly

conflicting with the company's promise of a seamless digital platform. As a neo-bank, delivering reliable and innovative technology is a cornerstone of success, and GloriFi's inability to meet these expectations further weakened its competitive position. Operational shortcomings also led to missed opportunities for customer retention and market expansion, eroding confidence in its ability to deliver value.

- **Financial Instability**: Despite securing over $50 million in initial funding, GloriFi faced significant financial challenges that it could not overcome. High operational costs quickly depleted its resources, while inefficient financial management exacerbated the problem. The company struggled to secure additional investment, leaving it without the capital needed to sustain its operations or scale effectively. Cost-cutting measures, including layoffs, were implemented but proved insufficient to address the root causes of its financial instability. Without a steady influx of funds, GloriFi could not invest in innovation, marketing, or customer acquisition, all of which are critical for growth in the neo-banking sector. The financial instability resulted from poor management and the company's overestimation of its market size and customer loyalty, which failed to generate the expected revenue streams.
- **Market Misjudgement**: GloriFi's market strategy was built on the assumption that political and cultural alignment could drive customer loyalty and acquisition. While this approach garnered initial attention, it did not translate into a sustainable customer base. The company failed to account for the competitive nature of the neo-banking space, where customers prioritise value, convenience, and innovation over ideological alignment. Traditional banks and other neo-banks offered broader and more inclusive value propositions, attracting a wider range of customers. GloriFi's narrow market focus not only limited its appeal but also left it vulnerable to competitors who could better meet the diverse needs of consumers. This misjudgement of market dynamics hindered its ability to scale and remain competitive.

Lessons Learned from the Failure of GloriFi

The failure of GloriFi serves as a poignant reminder of the complexities and challenges inherent in establishing a neo-bank in today's competitive and rapidly evolving financial landscape. While GloriFi entered the market with bold ambitions and a unique positioning, its collapse highlights critical lessons about the importance of balanced strategy, operational execution, and market adaptability. The key takeaways from GloriFi's short-lived tenure provide a roadmap for aspiring neo-banks to navigate the pitfalls that can undermine even the most promising ventures. One of the most glaring lessons from GloriFi's demise is the risk of overly niche branding. By positioning itself exclusively as a "pro-America" financial institution catering to conservative consumers, GloriFi alienated a significant portion of potential customers. In a sector where scale is essential for profitability, narrowing the target audience to a politically polarised demographic limited its growth potential. A successful neo-bank must appeal to a broad customer base, offering services and values that resonate universally rather than restricting itself to ideologically driven messaging. This misstep underscores the necessity of inclusivity and adaptability in branding strategies, particularly in industries as competitive and scale-dependent as banking. Operational inefficiencies played a critical role in GloriFi's failure. From the outset, the company struggled to deliver on its promises, with delays in launching key products and persistent technical issues undermining its credibility. Early adopters, expecting a seamless digital experience, were left disappointed, eroding trust in the brand before it could gain a strong foothold. For any neo-bank, operational excellence is not a luxury but a prerequisite. Customers demand reliable, intuitive platforms that simplify financial management, and any deviation from these expectations can be fatal. GloriFi's operational missteps highlight the importance of prioritising infrastructure development and quality assurance before launching into a competitive market.

Financial instability was another Achilles' heel for GloriFi. Despite raising significant initial funding, the company's high operational costs and poor financial management quickly drained its resources. The inability to secure additional investment, exacerbated by its niche market positioning, left GloriFi unable to sustain its operations. This outcome reinforces the critical importance of sound financial planning and efficient resource allocation. Neo-banks must secure adequate funding and ensure that their expenditure aligns with strategic priorities, enabling them to build momentum and withstand initial challenges. Diversifying revenue streams and maintaining financial discipline are key to ensuring long-term viability in an industry that demands sustained investment. Another critical insight from GloriFi's failure is the danger of misjudging market dynamics. While catering to a politically aligned demographic might have seemed innovative, it failed to account for the broader factors that drive customer choices in banking. Convenience, security, and value often precede ideological alignment when consumers select financial institutions. By prioritising cultural messaging over practical offerings, GloriFi overlooked the needs of a wider audience, effectively ceding ground to competitors with more inclusive and customer-focused value propositions. Understanding market preferences and aligning offerings with these needs is crucial for any neo-bank seeking relevance and growth. GloriFi's story is a cautionary tale about the perils of neglecting core business fundamentals. While its branding generated initial buzz, the lack of operational readiness, financial prudence, and market adaptability ensured its rapid decline. The lessons from GloriFi's failure are a potent reminder that success in the neo-banking sector requires more than a bold vision or a distinctive brand. It demands meticulous planning, a commitment to operational excellence, and an unwavering focus on delivering value to customers. Aspiring neo-banks can draw valuable insights from GloriFi's experience. Broadening market appeal, prioritising operational reliability, maintaining financial discipline, and aligning with customer needs are non-negotiable elements of a sustainable strategy.

c. How Impact Investing Can Support Digital Banking Services in Promoting Financial Inclusion

The convergence of impact investing and digital banking represents a transformative opportunity to enhance financial inclusion, particularly in underserved and marginalised communities. By aligning the goals of social impact and financial innovation, impact investing can serve as a catalyst for expanding access to essential financial services, reducing economic inequality, and fostering sustainable economic growth.

Mobilising Capital for Inclusive Growth: Impact investing channels capital into ventures that prioritise measurable social and environmental outcomes alongside financial returns. Digital banking services, with their ability to reach unbanked and underbanked populations, are an ideal target for such investments. In regions where traditional banking infrastructure is limited or inaccessible, digital banking platforms can bridge the gap by offering low-cost, technology-driven solutions. Impact investors play a critical role in providing the necessary funding for these platforms to scale their operations, develop innovative products, and expand their reach. For instance, investments in digital wallets and mobile banking applications can empower individuals in rural and remote areas to participate in the formal financial system. These tools enable users to save money, access credit, and make payments seamlessly, reducing reliance on informal financial mechanisms that are often costly and unreliable. By directing capital to such initiatives, impact investors help create an ecosystem that fosters economic participation and resilience.

Encouraging Innovation and Accessibility: Digital banking services rely on continuous innovation to address the diverse needs of their customer base. Impact investing supports this innovation by funding research and development efforts aimed at creating products tailored to the needs of low-income individuals. For example, microloan platforms and savings schemes designed for small-scale entrepreneurs can provide the financial flexibility needed to grow businesses and improve livelihoods. Similarly, investment in user-friendly interfaces and local language support ensures that digital banking services are accessible to a broad spectrum of users, including those with limited digital literacy. Furthermore, impact investors can incentivise digital banks to adopt inclusive practices, such as lower fees, flexible repayment terms, and partnerships with community-based organisations. These practices not only enhance the value proposition of digital banking services but also build trust among underserved populations, encouraging greater adoption and engagement.

Addressing Systemic Barriers: Impact investing can also help digital banking services overcome systemic barriers to financial inclusion. Issues such as inadequate digital infrastructure, regulatory challenges, and limited financial literacy often hinder the expansion of digital banking in underserved regions. By funding initiatives that address these barriers, impact investors contribute to creating an enabling environment for digital financial services to thrive. For example, investments in digital infrastructure—such as internet connectivity, mobile networks, and cybersecurity—are essential for ensuring that digital banking platforms operate reliably and securely. Similarly, impact-driven collaborations with governments and regulatory bodies can advocate for policies that promote open banking, data protection, and financial inclusion. Additionally, funding educational programmes focused on improving financial and digital literacy can empower individuals to use digital banking tools effectively, maximising their benefits.

Promoting Sustainable Business Models: A key challenge for digital banking services targeting underserved populations is achieving financial sustainability while maintaining affordability. Impact investing supports the development of sustainable business models by prioritising patient capital and long-term returns. Unlike traditional investors who may prioritise short-term profitability, impact investors are more likely to support ventures that balance social impact with financial performance. This approach allows digital banks to experiment with pricing models, cross-subsidisation, and partnerships that make their services more accessible without compromising viability. For instance, a digital bank might use impact capital to subsidise fees for low-income customers while generating revenue through premium services for higher-income users. Such models ensure that digital banking services remain inclusive and scalable, contributing to their long-term success.

Strengthening Partnerships and Ecosystems: Impact investing fosters collaboration among stakeholders in the financial inclusion ecosystem. By funding partnerships between digital banks, fintech companies, non-governmental organisations (NGOs), and local governments, impact investors help create synergies that amplify the reach and effectiveness of digital banking services. These collaborations can lead to innovative solutions, such as integrating digital banking with e-commerce platforms, agricultural value chains, or healthcare services, further enhancing their value to underserved communities. For example, a partnership between a digital bank and an agricultural cooperative partnership might enable farmers to access tailored financial products, such as crop insurance and seasonal loans, through their mobile devices. Such initiatives can promote financial inclusion and so address broader developmental goals, such as food security and rural economic development.

Impact investing holds immense potential to support digital banking services in advancing financial inclusion. By mobilising capital, fostering innovation, addressing systemic barriers, and promoting sustainable business models, impact investors can enable digital banks to transform the financial landscape for underserved populations. The intersection of impact investing and digital banking exemplifies the power of aligning financial resources with social purpose, creating a more inclusive and equitable financial ecosystem for all.

Digital Identity Solutions

Digital identity solutions refer to electronic systems that securely store and manage an individual's identity data, enabling online identity verification without reliance on traditional physical documents, such as passports or driver's licences. These systems leverage advanced technologies, including biometrics, cryptography, and blockchain, to ensure data security and empower users with control over how their information is shared. As the World Bank (2023) highlights, digital identity solutions hold the potential to provide over a billion people with access to essential services, such as finance and healthcare. By digitising identity verification, these solutions facilitate faster onboarding, improved service access, and enhanced user experiences, particularly in regions lacking widespread access to traditional identity documentation (White et al., 2019). The current state of digital identity solutions reflects their rapid adoption and integration across industries and societies, driven by the global push for digital transformation. Over 90 countries have implemented national digital identity programmes, with varying degrees of success, according. Developed economies increasingly integrate digital identity solutions into e-governance, banking, and healthcare systems, creating interconnected ecosystems that streamline service delivery. Estonia serves as a leading example, offering extensive public services through secure digital IDs that enhance government efficiency (White et al., 2019). In developing regions, digital identity solutions are instrumental in bridging gaps in financial inclusion by addressing the challenges posed by underdeveloped or inaccessible traditional identity systems. Programmes such as India's Aadhaar demonstrate the transformative potential of digital identities in improving socio-economic outcomes, enabling millions to access financial services and government support. The World Economic Forum (2023) reports that these solutions have played a pivotal role in extending financial services to over a billion unbanked individuals globally, particularly in Africa and South Asia. The private sector is also a key player in advancing digital identity solutions. Major technology companies, including Microsoft, IBM, and Mastercard, are investing heavily in secure and user-friendly identity management platforms (Deloitte, 2023). Innovations in decentralised identity systems, supported by blockchain technology, are gaining traction due to their emphasis on user control, privacy, and tamper-proof functionality. Digital identity solutions are not merely technological innovations; they represent a critical enabler of inclusive development in the digital age. By addressing the limitations of traditional identity systems and offering secure, user-centric, and privacy-enhancing approaches, these solutions empower individuals to access essential services seamlessly. As such, they play an increasingly vital role in promoting financial inclusion, fostering socio-economic development, and supporting the global agenda for equitable access to resources and opportunities (Kumar & Jie, 2023).

a. **Applications of Digital Identity Solutions in Financial Services to Promote Financial Inclusion**

Enabling Access to Financial Services: Digital identity solutions serve as a transformative mechanism to address the exclusion of underserved populations from formal financial systems. These systems eliminate the need for traditional identification documents, such as passports or driver's licenses, by enabling individuals to verify their identities electronically. In regions where documentation infrastructure is weak or inaccessible, digital identities provide a gateway to essential financial services, including savings accounts, credit facilities, and insurance products. For instance, India's Aadhaar system, which leverages biometric technology, has successfully brought millions into the formal financial ecosystem, highlighting its scalability and impact on financial inclusion. Moreover, digital identity solutions facilitate tailored financial products that cater to the unique needs of previously excluded demographics, fostering economic participation and empowerment.

Streamlining Onboarding Processes: Traditional onboarding in financial services is fraught with inefficiencies, often requiring significant time, cost, and effort from both service providers and customers. Digital identity solutions revolutionise this process by enabling instant, remote identity verification. Through biometric authentication and secure digital platforms, customers can access financial services without needing to visit physical branches. This is particularly beneficial in remote or underserved areas where financial institutions are scarce. Additionally, streamlined onboarding reduces operational costs for providers, allowing them to focus resources on expanding services and improving customer experiences. This approach enhances the scalability of financial services, ensuring that more individuals and small businesses can access and utilise formal financial systems.

Enhancing Security and Trust: In an era where identity fraud and cyber threats pose significant risks to financial systems, digital identity solutions provide a robust framework for enhancing security and trust. By employing advanced technologies such as biometrics, cryptography, and blockchain, these systems ensure secure and tamper-proof identity verification. This reduces the risk of fraudulent activities, such as identity theft and unauthorised transactions, thereby fostering greater confidence in digital financial services. Trust is a crucial factor in encouraging individuals, especially those previously excluded from financial systems, to engage with formal banking and payment platforms. Furthermore, enhanced security measures contribute to compliance with regulatory standards, strengthening the overall integrity of financial ecosystems.

Facilitating Digital Payments: The proliferation of digital payment systems is a cornerstone of financial inclusion, and digital identity solutions play a critical role in their success. By enabling secure and seamless identity verification, these solutions ensure that individuals can confidently engage in peer-to-peer transfers, e-commerce transactions, and mobile banking activities. In countries like Kenya, digital identity systems underpin mobile payment platforms such as M-Pesa, which have revolutionised financial access for unbanked populations. These systems not only facilitate day-to-day transactions but also enable users to participate in broader economic activities, thereby promoting financial empowerment and socio-economic development.

Promoting Microfinance and Credit Access: Access to credit is a key driver of economic development, yet many individuals and small businesses remain excluded due to the lack of traditional credit histories. Digital identity solutions address this gap by integrating alternative data sources, such as mobile phone usage, utility payments, and transactional behaviours, to assess creditworthiness. This enables financial institutions to extend microloans and credit products to those previously deemed unbankable. By empowering individuals to secure financing for education, healthcare, or entrepreneurial ventures, digital identity solutions foster economic resilience and mobility, contributing to poverty reduction and inclusive growth.

Enabling Cross-Border Remittances: For millions of families in developing regions, remittances from migrant workers serve as a lifeline. However, traditional cross-border payment systems are often expensive, slow, and encumbered by regulatory complexities. Digital identity solutions streamline these processes by simplifying compliance with Anti-Money Laundering (AML) and Know Your Customer (KYC) requirements. This enables faster, more affordable, and secure cross-border transactions, enhancing the accessibility of remittance services. By reducing transaction costs and ensuring the reliability of fund transfers, digital identity systems play a vital role in improving the financial well-being of migrant families and their communities.

Supporting Social Protection and Subsidy Programmes: Governments and financial institutions increasingly leverage digital identity solutions to optimise the delivery of social protection schemes and subsidies. By linking digital identities with financial accounts, these programmes ensure that benefits reach the intended recipients directly, reducing leakages and fraud. This approach enhances transparency and accountability while introducing beneficiaries to formal financial systems. For example, targeted cash transfer programmes in countries such as Brazil and India have demonstrated how digital identity systems can improve the efficiency and effectiveness of social welfare initiatives. These solutions not only address immediate needs but also lay the foundation for long-term financial inclusion by integrating marginalised populations into the formal economy.

b. Successes and Failures of Digital Solutions

i. BankID (Sweden)—Success Case

BankID, launched in Sweden in 2003, is one of the most successful digital identity solutions globally. Developed as a collaborative initiative between Swedish banks, it provides a secure and standardised electronic identification system that allows individuals and organisations to authenticate their identities and perform digital transactions. Widely used across both public and private sectors, BankID underpins a range of services, including banking, e-commerce, healthcare, and government interactions Robust encryption technologies are employed by BankID to ensure the security and privacy of user data. It operates through multiple mediums, including mobile apps, desktop applications, and physical devices, offering users flexibility and accessibility. Integration of biometric authentication, such as fingerprint or facial recognition, ensures a seamless and user-friendly experience while maintaining high security standards. Additionally, BankID adheres to stringent European Union regulations, including the General Data Protection Regulation (GDPR) and eIDAS (electronic IDentification, Authentication, and trust Services), ensuring compliance and trustworthiness. The solution has achieved widespread adoption in Sweden, with over 8 million active users in a population of approximately 10 million. This remarkable penetration highlights its role as a vital infrastructure in Sweden's digital economy. BankID facilitates millions of daily transactions, ranging from online banking and tax filing to signing legal documents and accessing healthcare services. Its seamless integration into various digital platforms has made it indispensable for both citizens and businesses.

BankID plays a significant role in promoting financial inclusion by enabling secure and efficient access to banking services. Onboarding processes for financial institutions are simplified, allowing individuals to open accounts and apply for loans without the need for physical documentation or in-person verification. This feature has been particularly impactful for younger populations and those residing in remote areas, as they can access financial services conveniently through digital channels. Furthermore, BankID's interoperability with other services amplifies its inclusivity, enabling users to access government benefits, apply for housing, and manage utilities, fostering participation in both financial and social systems. The economic and social impact of BankID is profound. Identity verification processes have been streamlined, reducing administrative costs for businesses and government agencies and enabling more efficient service delivery. Its secure infrastructure has enhanced trust in digital interactions, fostering greater adoption of online services and contributing to Sweden's reputation as a leader in digital transformation. During crises such as the COVID-19 pandemic, BankID demonstrated its adaptability, enabling remote access to essential services and underscoring its importance in ensuring resilience and continuity during disruptions. It exemplifies how a well-designed digital identity solution can drive financial inclusion, enhance service delivery, and foster a digitally inclusive society. Its success serves as a benchmark for other countries aiming to harness the potential of digital identity to transform their financial and social landscapes.

ii. **Key Factors Enabling the Success of BankID**

Strong Stakeholder Collaboration: The success of BankID is rooted in the effective collaboration between Sweden's financial institutions, government bodies, and private-sector entities. This tripartite partnership ensured that BankID addressed diverse needs while maintaining a focus on security and usability. The trust established between these stakeholders fostered widespread adoption and helped create a solution that is both scalable and sustainable. Such collaboration also facilitated interoperability between systems and services, allowing BankID to become an integral part of Sweden's digital ecosystem.

Comprehensive Regulatory Framework: Adherence to stringent regulatory standards, including the General Data Protection Regulation (GDPR) and eIDAS (electronic IDentification, Authentication, and trust Services), played a crucial role in BankID's success. These regulations provided a clear legal framework that prioritised user privacy and data security. By complying with these standards, BankID gained the trust of both users and service providers. The strong regulatory framework also served as a safeguard against misuse, ensuring that the system maintained its credibility and integrity over time.

Accessibility and User-Friendly Design: BankID's design prioritised accessibility, making it available through multiple platforms, including mobile applications, desktop interfaces, and physical devices. This flexibility allowed it to cater to a diverse user base, ranging from tech-savvy individuals to those with limited technological exposure. Features like biometric authentication enhanced the user experience, offering quick and secure access to services. By ensuring that the system was intuitive and easy to use, BankID minimised barriers to adoption, enabling even hesitant users to engage with digital services confidently.

Advanced Digital Infrastructure: Sweden's advanced digital infrastructure provided a strong foundation for the widespread adoption of BankID. With high levels of internet penetration, robust telecommunications networks, and a digitally literate population, the country was well-positioned to implement a national digital identity solution. These conditions not only facilitated the initial rollout of BankID but also ensured its sustained use across various sectors, from banking to healthcare and e-government services.

Interoperability Across Sectors: A key factor in BankID's success is its seamless integration with a wide range of public- and private-sector services. This interoperability allowed users to rely on a single digital identity for multiple purposes, including accessing government portals, signing contracts, and conducting financial transactions. By serving as a universal identification tool, BankID enhanced convenience for users and increased its value proposition. This widespread applicability solidified its role as an indispensable component of Sweden's digital infrastructure.

Proactive Promotion and Support: The promotion of BankID by public and private entities played a significant role in its widespread adoption. Awareness campaigns emphasised the benefits of using BankID, such as convenience, security, and access to essential services. These efforts demystified the technology for potential users, addressing concerns about privacy and ease of use. The ongoing support provided by banks and other institutions also ensured that users could navigate the system effectively, reinforcing trust and satisfaction.

iii. **BVN (Nigeria)—Success Case**

The Bank Verification Number (BVN), introduced in Nigeria in 2014 by the Central Bank of Nigeria (CBN) in collaboration with the Nigerian Interbank Settlement System (NIBSS), represents a transformative digital identity solution in the country's financial sector. Designed to uniquely identify bank customers, the BVN serves as a biometric-based digital identification system that enables secure and efficient banking transactions while addressing challenges such as fraud and financial exclusion. The BVN system was established to address the growing need for a unified identification mechanism within Nigeria's banking ecosystem. Each BVN is a unique identifier assigned to individuals upon enrolment, which involves capturing their biometric data, including fingerprints and facial images. This data is stored in a secure central database, allowing financial institutions to verify customer identities across multiple banks and platforms. The initiative aims to strengthen the financial system's security, promote transparency, and improve access to financial services for Nigeria's largely unbanked population. Since its launch, the BVN system has achieved widespread adoption, with over 57 million Nigerians enrolled as of 2023. This significant uptake reflects the success of the initiative in creating a robust and secure identity verification framework. BVN enrolment has become a mandatory requirement for accessing banking services in Nigeria, including opening accounts, obtaining loans, and conducting high-value transactions. The system's extension to support mobile banking and digital wallets has further increased its relevance in Nigeria's evolving financial landscape.

Advanced biometric technology underpins the BVN system, ensuring the accuracy and reliability of identity verification processes. Linking biometric data to unique identifiers minimises identity fraud and ensures that individuals cannot maintain multiple identities within the banking system. Integration with various financial services platforms enhances interoperability across banks, simplifying administrative processes and improving customer convenience. The BVN database is securely managed and routinely updated to maintain data integrity and safeguard sensitive customer information. The BVN has significantly contributed to financial inclusion in Nigeria, a country where a substantial portion of the population remains unbanked. The system provides individuals with a secure and universally accepted digital identity, lowering barriers to accessing formal financial services. Rural populations, who often lack traditional identification documents, have benefited considerably, gaining access to bank accounts and credit facilities with minimal documentation. Additionally, the BVN has facilitated the growth of microfinance and mobile banking services, empowering underserved communities to engage with the formal economy. The introduction of the BVN has had profound economic and social implications. Fraud within the banking sector has been curtailed through accurate customer identification and verification, enhancing trust in Nigeria's financial system and encouraging increased participation from individuals and businesses. The BVN has also improved the efficiency of government programmes, such as cash transfer schemes, ensuring benefits reach the intended recipients directly. During the COVID-19 pandemic, the BVN proved critical in facilitating the distribution of palliatives to vulnerable populations, underscoring its adaptability and importance in addressing societal challenges. The BVN initiative has reshaped Nigeria's financial landscape by addressing core issues of identity, security, and access. It has not only streamlined banking operations but also fostered greater inclusion and trust in the financial system. As Nigeria continues its journey towards digital transformation, the BVN stands as a vital instrument for creating a more inclusive and resilient financial ecosystem.

Key Factors Enabling the Success of BVN

Strong Regulatory Support: The Central Bank of Nigeria (CBN) played a pivotal role in ensuring the success of the BVN initiative by mandating its enrolment as a prerequisite for accessing banking services. This regulatory framework established uniformity across financial institutions, compelling both banks and customers to comply. Clear guidelines and timelines introduced by the CBN provided a structured approach to BVN implementation. Regulatory oversight assured stakeholders of the system's credibility and reliability, creating an environment of trust that encouraged widespread adoption. Furthermore, the mandate reduced resistance from institutions and individuals, ensuring the BVN's rapid integration into Nigeria's financial ecosystem.

Biometric Technology Integration: The BVN's incorporation of advanced biometric technology addressed the long-standing issue of identity fraud within Nigeria's financial sector. Biometric identifiers such as fingerprints and facial recognition provided a unique and tamper-proof method of verifying identities. This technological innovation eliminated the possibility of duplicate or fraudulent identities, a significant problem in the country's banking system. The reliability and accuracy of biometric systems fostered confidence among financial institutions, enabling them to streamline processes such as customer onboarding and transaction verification. Additionally, biometric systems offered users a secure way to interact with financial institutions, ensuring their personal information was protected.

Interoperability Across Financial Institutions: One of the standout features of the BVN system is its interoperability, which allows seamless integration across multiple banks and financial service providers. This unified approach eliminated inefficiencies associated with fragmented identification systems, enabling customers to use a single BVN for various banking and financial transactions. The system's interoperability simplified the customer experience and reduced administrative burdens for banks, enabling them to focus on improving service delivery. The ability to share data and collaborate among financial institutions created a cohesive ecosystem that amplified the BVN's effectiveness.

Public Awareness Campaigns: Effective public awareness campaigns were instrumental in driving the adoption of the BVN system. The

Central Bank of Nigeria, in partnership with financial institutions and media outlets, launched extensive educational programmes to inform citizens about the BVN's benefits. These campaigns targeted diverse demographics, including rural and semi-literate populations, through radio broadcasts, community outreach, and social media platforms. Addressing common misconceptions and emphasising the BVN's role in enhancing financial security and inclusion were key components of these efforts. Highlighting the long-term advantages of digital identity systems fostered trust and encouraged widespread participation.

Focus on Financial Inclusion: The BVN initiative was explicitly designed to address Nigeria's significant financial inclusion gap. Recognising the challenges faced by underserved and unbanked populations, the programme implemented mobile enrolment units and outreach programmes to bring BVN registration to rural and remote areas. This proactive approach reduced barriers such as physical distance and lack of infrastructure, enabling marginalised communities to participate in the formal financial sector. Providing a secure and accessible digital identity empowered individuals to open bank accounts, access credit facilities, and engage with financial services, thereby driving broader economic participation.

Scalable Infrastructure The infrastructure supporting the BVN system was engineered to accommodate Nigeria's large and diverse population. A centralised database combined with advanced biometric technology allowed the system to securely process and store vast amounts of data. The scalability of this infrastructure ensured the system could handle increasing enrolment rates and integrate with emerging financial technologies such as digital wallets and mobile banking platforms. The robust design of the system also guaranteed resilience against cyber threats and technical failures, ensuring long-term reliability and efficiency.

Government and Private-Sector Collaboration: Collaboration between the government and private-sector entities, particularly financial institutions, was a cornerstone of the BVN's success. Banks played a critical role in operationalising the system, ensuring their customers were enrolled and that the BVN was integrated into their services. This partnership ensured the BVN addressed both regulatory requirements and practical banking needs. The synergy between stakeholders created a unified vision for the BVN's implementation, fostering innovation and streamlining its rollout.

Adaptability and Responsiveness: The BVN system's ability to adapt to evolving challenges and opportunities significantly contributed to its success. For example, during the COVID-19 pandemic, the BVN facilitated the distribution of government palliatives to vulnerable populations. This adaptability demonstrated the system's utility beyond traditional banking applications, reinforcing its importance in addressing broader socio-economic issues. The BVN's responsiveness to emerging needs has solidified its role as a critical tool in Nigeria's financial and social development.

c. **Impact Investing and the Growth of Digital Identity Solutions**

The demand for funding in the digital identity solutions segment is growing, particularly in regions where formal identification is inaccessible. In developing economies, an estimated 1 billion people lack legal identity, preventing them from accessing essential services, such as banking and healthcare (World Bank, 2023). This presents a significant opportunity for impact investors to address the funding gap and enable large-scale deployment of digital identity solutions that can enhance financial inclusion and socio-economic development. Impact investing can support the growth of digital identity solutions by addressing specific funding demands, including infrastructure development, capacity-building, decentralised solutions, and Public–Private Partnerships. These areas of funding can directly enhance the adoption and scalability of digital identity systems, ultimately promoting financial inclusion.

d. **Demand for Impact Capital to Support Digital Identity Solutions**

Infrastructure Development: Digital identity solutions require a robust technological infrastructure, including reliable internet access and biometric devices. In many emerging markets, this infrastructure is underdeveloped, limiting the implementation of digital identity programmes. Impact investments can provide funding to build and expand the necessary digital infrastructure, ensuring that even the most remote communities have access to digital identity solutions. By investing in infrastructure, impact investors can help bridge the digital divide, making digital identities accessible to underserved populations and enhancing their ability to access essential services, including financial services.

Capacity-Building and Awareness: The successful adoption of digital identity solutions depends on the ability of individuals to understand and use them effectively. There is a significant need for funding educational campaigns and training programmes that raise awareness about the benefits of digital identities and teach individuals how to use them. This is particularly important in rural areas, where digital literacy levels are often low. Impact investors can fund initiatives that educate users, ensuring that they can fully benefit from digital identity solutions and access services such as banking, healthcare, and government support.

Scaling Decentralised Solutions: Decentralised identity solutions, such as those using blockchain technology, have the potential to address privacy concerns and provide individuals with more control over their data. However, these solutions are still in the early stages of adoption and require substantial investment to scale. Impact investors can support pilot projects and the development of decentralised identity frameworks that prioritise user privacy and control. By scaling these solutions, impact investors can help create secure, user-centric identity systems that are resilient to data breaches and give individuals greater autonomy over their personal information.

Promoting Financial Inclusion: By providing funding to establish digital identities for marginalised populations, impact investors can help bring millions of people into the formal financial system. With a verifiable digital identity, individuals can open bank accounts, access loans, and receive government benefits more easily. According to McKinsey (2019), providing digital identities to underserved populations can improve financial access for up to 500 million people globally. Impact investments can thus play a crucial role in expanding financial inclusion and empowering individuals who lack access to traditional identification.

Supporting Innovative Technologies: Impact investors can help drive innovation in the digital identity space by funding the development of privacy-preserving technologies, such as zero-knowledge proofs and secure multi-party computation. These technologies enhance the security of digital identity solutions while ensuring user privacy, addressing one of the key challenges facing the adoption of digital identities. Investment in such technologies will help build user trust, which is crucial for widespread adoption.

Public–Private Partnerships: Collaboration between governments and the private sector is essential for scaling digital identity solutions. Impact investors can play a role by funding Public–Private Partnerships that aim to develop and implement national digital identity systems. These partnerships can help governments leverage private-sector expertise and technology while ensuring that digital identity solutions are inclusive and accessible to all citizens. Such partnerships can also help in creating interoperable frameworks, making digital identities usable across various platforms and services.

Alternative Credit Scoring

Alternative credit scoring leverages non-traditional data and advanced analytical technologies such as machine learning, artificial intelligence, and big data analytics, to assess the creditworthiness of individuals who may lack a formal credit history. Unlike conventional models that rely primarily on financial records, alternative credit scoring integrates a broader array of data points, including social media activity, utility and rent payments, mobile phone usage patterns, e-commerce transactions, employment history, and psychometric assessments. By capturing a more comprehensive picture of financial behaviour, this approach offers an inclusive mechanism for evaluating credit risk, particularly for underserved populations. The adoption of alternative credit scoring has expanded rapidly across both developed and emerging markets. In developed economies, fintech companies increasingly employ alternative data and machine learning to enhance lending decisions, extending credit access to demographics traditionally excluded from the financial system—such as young adults, immigrants, and gig economy workers, who often lack extensive credit histories (White et al., 2021). By integrating non-traditional data sources, lenders can expand their customer base while effectively managing risk. In emerging markets, alternative credit scoring has proven transformative in promoting financial inclusion. Across Africa, Asia, and Latin America, where formal financial records are often scarce, fintech firms leverage mobile phone data, utility payment histories, and e-commerce transactions to assess creditworthiness. Companies such as Tala and Branch have successfully deployed these models to provide microloans to millions of unbanked individuals, enabling them to establish credit profiles and gain access to essential financial services (World Bank, 2019). This approach is particularly impactful in regions with weak or inaccessible traditional banking infrastructure, allowing previously excluded populations to participate in the formal financial system for the first time. Beyond individual consumers, alternative credit scoring also holds significant potential for Small and Medium-sized Enterprises (SMEs), which frequently struggle to access credit due to a lack of collateral or formal financial history. By analysing business cash flows, customer relationships, and other non-financial indicators, alternative models enhance SME financing opportunities, fostering entrepreneurship and economic growth. This is particularly critical in developing economies, where SMEs serve as a major driver of job creation and economic development (Deloitte, 2021).

As financial institutions and fintech innovators continue to refine alternative credit assessment methodologies, the approach is poised to play a pivotal role in advancing financial inclusion worldwide. Access to credit remains a fundamental enabler of economic empowerment, allowing individuals and businesses to invest in opportunities that drive sustainable growth. By incorporating a diverse range of data sources and harnessing technological advancements, alternative credit scoring presents a more inclusive framework for evaluating creditworthiness, bridging the financial divide, and fostering long-term economic resilience.

a. **Applications of Alternative Credit Scoring**

Financial Inclusion for the Unbanked: Traditional credit-scoring systems are often a barrier to financial services for individuals without formal financial records. Alternative credit scoring allows lenders to assess the creditworthiness of the unbanked by analysing non-traditional data points, providing them with access to credit (World Bank, 2019). Companies like Tala and Branch use mobile phone data to provide microloans to individuals in countries like Kenya, Tanzania, and the Philippines, promoting financial inclusion.[8]

Credit for Gig Economy Workers: Gig economy workers, freelancers, and self-employed individuals often face challenges when accessing credit due to irregular income patterns and lack of formal employment records.

Alternative credit-scoring models consider other factors, such as transaction history and income streams from different platforms, to determine their creditworthiness (White et al., 2021). Companies like Kiva and Upstart are using alternative data to help gig workers obtain loans and build a credit profile.[9]

Small Business Lending: Small businesses, particularly those in developing markets, may not have an established credit history, making it difficult to secure loans from traditional financial institutions. Alternative credit scoring can be used to evaluate business cash flow, customer reviews, supplier relationships, and transaction data to assess the creditworthiness of SMEs (Deloitte, 2021). Lenddo and CredoLab are examples of companies using alternative data to help small businesses access credit.[10]

b. **Impact Investing and the Growth of Alternative Credit Scoring**

The demand for funding for alternative credit-scoring segment is substantial, as fintech companies and lenders seek to expand their reach and develop innovative models to assess creditworthiness. This section discusses the biggest gaps in innovation in alternative credit scoring and explore how impact capital can support to bridge these gaps to enhance financial inclusion.

Limited Access to Quality Data: One of the biggest gaps in alternative credit scoring lies in the availability and quality of non-traditional data. In many regions, particularly in parts of Africa and Southeast Asia, the necessary data to effectively assess creditworthiness is either unavailable or of low quality. This limits the ability of alternative credit-scoring models to accurately assess risk. Impact investors can help address this gap by funding initiatives that focus on building digital infrastructure, such as improving internet access and mobile network coverage, which are crucial for data collection. Investments can also support partnerships between fintech companies and telecommunications or utility companies to access relevant data, ensuring that the data collected is accurate and reliable.

Scalability Challenges: Scaling alternative credit-scoring solutions to reach large populations remains a significant challenge. Many fintech companies lack the resources to expand beyond pilot projects and local markets, particularly in regions like Latin America and Sub-Saharan Africa.

Impact investments can provide the capital needed to scale these solutions, enabling fintech companies to expand their operations and reach more underserved populations. By investing in infrastructure, technology, and talent, impact investors can help ensure that alternative credit-scoring solutions can be deployed at scale and have a broader impact on financial inclusion. For instance, impact investors can fund the expansion of mobile lending platforms in rural areas, where traditional banking services are limited.

Algorithmic Bias and Fairness: Machine learning algorithms used in alternative credit scoring are susceptible to bias, which can result in unfair lending decisions and perpetuate inequality. Addressing this gap requires significant investment in research and development to create unbiased and fair algorithms. Impact investors can support initiatives that focus on developing ethical AI models, ensuring that alternative credit-scoring systems promote inclusivity and do not unintentionally discriminate against certain groups. This can include funding projects that develop diverse and representative training datasets, as well as supporting regular audits of credit-scoring algorithms to identify and rectify biases.

Regulatory and Compliance Barriers: The regulatory environment for alternative credit scoring is still evolving, and compliance with data privacy laws varies across countries. This creates challenges for fintech companies looking to expand their services internationally, particularly in regions like the Middle East and North Africa (MENA), where regulatory frameworks are less developed. Impact investors can fund efforts to navigate these regulatory landscapes by supporting fintech companies in building compliant data management systems and working with regulators to create frameworks that support innovation while protecting consumer rights. Additionally, impact investors can advocate for regulatory harmonisation across regions, making it easier for fintech companies to operate in multiple markets.

Supporting Technological Innovation: The development of advanced analytics, AI, and machine learning is critical to the success of alternative credit scoring. Impact investors can support the growth of these technologies by funding research and development initiatives that improve the accuracy and reliability of credit assessments. This investment in technology will also help address challenges related to data quality, algorithmic bias, and scalability. By supporting innovation in AI-driven credit scoring, impact investors can help create more sophisticated models that are capable of accurately assessing credit risk for individuals without formal financial histories.

Addressing Data Privacy and Compliance: Data privacy is a major concern in alternative credit scoring, especially when using sensitive personal information from unconventional sources. Impact investors can play a role in ensuring that alternative credit-scoring models are developed responsibly by funding initiatives that prioritise data privacy, security, and compliance with regulatory standards. This includes supporting companies in developing transparent data usage policies, obtaining informed consent from borrowers, and implementing robust data protection measures. By funding privacy-preserving technologies, such as encryption and anonymisation, impact investors can help build trust among users and regulators, facilitating wider adoption of alternative credit scoring.

Digital Savings and Insurance Products

Digital savings leverage technology to facilitate secure, accessible, and flexible savings mechanisms, offering individuals the opportunity to accumulate financial assets without requiring a physical bank presence. The evolution of digital savings has been propelled by innovations in mobile banking, fintech applications, and seamless payment integrations, transforming the way people manage their financial resources. The nature of digital savings products varies widely, with many platforms offering features tailored to the needs of diverse user groups. **Automated savings** functionalities allow individuals to schedule recurring deposits or round up transactions, fostering consistent saving habits. **Goal-based savings** structures enable users to allocate funds towards specific financial objectives, while **higher interest rates**—compared to traditional banks—enhance the appeal of digital platforms. Additionally, integration with mobile money services streamlines transactions, ensuring that individuals, including those in underserved regions, can seamlessly transfer and access funds. The expansion of digital savings products has been significant across both developed and emerging economies. In **developed countries,** challenger banks such as Chime (United States) and Monzo (United Kingdom) have redefined savings experiences through innovative, fee-minimal structures, often augmented by artificial intelligence to encourage disciplined financial behaviour. These platforms provide intuitive user interfaces, real-time spending insights, and personalised savings recommendations, making them attractive to digitally savvy consumers.

In **emerging markets,** the role of digital savings products in financial inclusion is even more pronounced. Mobile money platforms like **M-Pesa in Kenya and Tanzania** have integrated savings products that empower millions of previously unbanked individuals to store funds securely. Digital savings solutions are particularly crucial in regions where traditional banking services are inaccessible due to geographic, economic, or infrastructural constraints. The success of such initiatives underscores the transformative potential of digital savings in enabling financial resilience among low-income populations. Several case studies highlight the effectiveness of digital savings products in bridging financial gaps:

- **M-Pesa and M-Shwari (Kenya)**: The collaboration between M-Pesa and Commercial Bank of Africa led to the creation of M-Shwari, a mobile-based savings and microloan platform. With over 30 million users, M-Shwari has significantly contributed to the formalisation of savings behaviour among previously unbanked populations, exemplifying the potential of digital savings in financial empowerment.
- **Tala (Multiple Emerging Markets)**: Tala operates in Kenya, the Philippines, Mexico, and India, offering financial services that leverage alternative credit-scoring methods. By integrating mobile-based savings and lending solutions, Tala extends financial access to underserved communities, demonstrating the scalability of digital savings models in diverse economic environments.
- **Alipay Yu'e Bao (China)**: As one of the largest digital savings platforms globally, Yu'e Bao by Alipay allows users to deposit spare funds into a money market fund, providing superior returns compared to conventional bank accounts. Its immense adoption reflects the demand for digital savings alternatives that offer both accessibility and competitive financial returns.

How Impact Investing Can Address the Gaps in Digital Savings

Several barriers continue to limit their reach and impact of digital savings, particularly in underserved regions. Impact investing can bridge the gaps by providing the necessary capital to develop and expand digital savings solutions, ultimately fostering financial inclusion.

Limited Awareness and Financial Literacy: One of the most significant obstacles to the adoption of digital savings products is the lack of awareness and financial literacy among potential users, especially in rural and low-income communities. Many individuals are unfamiliar with digital savings mechanisms or remain sceptical due to past negative experiences with financial institutions. Impact investors can support initiatives that enhance financial literacy through educational campaigns, community outreach programmes, and partnerships with local organisations. By funding these efforts, investors can build trust and encourage the adoption of digital savings, particularly in regions like Sub-Saharan Africa and South Asia, where financial literacy levels remain low.

Access to Digital Infrastructure: Digital savings platforms require reliable digital infrastructure, including mobile network coverage, internet access, and efficient digital payment systems. In many remote areas, the absence of such infrastructure prevents individuals from accessing digital financial services. Impact investments can help bridge this gap by funding projects that expand mobile network coverage and improve internet connectivity in underserved regions. Strengthening digital infrastructure is essential for ensuring that digital savings products can reach and benefit marginalised populations.

Affordability and Product Design: Many digital savings products are not designed to meet the specific needs of low-income individuals. Minimum deposit requirements, rigid withdrawal policies, and complex terms often create barriers to adoption. Impact investors can help address these issues by funding the development of affordable, user-centric savings products. This could include flexible savings plans, pay-as-you-go models, and features that align with the irregular income patterns of gig economy workers and smallholder farmers. By supporting inclusive product design, impact investors can ensure that digital savings platforms cater to the financial realities of underserved communities.

Regulatory and Compliance Challenges: The regulatory landscape for digital savings varies significantly across countries, creating challenges for scaling these solutions. Some regions lack clear regulatory frameworks that support digital financial services, while others impose stringent compliance requirements that make it difficult for fintech firms to operate.

Impact investors can play a pivotal role in advocating for regulatory reforms that encourage innovation while safeguarding consumer interests. By funding policy research, providing technical assistance, and collaborating with regulators, investors can help create an enabling environment for digital savings to thrive.

Scalability and Distribution Channels: Expanding digital savings products to reach a broader population remains a challenge, particularly in regions where traditional financial distribution channels are limited. Innovative distribution models are needed to scale these solutions effectively. Impact investors can support partnerships with mobile network operators, agent networks, and community-based organisations to enhance the accessibility of digital savings platforms. Investing in scalable distribution strategies ensures that these financial solutions reach the individuals who need them most.

Building Trust and Security: Many potential users are hesitant to adopt digital savings products due to concerns about security and trust. Fears of fraud, data breaches, and unstable financial service providers deter participation. Impact investors can fund initiatives that focus on building consumer confidence, such as financial literacy programmes, transparent operational practices, and strengthened cybersecurity measures. By addressing these concerns, digital savings providers can foster greater trust and encourage broader adoption.

Investing in Product Innovation and Technology: Reliable digital infrastructure and technological advancements are critical for the success of digital savings products. Impact investors can provide funding to develop and implement innovative solutions that enhance user experience and financial security. This includes leveraging artificial intelligence for personalised savings recommendations, integrating blockchain for secure transactions, and enhancing mobile-friendly interfaces for seamless navigation. Investing in technological innovation can significantly enhance the usability and effectiveness of digital savings solutions.

By addressing these challenges and providing the necessary funding for growth, impact investing can play a transformative role in expanding digital savings solutions. These efforts will enhance financial inclusion, empower individuals to accumulate assets, and create economic opportunities in regions that have historically lacked access to formal financial services. As digital savings continue to evolve, strategic investments in infrastructure, education, policy advocacy, and innovation will be crucial in ensuring their sustainability and widespread adoption.

Innovative Insurance Products

Insurance products are fundamental to financial inclusion, offering mechanisms that mitigate economic shocks and foster resilience among low-income and marginalised populations. Traditional insurance models, often characterised by high premiums, complex documentation, and limited outreach, have failed to address the needs of underserved communities. However, the emergence of innovative insurance solutions, such as micro-insurance and parametric insurance, has reshaped the sector. These models leverage technology, alternative data sources, and digital platforms to create accessible, cost-effective, and scalable financial protection systems. By reducing vulnerability to unexpected financial losses, they enable individuals and small enterprises to engage more confidently in economic activities, fostering sustainable development and poverty reduction. Beyond providing a safety net, innovative insurance products play a crucial role in enhancing economic participation and stability. By reducing exposure to unforeseen financial losses, these solutions empower individuals and businesses to engage in productive economic activities with greater confidence. Smallholder farmers, for instance, can invest in improved agricultural inputs without the fear of losing their livelihoods to climate shocks. Similarly, micro-entrepreneurs are more likely to expand their businesses when they have the assurance of financial protection against potential losses. Furthermore, insurance solutions tailored to small-scale farmers and micro-enterprises enhance resilience against economic volatility. This stability ensures business continuity and fosters economic diversification, reducing the long-term risks associated with financial shocks. Additionally, the accessibility, transparency, and efficiency of digital insurance models contribute to greater trust in financial institutions, fostering broader engagement with formal financial services. The convergence of fintech, big data, and mobile networks has further streamlined insurance delivery, lowering operational costs and making risk protection more affordable and scalable. The integration of technology enhances risk assessment models, expedites claims processing, and facilitates seamless policy distribution. This shift is particularly vital for advancing climate adaptation strategies, as timely financial relief enables communities affected by climate change to implement sustainable resource management practices.

Micro-Insurance: These are affordable, simplified insurance products specifically designed to protect low-income individuals against common risks such as health emergencies, agricultural failures, and property loss. Unlike conventional insurance, micro-insurance operates on a low-margin, high-volume model, enabling insurers to scale services across vast underserved populations. These products often feature flexible payment options, minimal underwriting requirements, and mobile-based claim processing, ensuring accessibility and affordability for those outside the formal financial system. The impact of micro-insurance is particularly evident in agricultural economies, where smallholder farmers face persistent climate-related vulnerabilities. By safeguarding against income shocks caused by droughts, floods, or pest infestations, micro-insurance not only protects livelihoods but also encourages greater investment in productivity-enhancing agricultural inputs, such as improved seeds, fertilisers, and irrigation systems.

Parametric Insurance: Parametric insurance departs from traditional indemnity-based models by offering payouts based on predefined, independently verifiable parameters such as rainfall levels, seismic activity, or temperature fluctuations rather than actual financial loss assessments. This innovation eliminates the need for lengthy claims verification processes, enabling immediate disbursement of funds to affected policyholders. Parametric insurance is particularly relevant for addressing climate risks in agriculture, infrastructure, and disaster-prone regions. By providing rapid liquidity in the aftermath of adverse events, it allows businesses and communities to recover more efficiently, reducing long-term economic disruptions. Moreover, the integration of satellite imagery, weather sensors, and blockchain technology enhances the accuracy, transparency, and efficiency of these insurance models, bolstering trust and adoption among users.

Case Study: Kilimo Salama (Kenya).

Kilimo Salama, a weather-indexed crop insurance programme in Kenya, exemplifies the effectiveness of parametric insurance in strengthening agricultural resilience. Developed by the Syngenta Foundation in collaboration with local financial institutions and telecom providers, Kilimo Salama provides smallholder farmers insurance coverage against erratic weather conditions, leveraging mobile technology to enhance accessibility and efficiency. Key features and impact of the programme include the following[11]:

Features:

- **Weather-Indexed Model:** Uses data from satellite and ground-based weather stations to determine insurance payouts automatically when predefined climate conditions (e.g., inadequate rainfall) are met.
- **Mobile Money Integration:** Farmers can enrol in the programme, pay premiums, and receive payouts directly through mobile money platforms such as M-Pesa, eliminating the need for physical banking infrastructure.
- **Affordable and Scalable:** The initiative ensures inclusivity and widespread adoption among smallholder farmers by offering low-cost premiums and flexible payment options.

Impact:

- **Risk Mitigation for Vulnerable Farmers:** Enables small-scale farmers to recover from weather-related income shocks, ensuring food security and agricultural sustainability.
- **Encouraging Productive Investments:** Farmers insured under Kilimo Salama demonstrate a higher likelihood of investing in high-yield crops, fertilisers, and irrigation technologies, driving overall agricultural productivity.
- **Promoting Financial Inclusion:** Integrating insurance with mobile technology introduces financially excluded populations to broader financial services, including credit and savings products.

The integration of micro-insurance and parametric insurance into financial inclusion strategies represents a transformative shift in risk management for low-income populations. These products provide essential financial protection while fostering broader economic participation by leveraging technology, alternative risk assessment models, and mobile platforms. The success of programmes like Kilimo Salama in Kenya underscores the potential of innovative insurance to drive sustainable development, build economic resilience, and create pathways for inclusive growth in emerging economies. As the financial ecosystems continue to evolve, strengthening regulatory frameworks, expanding technological integration, and enhancing consumer education will be critical to maximising the impact of insurance-driven financial inclusion.

How Impact Investing Can Address the Gaps in Insurance

Limited awareness, affordability constraints, digital infrastructure deficiencies, regulatory barriers, and challenges in scalability are some of the obstacles limiting the sector from driving financial inclusion. Impact investing can address these issues by providing the necessary capital, technical expertise, and strategic partnerships to enhance the reach and efficiency of micro-insurance and parametric insurance solutions. By directing investments towards financial inclusion-oriented insurance products, impact investors can help build resilient economic systems that empower vulnerable populations and promote long-term stability.

Bridging the Awareness and Financial Literacy Gap: One of the primary obstacles to the adoption of innovative insurance solutions is the lack of financial literacy and awareness among potential beneficiaries, particularly in low-income and rural communities. Many individuals either lack trust in insurance products or do not fully understand their benefits. Impact investing can be vital in funding initiatives that promote financial education through community-based training programmes, digital literacy campaigns, and partnerships with local organisations. By integrating technology-driven awareness campaigns and user-friendly digital interfaces, investors can contribute to demystifying insurance concepts and improving consumer confidence in insurance mechanisms.

Enhancing Affordability and Accessibility: Affordability remains a key concern for low-income individuals who are often excluded from traditional insurance markets due to high premiums and complex underwriting processes. Impact investors can support the development of inclusive insurance models that feature low-cost, flexible premium structures tailored to the financial capabilities of underserved populations. Investments in micro-insurance schemes that allow policyholders to pay premiums in small, manageable instalments—potentially linked to mobile money platforms—can ensure greater accessibility. Additionally, investors can encourage the adoption of pay-as-you-go (PAYG) models, where coverage is dynamically adjusted based on usage patterns, making insurance more responsive to the financial realities of low-income consumers.

Strengthening Digital Infrastructure for Scalability: For insurance products to reach their full potential, they must be seamlessly integrated into digital ecosystems that ensure efficiency, transparency, and ease of use. In many regions, limited mobile network coverage, weak digital payment infrastructure, and inadequate data analytics capabilities hinder the scalability of digital insurance solutions. Impact investing can address these gaps by funding initiatives that expand digital connectivity, improve mobile payment systems, and integrate advanced analytics into insurance platforms. For instance, investments in blockchain-based smart contracts can automate claim processing, reducing fraud risks and enhancing trust in digital insurance platforms.

Overcoming Regulatory and Compliance Barriers: Regulatory environments vary significantly across regions, posing challenges for scaling micro-insurance and parametric insurance products. Some jurisdictions lack clear guidelines for digital insurance, while others impose regulatory burdens that increase operational costs. Impact investors can facilitate collaboration between insurance providers, policymakers, and financial regulators to develop conducive regulatory frameworks that balance consumer protection with industry growth. By funding research and advocacy efforts aimed at policy innovation, investors can help create an enabling environment that fosters responsible insurance expansion while safeguarding the interests of vulnerable consumers.

Scaling Distribution Networks Through Strategic Partnerships: Reaching underserved populations requires innovative distribution channels that go beyond traditional banking and insurance institutions. Impact investors can support partnerships between fintech firms, mobile network operators, agribusinesses, and community-based organisations to expand insurance outreach. Leveraging mobile money platforms and agent networks can ensure that insurance products reach the last mile, particularly in rural and informal economies. Additionally, investments in alternative risk-sharing mechanisms, such as cooperative insurance models, can create sustainable distribution frameworks that align with community needs and cultural norms.

Case: Leveraging Impact Investing for Agricultural Insurance

Agriculture and Climate Risk Enterprise (ACRE) Africa provides index-based agricultural insurance solutions to smallholder farmers in Kenya, Rwanda, and Tanzania. ACRE Africa utilises mobile technology and remote sensing data to assess weather risks and trigger automated payouts. Impact investment in ACRE Africa has facilitated the expansion of weather-indexed insurance products for small-scale farmers, reducing their vulnerability to climate change, integration of mobile payment systems to streamline premium collection and claims disbursement, and partnerships with agribusinesses and cooperatives to promote large-scale adoption of insurance solutions. ACRE underscores the potential of targeted investments in strengthening insurance ecosystems that empower marginalised communities while ensuring financial sustainability. Moving forward, strategic collaboration between investors, policymakers, insurers, and technology providers will be essential to unlocking the full potential of insurance as a tool for sustainable development and financial empowerment.

6.4.2 Business Models and initiatives

Microfinance

Microfinance Institutions (MFIs) are pivotal players in enhancing financial inclusion for low-income individuals and micro-entrepreneurs who often lack access to formal banking services. They provide small loans, savings products, insurance, and other financial services to empower economically marginalised groups, enabling them to improve their livelihoods, invest in income-generating activities, and build financial resilience (Babalola et al., 2023; Pattnaik et al., 2024). Microfinance gained global attention in the 1970s with Dr. Muhammad Yunus' pioneering work and the Grameen Bank's establishment in Bangladesh. Yunus envisioned a financial model that would empower the poor, particularly women, through small loans that could help start or expand small businesses.[12] Since then, the microfinance sector has grown globally, with diverse institutions including non-governmental organisations (NGOs), cooperatives, and even commercial banks providing microfinance services. MFIs today operate across various regions, with notable growth in South Asia, Sub-Saharan Africa, and Latin America. The financial products offered by MFIs expanded beyond microcredit to include savings accounts, micro-insurance, remittance services, and capacity-building programmes. The aim is to provide access to credit, help individuals manage financial risks and improve their overall economic well-being.

Microfinance has demonstrated significant potential in alleviating poverty and empowering disadvantaged communities. Studies have shown that access to microfinance can lead to increased household income, improved access to education and healthcare, and enhanced social empowerment, particularly for women.[13] In many countries, MFIs have played a critical role in providing financial services to those who remain outside the reach of formal banking institutions. Recent data suggest that MFIs are serving over 140 million clients worldwide, with South Asia and Sub-Saharan Africa leading in terms of outreach (CGAP, 2020). These regions have seen significant expansion in microfinance services, with countries like India, Bangladesh, and Kenya becoming prominent players. In India, MFIs have grown substantially, with an active client base of over 60 million borrowers, contributing to improved income stability and increased entrepreneurship among low-income groups (Sa-Dhan, 2022). Women have been a key target demographic for MFIs, with the idea that empowering women through access to finance has a ripple effect on household welfare and community development. Approximately 80 per cent of microfinance clients globally are women, who use microloans to start small businesses, invest in livestock, or fund their children's education, thereby contributing to economic growth and community well-being (Hussain et al., 2025). Women's access to financial services has also been linked to enhanced decision-making power within households, resulting in better outcomes for families (Kochar et al., 2022).

However, the impact of microfinance is not uniform, and several studies have highlighted limitations and challenges. Some research suggest that the effects of microfinance on poverty alleviation can be modest or mixed, with instances where microloans have led to over-indebtedness or failed businesses (Banerjee & Jackson, 2017). In certain cases, microfinance has not always translated into sustainable improvements in living standards, prompting debates on the need for more targeted interventions and better product design (Bauchet et al., 2011). The challenge of ensuring that microfinance reaches the poorest of the poor, rather than those who are already relatively economically stable, remains a topic of concern (Duvendack et al., 2011). Another key aspect of microfinance is the growing use of digital technologies to enhance service delivery. The integration of digital solutions such as mobile banking, digital payment platforms, and fintech partnerships has revolutionised how MFIs operate, making financial services more accessible, especially in rural and remote areas. Digital microfinance has helped reduce operational costs, improved efficiency, and expanded the reach of MFIs to previously underserved populations (CGAP, 2023).

Challenges persist with MFIs—many struggle with sustainability, as they often operate in complex environments with clients with limited repayment capacity. External funding and subsidies are crucial for many MFIs, particularly those that serve the most vulnerable communities (Cull et al., 2021). Furthermore, the issue of over-indebtedness remains a significant problem, with borrowers sometimes taking multiple loans from different providers, leading to a debt cycle that is difficult to escape (Guérin and Venkatasubramanian, 2022). Overcoming these challenges requires a combination of responsible lending practices, effective regulation, and the use of technology to improve credit assessment and monitoring.

Challenges Faced by Microfinance Institutions

Sustainability Challenges: One of the primary challenges MFIs face is achieving financial sustainability while maintaining their social mission of serving low-income populations. Many MFIs struggle to balance profitability with their commitment to providing financial services to high-risk, low-income clients. This challenge is exacerbated by the reliance on donor funding and concessional loans, which, while beneficial in the short term, raise concerns about long-term sustainability (Cull et al., 2009; CGAP, 2023). Additionally, high operational costs associated with serving remote and underserved communities further strain the financial viability of MFIs. While digital innovations such as mobile banking and digital field applications have the potential to reduce costs, their adoption requires significant upfront investment (IFC, 2023).

Over-Indebtedness: Another critical challenge is the issue of over-indebtedness among borrowers. Many individuals take multiple loans from different MFIs, leading to unsustainable debt cycles. This problem is particularly acute in regions where competition among MFIs is intense, leading to aggressive lending practices and insufficient credit assessments (Bylander et al., 2019; Manford, 2022). Recent data from India and Cambodia highlight the severity of the issue, showing that borrowers often struggle to meet their repayment obligations, resulting in elevated default rates. MFIs must adopt responsible lending practices and enhance their credit assessment mechanisms to mitigate this risk.

Regulatory Challenges: The regulatory environment presents both opportunities and obstacles for MFIs. In some countries, weak regulatory frameworks enable predatory lending practices and lack of transparency, undermining the sector's credibility. Conversely, overly stringent regulations such as interest rate caps can hinder the ability of MFIs to cover operational costs and expand their services, particularly for high-risk clients (Christen et al., 2012). Striking a balance between consumer protection and operational flexibility is essential for fostering an environment conducive to microfinance growth.

Climate Change and Environmental Risks: Climate change poses a growing challenge for MFIs, particularly in agrarian communities where smallholder farmers rely on microfinance to fund agricultural activities. Climate-related shocks, such as floods and droughts, can severely impact borrowers' ability to repay loans. To address this issue, MFIs need to develop resilience-focused financial products, such as micro-insurance and climate adaptation loans, to help clients manage environmental risks (Helwig et al., 2020). Impact investors can play a crucial role in supporting the development and deployment of these products.

Operational Inefficiencies: Many MFIs face operational inefficiencies due to outdated processes, lack of technological infrastructure, and limited data analytics capabilities. Manual operations contribute to high transaction costs and slow service delivery, reducing the efficiency of microfinance services. The adoption of fintech solutions such as digital loan processing, automated credit scoring, and blockchain for secure transactions can enhance efficiency and scalability. However, transitioning to these technologies requires significant capital investment and capacity-building (IFC, 2024a).

Impact Investing for Microfinance

Patient Capital and Subsidised Funding: MFIs often operate in high-risk environments where loan repayment can be irregular, and clients may not have formal credit histories. Impact investors can play a crucial role by providing patient capital to allow MFIs to prioritise social impact without the immediate pressure of generating high financial returns. By offering concessional loans or blending finance structures that combine commercial and philanthropic funding, impact investors can create financial buffers that enable MFIs to expand outreach while maintaining operational stability. Additionally, these investments allow MFIs to invest in technological innovations and staff training without passing excessive costs onto borrowers, ensuring financial services remain accessible to the most vulnerable populations.

Capacity-Building and Institutional Strengthening: Sustainable growth in the microfinance sector depends on financial resources as well as institutional strength and operational efficiency. Many MFIs, particularly in emerging markets, struggle with limited technical expertise, outdated management systems, and inefficient loan processing structures. Impact investors can support MFIs by funding capacity-building initiatives that enhance institutional governance, risk management, and client service delivery. Through technical assistance, training programmes, and mentorship, investors can help MFIs adopt best practices in financial planning, data management, and impact assessment. Strengthening these core functions ensures that MFIs can manage expansion effectively, improve client retention, and maintain high repayment rates while upholding ethical lending standards.

Technology Integration and Digital Transformation: The integration of digital technology in microfinance is essential for reducing operational costs, increasing outreach, and improving service delivery. However, many MFIs face significant barriers in adopting digital tools due to limited capital and technological expertise. Impact investors can bridge this gap by financing the development and deployment of mobile banking solutions, digital loan processing systems, and automated credit-scoring algorithms. Mobile banking, for instance, can facilitate seamless transactions in remote areas where traditional banking infrastructure is absent, reducing dependency on cash-based transactions and minimising fraud risks. Similarly, artificial intelligence-powered credit assessment tools can help MFIs make more accurate lending decisions, improving financial sustainability while reducing default rates. Investing in technology-driven solutions allows MFIs to scale efficiently and offer services that are both cost-effective and accessible.

Development of Inclusive Financial Products: A one-size-fits-all approach to financial services often fails to meet the diverse needs of low-income borrowers. Many microfinance clients, such as smallholder farmers and informal entrepreneurs, face seasonal income fluctuations and unexpected financial shocks. Impact investors can support the development of tailored financial products that address these specific challenges. For example, micro-insurance products can provide financial protection against health emergencies or crop failures, ensuring borrowers do not fall into debt traps due to unforeseen crises. Climate adaptation loans, structured to align with agricultural cycles, can help farmers invest in resilient crops and irrigation systems without worrying about rigid repayment schedules. Flexible loan structures that allow for grace periods or variable instalment plans can further enhance financial stability for borrowers with irregular income streams. By fostering product innovation, impact investors ensure that MFIs can better serve their clients' evolving financial needs.

Promoting Responsible Lending and Consumer Protection: While microfinance has been praised for its role in financial inclusion, it has also been criticised for instances of over-indebtedness and predatory lending practices. Unchecked expansion and competition among MFIs in some regions have led to aggressive lending, high interest rates, and unsustainable client borrowing. Impact investors can be pivotal in promoting responsible finance by supporting MFIs in implementing rigorous credit assessment frameworks, borrower education programmes, and ethical lending guidelines. Funding initiatives that promote financial literacy and responsible borrowing behaviour can help clients make informed decisions, reducing the risk of default and financial distress. Furthermore, impact investors can encourage transparency and accountability in the sector by advocating for stronger consumer protection regulations and supporting the establishment of grievance redressal mechanisms that empower borrowers to report unfair practices.

Financial Literacy and Client Empowerment: Access to financial services alone is not enough to drive economic empowerment; clients must also have the knowledge and skills to use these services effectively. Many microfinance clients have limited experience with formal financial systems and may lack the understanding necessary to make informed financial decisions. Impact investors can contribute to financial literacy initiatives that equip borrowers with essential knowledge on budgeting, credit management, and long-term financial planning. This can be achieved through community-based education programmes, mobile-based financial literacy apps, and partnerships with local organisations that provide on-the-ground training. By investing in financial literacy, impact investors help ensure that clients can maximise the benefits of microfinance services, improve their financial resilience, and break the cycle of poverty.

Strengthening Climate Resilience Through Impact Investing: Climate change poses a significant threat to microfinance clients, particularly small-scale farmers and rural entrepreneurs who depend on stable environmental conditions for their livelihoods. Droughts, floods, and erratic weather patterns can lead to income losses, making it difficult for borrowers to repay loans and sustain their businesses. Impact investors can support climate resilience by funding micro-insurance programmes that protect against climate-related risks, financing climate adaptation projects, and promoting sustainable agricultural practices. By integrating climate risk considerations into microfinance strategies, investors can help MFIs offer products that enable borrowers to adapt and thrive in changing environmental conditions. This protects the financial stability of microfinance clients and contributes to broader efforts in sustainable development and poverty reduction.

Agent Banking and Banking Correspondents

Agent banking and banking correspondents have become vital enablers of financial access for underserved and unbanked populations. In regions where traditional banking infrastructure is either non-existent or insufficient, these models enable financial institutions to overcome barriers related to geography, operational costs, and infrastructural constraints. It leverages local businesses and trusted community members as intermediaries to expand financial accessibility, enhance financial inclusion, reduce transaction costs, and foster economic participation at the grassroots level. Financial institutions outsource key banking operations to third-party agents who facilitate essential services such as cash deposits, withdrawals, bill payments, and money transfers. These agents, embedded within their local communities as small business owners or retail operators, utilise mobile banking technologies, point-of-sale (POS) devices, and biometric authentication tools, all of which are securely connected to the bank's system. This model enables individuals, particularly in remote and rural areas, to access financial services without the need to travel long distances to physical bank branches. The adoption of digital financial tools further enhances efficiency, allowing for real-time transactions and reducing service delays while also fostering the transition from cash-based transactions to mobile and electronic banking solutions.

A subset of agent banking, banking correspondents—also called Business Correspondents (BCs)—play a more structured role in facilitating financial services, particularly in areas where banks do not have branches. Unlike informal agents, BCs are formally appointed by banks under stringent regulatory frameworks and are authorised to perform a broader range of banking functions. These include account opening, loan disbursement, collection of repayments, and customer due diligence under financial inclusion programmes. In addition to transactional services, banking correspondents play a crucial role in promoting financial literacy, guiding individuals on responsible financial practices such as savings, credit management, and investment planning. Their proximity to local communities fosters trust in formal banking institutions and encourages individuals to transition from informal financial systems to structured banking. The agent banking model, complemented by banking correspondents, offers multiple advantages for financial institutions and consumers alike. By extending banking services to previously unbanked populations, these models enhance financial inclusion, enabling millions of individuals to participate in the formal financial system. Furthermore, they present cost-efficiency benefits, as financial institutions can leverage existing retail infrastructure instead of incurring high capital expenditures for brick-and-mortar branches. Customers also benefit from improved convenience, as they can access banking services within their communities, reducing travel costs and wait times. Additionally, agent banking serves as a gateway for digital financial services, fostering the adoption of cashless transactions and driving financial technology integration in underserved markets. Given that many agents and banking correspondents are trusted members of the community, they play a pivotal role in educating customers about formal banking services, fostering trust and mitigating scepticism towards banks.

The agent banking model faces several challenges that must be addressed to ensure long-term sustainability. Regulatory and compliance risks remain significant, as adherence to Know Your Customer (KYC) and Anti-Money Laundering (AML) requirements is essential for safeguarding the financial system. The decentralised nature of agent banking increases exposure to fraud, cybersecurity threats, and operational inefficiencies, necessitating the implementation of strong oversight mechanisms and digital security measures. Liquidity management is another challenge, as agents often struggle to maintain sufficient cash flow to meet customer demands. Solutions such as real-time liquidity monitoring and liquidity support programmes can help mitigate these issues. Infrastructure limitations, particularly in regions with poor internet connectivity or unreliable power supply, also pose barriers to efficient service delivery. Furthermore, consumer protection and dispute resolution mechanisms must be strengthened to build customer confidence in agent banking services and ensure the safeguarding of consumer rights. Investments in technology, regulatory frameworks, and financial literacy initiatives will be key to unlocking the full potential of agent banking and its role in fostering global financial inclusion.

Successful Case Studies and Examples

Brazil's Banking Correspondent Model: Brazil is one of the pioneers of the banking correspondent model, which has been instrumental in expanding financial inclusion across the country. Banking correspondents, often located in supermarkets, post offices, and lottery shops, offer various services, including account opening, deposits, withdrawals, and bill payments. The success of Brazil's model is attributed to supportive regulations and partnerships between banks and non-financial entities, which have made banking services accessible to millions of unbanked individuals.[14]

India's Business Correspondent Network: India has also seen significant success with its Business Correspondent (BC) model, particularly under the Pradhan Mantri Jan Dhan Yojana (PMJDY) financial inclusion scheme. BCs have been crucial in providing banking services to rural and underserved populations, enabling the opening of millions of no-frills savings accounts (Uzma & Pratihari, 2019). The use of Aadhaar based biometric authentication has further enhanced the efficiency and security of transactions conducted through BCs, ensuring that services reach the intended beneficiaries.

Kenya's Agent Banking Model: In Kenya, agent banking has been instrumental in increasing financial inclusion, complementing the success of mobile money services like M-Pesa. Banks like Equity Bank and KCB Bank have established extensive agent networks that provide a range of services, from deposits and withdrawals to loan disbursements and bill payments (Ndungu, 2021). These agents, often located in rural shops and kiosks, have brought banking services closer to people previously excluded from the formal financial system (Equity Bank Kenya, 2023).

How Impact Capital Can Support Agent Banking

The demand for funding in agent banking is substantial, particularly in regions where financial infrastructure is weak and where agents face challenges in liquidity management, training, and technology integration. Countries in Sub-Saharan Africa, South Asia, and Latin America have shown a high demand for funding to support the expansion of agent networks and improve service quality. In these regions, smaller agents often struggle with limited access to capital, which restricts their ability to compete with larger players and provide consistent services to underserved communities. Impact investing can play a significant role in expanding agent banking networks, particularly in underserved regions where financial infrastructure is lacking. Impact investors can provide funding for capacity-building, enabling banks to train agents and equip them with the necessary tools to deliver quality services. This is particularly relevant in regions like rural Kenya and India, where agents require ongoing training to handle complex financial products and provide advisory services to customers effectively. Investments can also support the development of digital infrastructure, ensuring that agents have reliable connectivity to conduct transactions efficiently. In areas such as Tanzania and rural Brazil, where connectivity issues persist, funding for digital infrastructure can make a significant difference in service availability and reliability.

Impact investments can also be directed towards initiatives that promote agent liquidity, such as revolving funds or partnerships with local financial institutions that help agents maintain adequate cash balances. This is especially critical in regions like Nigeria, where liquidity issues frequently disrupt agent services, leading to customer dissatisfaction and reduced trust in agent banking. By supporting liquidity management solutions, impact investors can ensure that agents are able to meet customer needs consistently, thereby enhancing the overall effectiveness of agent banking. Furthermore, impact investors can support innovations in agent banking technology, such as mobile applications, digital wallets, and biometric authentication systems. In countries like India and Kenya, these technological advancements can help agents provide faster, more secure services, thereby increasing customer confidence and expanding financial inclusion. These technologies can improve the efficiency, security, and transparency of transactions, reducing the risks associated with agent banking and enhancing customer trust.

Remittance Solutions

Remittances serve as a financial lifeline for millions of families worldwide, particularly in developing economies, providing essential income for necessities such as food, education, healthcare, and housing. Traditionally dominated by banks and money transfer operators like Western Union and MoneyGram, remittance solutions have evolved significantly with technological advancements. These innovations have made transactions faster, more affordable, and accessible to underserved populations, driving financial inclusion and economic resilience. Initially, remittance services relied on bank branches and physical outlets, often requiring both senders and recipients to have formal bank accounts. However, the rise of fintech solutions has introduced digital platforms, mobile money services, and blockchain-based transactions, improving efficiency and reducing costs. Users can now send and receive remittances through mobile apps, digital wallets, agent networks, and even social media platforms, reducing dependency on traditional banking infrastructure. The adoption of mobile money services has been particularly transformative, enabling instant cross-border transfers without requiring a bank account, significantly increasing financial inclusion in rural and remote areas.

The global remittance market has experienced sustained growth, driven by increasing migration from developing to developed nations. According to the World Bank (2023), remittances to low- and middle-income countries reached a record $630 billion in 2022, highlighting their importance in household income support. The COVID-19 pandemic accelerated the shift towards digital remittances, as lockdowns and restrictions underscored the need for contactless and efficient financial solutions. Digital remittance platforms have disrupted the market by offering faster transactions, lower fees, and greater transparency. Companies such as Wise (formerly TransferWise), Remitly, and WorldRemit leverage digital technology to bypass intermediary banks, reducing costs and ensuring that recipients receive a larger share of the transferred funds. Additionally, real-time exchange rate comparisons provided by these platforms enable users to make more informed financial decisions. Blockchain technology is playing an increasingly significant role in cross-border remittances, with platforms like Ripple utilising decentralised ledger systems to facilitate secure, near-instant transfers at minimal fees. However, regulatory concerns and varying adoption levels across different countries present ongoing challenges to widespread implementation. Overall, remittance solutions have undergone a significant transformation, with digital and blockchain-based innovations reshaping the way people transfer money across borders. While traditional methods remain relevant, the transition to mobile and digital platforms has vastly improved affordability, efficiency, and accessibility. Overcoming regulatory barriers, enhancing digital infrastructure, and fostering consumer trust will be critical in unlocking the full potential of remittance solutions and furthering financial inclusion worldwide.

Challenges in Remittance Solutions and the Role of Impact Investment

Several challenges hinder their remittance solutions efficiency, accessibility, and affordability. These barriers affect both senders and recipients, particularly in developing countries where remittances serve as a crucial economic lifeline. Addressing these challenges is essential for improving financial inclusion and maximising the benefits of remittance flows.

High Transaction Costs: One of the most persistent challenges in remittance services is the high cost of transactions. Traditional money transfer operators and banks often impose high fees, which can significantly reduce the actual amount received by beneficiaries. These fees typically include service charges, foreign exchange markups, and correspondent banking costs, all of which add up to make cross-border transfers expensive. While digital and blockchain-based solutions have introduced lower-cost alternatives, they still face regulatory costs, compliance expenses, and network maintenance charges that prevent widespread affordability. According to the World Bank, the global average cost of sending remittances remains above the UN Sustainable Development Goal target of 3 per cent, disproportionately affecting low-income individuals who rely on these funds for basic necessities. Reducing transaction costs requires innovative approaches, including the use of decentralised finance, direct peer-to-peer payment systems, and streamlined cross-border financial regulations. Impact investors can support fintech companies and startups working on innovative technologies that reduce transaction fees, such as blockchain-based platforms that provide low-cost, secure remittance services. By funding these solutions, impact investors can help make remittance services more affordable for everyone, especially in high-cost corridors like Sub-Saharan Africa.

Regulatory and Compliance Barriers: Cross-border remittance services are heavily regulated to prevent money laundering, fraud, and financing of illicit activities. While these regulations are necessary, they often create bureaucratic obstacles that slow down transactions and increase costs. Strict Know Your Customer (KYC) and Anti-Money Laundering (AML) policies can make it difficult for unbanked individuals to access remittance services, particularly in regions where obtaining formal identification is challenging. Many low-income populations lack the necessary documentation to comply with KYC requirements, leading to financial exclusion. Additionally, varying regulatory frameworks across countries complicate international remittance flows, limiting interoperability and increasing compliance burdens for service providers. Achieving regulatory harmonisation and creating more inclusive identification frameworks can help address these challenges and improve access to remittance services. Impact investors can support policy advocacy efforts to create a more harmonised regulatory environment for remittances. They can work with regulators and industry stakeholders to help establish clear guidelines that protect consumers while encouraging innovation in the remittance sector.

Limited Digital Infrastructure: Although digital remittance platforms have significantly improved access to financial services, infrastructure limitations in many developing countries hinder their effectiveness. Poor internet connectivity, inadequate mobile penetration, and a lack of reliable financial service providers create barriers for recipients who rely on these solutions. Many remote communities do not have access to stable digital networks, making it difficult to send or receive remittances through mobile money platforms or digital wallets. Even when digital remittance solutions are available, the lack of widespread mobile device ownership and financial literacy can prevent individuals from fully utilising these services. Expanding internet connectivity, investing in mobile banking infrastructure, and improving digital literacy programmes are essential to making digital remittance solutions more accessible. Impact investments can help bridge this gap by funding infrastructure projects that expand mobile network coverage and improve internet connectivity, to ensure that remittance solutions are accessible to underserved populations, enabling them to receive funds quickly and securely.

Currency Volatility and Exchange Rate Fluctuations: Exchange rate volatility presents another major challenge in remittance transactions. Fluctuations in currency values can significantly reduce the purchasing power of recipients, particularly in economies with unstable exchange rates. Traditional remittance providers and banks often offer less favourable exchange rates, increasing the hidden costs of cross-border transfers. Additionally, unpredictable currency depreciation can erode the real value of remitted funds, making it difficult for recipients to plan their finances effectively. Blockchain-based solutions and cryptocurrency-backed remittances offer potential alternatives, as they can provide more stable and transparent exchange mechanisms. However, regulatory uncertainty, low adoption rates, and price volatility in cryptocurrencies pose significant hurdles to their mainstream use.

Fraud and Security Concerns: The remittance industry is a frequent target for fraud, cyber threats, and identity theft. Digital platforms, in particular, face challenges in ensuring secure transactions, as hackers and fraudsters attempt to exploit system vulnerabilities. Phishing scams, SIM card fraud, and account takeovers are common issues that can result in the loss of remitted funds. Additionally, informal remittance channels, such as unregulated hawala networks, pose risks of illicit activity and financial losses for users. Many low-income individuals are particularly vulnerable to scams due to limited financial literacy and a lack of awareness about cybersecurity best practices. Ensuring robust cybersecurity measures, consumer protection policies, and fraud prevention strategies is critical to maintaining trust in digital remittance solutions. Investment in biometric authentication, artificial intelligence-driven fraud detection, and encrypted transactions can enhance security and reduce financial risks.

Accessibility and Financial Literacy: While digital remittance services offer convenience, many recipients, particularly in rural areas, face challenges in accessing and using these platforms due to low financial literacy. Limited understanding of digital payments, banking systems, and security risks can prevent individuals from fully utilising available services. In many cases, remittance recipients continue to rely on cash-based transactions due to unfamiliarity with digital banking. Educational initiatives aimed at improving financial literacy, particularly for first-time users, are essential for bridging this gap. Government agencies, NGOs, and financial institutions can collaborate to provide tailored financial education programmes, including workshops, mobile-based learning tools, and peer-to-peer support networks. Impact investors can support financial literacy initiatives that educate individuals on the benefits of digital remittance solutions, how to use them, and how to manage their funds effectively.

Investment in educational campaigns, community outreach programmes, and partnerships with local organisations can help increase the adoption of digital remittance solutions, particularly in underserved regions like Sub-Saharan Africa and Latin America.

Dependence on Traditional Banking Systems: Despite innovations in fintech and blockchain, traditional banking systems still play a dominant role in processing remittances. Many digital solutions rely on existing banking networks for settlements and withdrawals, which can delay transactions and maintain high costs. Additionally, in countries with underdeveloped banking infrastructure, access to formal financial services remains a significant challenge. The lack of banking penetration forces many individuals to rely on intermediaries for cash-out services, adding additional costs and inefficiencies to the remittance process. Expanding alternative banking models, such as agent banking and mobile-based financial services, can help reduce dependency on traditional banks and improve remittance accessibility.

Platform-Based Ecosystems

Platform-Based Ecosystems, such as crowdfunding platforms, democratise access to finance. These platforms connect micro-entrepreneurs directly with investors, bypassing traditional financial intermediaries. Crowdfunding platforms is a powerful tool for raising funds, enabling individuals, startups, small businesses, and social causes to access capital from a large pool of people, typically via the internet. It leverages the power of digital technology and social networks to democratised access to finance, offering an alternative to traditional funding sources like banks, venture capital, and angel investors. These platforms have become especially popular for raising funds for innovative projects, creative endeavours, and social causes, as they allow anyone with a compelling idea to pitch directly to potential backers. Crowdfunding has evolved significantly since its inception, moving from small-scale community fundraising to a sophisticated global industry. It has been driven by advancements in technology, particularly the rise of social media, which has made it easier for project creators to reach a wider audience and build communities of supporters. Social media platforms like Facebook, Twitter, and Instagram have become essential tools for promoting crowdfunding campaigns, enabling creators to generate buzz, share updates, and engage with backers in real time. The introduction of blockchain technology and decentralised finance (DeFi) is transforming the financial landscape. In particular, blockchain-based crowdfunding platforms are emerging as innovative alternatives that enhance transparency, improve security, and reduce transaction costs. Additionally, the COVID-19 pandemic accelerated the adoption of digital crowdfunding as traditional fundraising methods became less feasible.

According to the Fortune Business Insights (2025), The global crowdfunding market size was valued at USD 1.60 billion in 2024 and is projected to grow to USD 4.50 billion by 2032, exhibiting a CAGR of 13.8 per cent during the forecast period. The proliferation of internet access, coupled with the increasing willingness of individuals to support projects online, has contributed to the expansion of crowdfunding platforms as a popular financing option. Crowdfunding offers a range of funding models to suit different needs. It is used by startups and creators and social enterprises, charitable organisations, and individuals in need of financial assistance. The growing popularity of crowdfunding reflects a broader trend towards democratising access to finance, empowering individuals and communities to take control of their financial futures. There are several types of crowdfunding models, each serving different purposes.

Reward-Based Crowdfunding: In reward-based crowdfunding, backers contribute money to a project in return for a non-financial reward. The reward often depends on the level of contribution and can range from a simple thank-you note to a prototype or an exclusive product. This model is popular for creative projects, new product launches, and artistic endeavours. Platforms like Kickstarter and Indiegogo are well-known for using this model. It is an effective way for creators to raise funds while building a community of early adopters and supporters. The concept relies heavily on emotional connection and excitement about a product or project. However, managing backers' expectations is crucial, as delays or changes in rewards can lead to dissatisfaction and harm the creator's reputation.

Equity Crowdfunding: Equity crowdfunding allows individuals to invest in a startup or business in exchange for equity or ownership shares. Unlike reward-based crowdfunding, backers receive a financial return if the company succeeds. This model has become a popular alternative to venture capital for early-stage startups. Platforms such as Seedrs, Crowdcube, and Republic facilitate equity crowdfunding, making it accessible to retail investors who were previously excluded from startup investments. The model democratises investing, allowing everyday individuals to own a stake in potentially high-growth businesses. While it offers significant upside potential, it also involves substantial risks, as many startups fail to succeed. Regulatory requirements in different countries add complexity to this model, often requiring platforms to comply with stringent securities laws to protect investors.

Debt-Based Crowdfunding (Peer-to-Peer Lending): Debt-based crowdfunding, also known as peer-to-peer (P2P) lending, allows individuals to lend money to others or to small businesses in return for interest payments. Platforms like LendingClub and Funding Circle connect borrowers directly with lenders, providing an alternative to traditional bank loans. This model benefits borrowers by offering more flexible terms and lower interest rates compared to traditional lenders. Lenders, on the other hand, receive interest payments, often with better returns than traditional savings accounts. Debt-based crowdfunding has been particularly useful in regions with limited access to formal credit. However, it comes with risks, including potential defaults. To mitigate these risks, platforms usually conduct credit assessments to determine the borrower's creditworthiness.

Donation-Based Crowdfunding: In donation-based crowdfunding, individuals donate money to support a cause or project without expecting any financial return. This model is used extensively for charitable projects, medical expenses, disaster relief, community initiatives, and other social causes. GoFundMe and JustGiving are leading platforms in this space. Donation-based crowdfunding relies on emotional appeal, with campaigns often gaining traction when they resonate personally with donors. It is highly effective for raising money quickly for emergencies or impactful community projects. However, it requires a compelling story and significant outreach to reach potential donors, as people are often motivated by empathy and a desire to make a difference.

Hybrid Crowdfunding: Hybrid crowdfunding combines elements of multiple crowdfunding models to create greater flexibility. For instance, a campaign may offer backers both reward-based incentives and equity, allowing them to choose based on their preferences. Hybrid models are beneficial for projects that aim to create both financial and social impact, making them attractive to different types of supporters. Fundable is an example of a platform that provides both reward-based and equity crowdfunding. Hybrid crowdfunding offers versatility, making it suitable for a wide range of projects, but it also adds complexity, requiring project creators to manage different types of backers with varying expectations.

How Crowdfunding Promotes Financial Inclusion

Crowdfunding has significant potential to promote financial inclusion by offering an alternative means of accessing capital, particularly for individuals and businesses that are underserved by traditional financial institutions. The accessibility of crowdfunding platforms allows entrepreneurs, small businesses, and social enterprises to reach a wide range of potential supporters without the barriers typically associated with banks and venture capital.

Access to Capital for the Underserved: Crowdfunding enables individuals and businesses that may be excluded from traditional financing options—such as those without a credit history or collateral—to raise funds. This is particularly important for marginalised groups, including women, rural entrepreneurs, and minority-owned businesses, who often face challenges in securing bank loans. By providing a platform where anyone can pitch their ideas and receive funding, crowdfunding democratises access to finance, allowing a wider range of people to participate in economic opportunities.

Supporting Micro and Small Enterprises: Micro, Small, and Medium Enterprises (MSMEs) are vital to economic growth and job creation, yet they often struggle to secure financing from traditional sources. Crowdfunding platforms provide MSMEs with an alternative way to access capital, helping them grow and expand their operations. By enabling small businesses to bypass conventional lenders and connect directly with supporters, crowdfunding contributes to the growth of local economies and job creation, promoting financial inclusion at the grassroots level.

Financial Empowerment Through Community Support: Crowdfunding also promotes financial inclusion by fostering a sense of community and collective financial empowerment. Individuals can support causes and projects they believe in, contributing to local development initiatives and social enterprises. This community-driven approach helps channel funds to areas where traditional financial institutions may not see profitability, thereby addressing social and economic challenges in underserved communities.

Financial Literacy and Awareness: Participating in crowdfunding campaigns, either as a creator or a backer, can also contribute to financial literacy. Project creators learn how to budget, manage campaigns, and communicate with supporters, while backers gain exposure to different types of investment opportunities. This increased financial awareness helps foster a more financially informed population, which is an essential component of financial inclusion.

Leveraging Technology for Inclusion: Crowdfunding platforms leverage technology to provide financial services to people who may not have access to physical banking infrastructure. This is particularly relevant in developing countries where access to banks is limited. Mobile-friendly platforms, social media integration, and digital payment options make it easier for individuals to participate in crowdfunding, regardless of their location. This technological reach allows crowdfunding to serve as a bridge for those who are otherwise excluded from the formal financial system.

Successful Case Studies and Examples

M-Changa (Kenya): M-Changa is a mobile-based crowdfunding platform in Kenya that allows individuals and communities to raise funds for social causes, medical expenses, and other community needs. M-Changa leverages mobile money, which is widely adopted in Kenya, making it easy for users to contribute and manage campaigns using their mobile phones. The platform's integration with mobile money services like M-Pesa has made it accessible to a broad population, particularly those who are unbanked. M-Changa has enabled thousands of Kenyans to raise funds for emergencies, education, and social initiatives, promoting financial inclusion by allowing people to access funds without traditional banking infrastructure.

Kickstarter (Global): Kickstarter is one of the most well-known reward-based crowdfunding platforms, having raised over $6 billion for creative projects since its inception in 2009. Kickstarter has supported a wide range of projects, from tech gadgets to films, art, and music. The platform's success lies in its ability to connect creators with a global audience, allowing them to validate their ideas, gather feedback, and build a community of supporters before launching their products. Notable projects funded on Kickstarter include the Pebble smartwatch, Oculus Rift, and Exploding Kittens card game.

Kiva (Global): Kiva is a donation-based and debt-based crowdfunding platform that provides microloans to underserved communities around the world. Kiva allows individuals to lend small amounts of money to entrepreneurs and small businesses in developing countries, helping them grow their enterprises and improve their livelihoods. Since its launch in 2005, Kiva has facilitated over $1.5 billion in loans to more than 3 million borrowers, demonstrating the power of crowdfunding to promote financial inclusion and empower communities in need.

GoFundMe (Global): GoFundMe is a donation-based crowdfunding platform that has become a go-to resource for individuals seeking to raise money for personal emergencies, medical expenses, and charitable causes. Since its launch in 2010, GoFundMe has helped raise over $15 billion from more than 200 million donations. The platform's success is largely due to its ease of use, emotional appeal, and ability to connect people with causes they care about.

Seedrs (UK/Europe): Seedrs is an equity crowdfunding platform that allows startups to raise capital by offering equity to individual investors. Since its launch, Seedrs has helped hundreds of startups raise millions of dollars in funding, democratising access to investment opportunities for retail investors. Seedrs has been particularly successful in Europe, where regulatory frameworks are more favourable for equity crowdfunding than other regions. The platform has enabled everyday investors to participate in early-stage funding rounds, contributing to the growth of the startup ecosystem.

Addressing Gaps in Crowdfunding Platforms Through Impact Investment

Crowdfunding platforms have revolutionised access to finance by enabling individuals, startups, and social enterprises to raise capital directly from a broad base of contributors. However, despite these advancements, several challenges limit their effectiveness, particularly in underserved regions and among marginalised groups. Impact investing can play a transformative role in addressing these gaps by providing the necessary capital and strategic support to scale crowdfunding platforms and enhance their inclusivity.

Limited Access in Underserved Regions: Crowdfunding platforms remain inaccessible in many underserved regions, including parts of Africa, South Asia, and Latin America, where internet penetration is low, digital payment systems are underdeveloped, and financial literacy is limited. Many potential campaign creators and backers in these areas lack the infrastructure or knowledge to engage effectively with crowdfunding, limiting the reach of these platforms.

Impact investors can bridge this gap by funding initiatives that expand digital infrastructure and mobile banking solutions, making it easier for people in remote areas to participate in crowdfunding campaigns. Investments in internet connectivity, local fintech partnerships, and digital financial literacy programmes can significantly increase participation rates. Additionally, impact investors can facilitate partnerships between crowdfunding platforms and local organisations to enhance awareness, trust, and engagement in these regions.

Financial Literacy and Campaign Support: Many entrepreneurs and individuals lack the financial literacy and skills needed to run a successful crowdfunding campaign. Crafting a compelling campaign requires knowledge of marketing, budgeting, storytelling, and audience engagement, which can be challenging for first-time fundraisers or small businesses with limited resources. Impact investors can fund educational initiatives, including online training programmes, mentorship opportunities, and localised workshops, to equip individuals with the skills needed to create and manage successful campaigns. By supporting capacity-building programmes, investors can empower marginalised communities to leverage crowdfunding effectively, leading to greater financial inclusion and economic empowerment.

Regulatory Barriers: Regulatory challenges hinder the growth of crowdfunding, particularly equity crowdfunding, which allows individuals to invest in startups in exchange for shares. In many countries, restrictive financial regulations limit the participation of retail investors, making it difficult for startups to access funding through these platforms. The absence of clear regulatory frameworks also creates uncertainty for crowdfunding platforms, discouraging potential investors and limiting market growth. Impact investors can play a critical role in policy advocacy by working with governments and regulatory bodies to develop balanced frameworks that protect investors while encouraging innovation. By funding research, stakeholder engagement, and legal advocacy initiatives, investors can contribute to a more enabling environment for crowdfunding platforms to thrive. Additionally, investments in compliance technology can help platforms navigate regulatory complexities more efficiently.

High Platform Fees and Cost Barriers: Platform fees can be a major barrier for project creators, particularly for those raising small amounts of capital. Transaction fees, processing charges, and service costs can reduce the net funds raised, discouraging small businesses and social enterprises from using crowdfunding as a financing option. Impact investors can support the development of low-cost or non-profit crowdfunding platforms that reduce fees for creators, ensuring more of the funds raised go directly to the intended projects. Investments in technological innovations such as blockchain-based payment processing can also help reduce operational costs, allowing platforms to lower fees without compromising service quality.

Risk Mitigation and Trust Issues: Trust remains a major challenge in crowdfunding, as backers often hesitate to contribute due to concerns about fraud, project failure, or mismanagement of funds. The lack of effective due diligence and transparency mechanisms makes it difficult for backers to assess the credibility of campaigns, limiting participation and funding potential. Impact investors can fund the development of advanced risk assessment tools that use artificial intelligence and blockchain technology to enhance transparency and accountability. Investments in fraud detection systems, escrow services, and impact verification models can strengthen trust in crowdfunding platforms. By improving due diligence mechanisms, investors can help create a more secure and reliable crowdfunding ecosystem that attracts more backers and larger investments.

While crowdfunding has democratised access to capital, key challenges continue to limit its impact, particularly in underserved regions. Impact investing presents a powerful solution by addressing gaps in digital infrastructure, financial literacy, regulatory frameworks, cost efficiency, and trust mechanisms. By strategically deploying capital to scale and enhance crowdfunding platforms, impact investors can unlock new opportunities for entrepreneurs, social enterprises, and marginalised communities, driving financial inclusion and economic development.

Pay-As-You-Go (PAYG) Models

Pay-As-You-Go (PAYG) models are transforming financial inclusion by enabling low-income consumers to access essential goods and services through incremental payments. By removing the requirement for large upfront costs, these models democratise access to products such as solar energy, clean cooking solutions, and mobile technology. The integration of digital payment platforms, particularly mobile money, has further expanded the reach of PAYG solutions, making them a powerful tool for addressing financial and energy exclusion in underserved communities. The PAYG models allow consumers to pay for products or services in small, manageable instalments, often using mobile payment platforms. The structure of these payments varies some models follow a rent-to-own system where users eventually gain full ownership, while others operate on a subscription basis. For example, in the PAYG solar energy sector, companies provide households with solar home systems that customers activate through regular payments. If payments stop, access to the service is temporarily restricted until payments resume. Digital financial services play a critical role in facilitating PAYG models. With mobile penetration increasing across developing regions, mobile money platforms such as M-Pesa, Airtel Money, and EcoCash have enabled seamless transactions for PAYG services. By integrating digital finance with product accessibility, PAYG models bridge financial barriers while fostering digital payment adoption among low-income users.

a. **Impact of PAYG Models on Financial Inclusion**

   Expanding Access to Essential Services: One of the most significant contributions of PAYG models is their ability to extend access to essential services, particularly for low-income populations. Traditional financing methods often exclude individuals with limited credit history or irregular income. PAYG models remove this barrier by offering flexible payment structures, allowing consumers to pay as they use the service without requiring collateral or formal banking relationships. For example, PAYG solar solutions have been instrumental in promoting energy inclusion. In many rural areas, unreliable grid access limits economic opportunities and quality of life. PAYG solar companies such as M-KOPA, Off-Grid Electric, and BBOXX provide affordable, off-grid energy solutions that customers can pay for over time. As a result, families gain access to clean electricity, reducing reliance on expensive and polluting alternatives like kerosene lamps.

Lowering Financial Barriers: Upfront costs often prevent low-income consumers from acquiring essential goods. PAYG models eliminate this financial hurdle by spreading payments over time, making products and services more affordable. This approach allows consumers to access life-enhancing goods without the need for traditional credit or loans, which are often unavailable to unbanked individuals. Additionally, PAYG transactions create digital financial records that can help customers build a financial identity. By establishing a history of consistent payments, PAYG users may become eligible for other financial services, including microloans, savings products, and insurance, thereby fostering broader financial inclusion.

Promoting Digital Financial Services Adoption: PAYG models encourage the adoption of digital financial services, particularly mobile money. Since most PAYG platforms integrate mobile payments, users become familiar with digital transactions, increasing financial literacy and participation in the formal financial system. This shift is especially crucial in rural areas, where mobile-based financial tools serve as a gateway to banking services for previously excluded populations. For example, PAYG solar providers often partner with mobile network operators to streamline payment processing. As users engage in frequent digital transactions, they become more comfortable using mobile banking for other financial needs, such as savings and remittances. This behavioural shift contributes to the long-term integration of digital finance into everyday economic activities.

Enhancing Credit Access for Underserved Populations: Since PAYG transactions generate reliable payment records, they provide an alternative form of credit assessment for individuals without traditional credit histories. Lenders and financial institutions can analyse PAYG repayment data to determine a customer's creditworthiness, enabling access to microloans and other financial products. In some cases, PAYG providers themselves extend credit to customers who maintain good repayment records, further integrating low-income individuals into the financial system. For instance, M-KOPA uses PAYG transaction data to offer additional financial products, including loans for smartphones and agricultural inputs. By leveraging technology to assess risk, PAYG models create new opportunities for financial inclusion beyond their immediate service offerings.

b. **Addressing Challenges with PAYG Models Through Impact Investment**

Affordability of PAYG Services: While PAYG models reduce the need for large upfront payments, they often lead to higher total costs over time compared to outright purchases. The financing model includes interest or service fees that can make essential products more expensive for low-income consumers. If not carefully structured, PAYG pricing can undermine affordability, limiting its intended impact on financial inclusion. Impact investors can support PAYG providers in developing fair and transparent pricing structures, ensuring that repayment terms are aligned with consumers' income levels. Investments can subsidise for low-interest credit facilities to help make PAYG solutions more affordable. Additionally, investments in cost-reducing technologies, such as more efficient payment processing systems and supply chain optimisation, can lower overall operational expenses and reduce consumer costs.

Regulatory Barriers: In many regions, regulatory frameworks for digital payments, mobile money, and PAYG financing remain underdeveloped or overly restrictive. This uncertainty can hinder the expansion of PAYG services, limiting access to essential goods and financial inclusion for underserved populations. Impact investors can support policy advocacy efforts to develop enabling regulatory environments for PAYG models. By funding research, stakeholder engagement, and legal support, investors can help create balanced regulations that protect consumers while allowing PAYG innovation to thrive. Additionally, investments in compliance technology can help providers navigate complex regulatory landscapes and scale their services responsibly.

Payment Default Risks: Low-income consumers often experience seasonal income fluctuations and financial instability, making it difficult to maintain consistent payments. Default risks are particularly high in rural areas where employment opportunities are irregular. When customers default, they lose access to essential services such as electricity, affecting their quality of life. Funding flexible repayment solutions that align with customers' cash flow patterns can mitigate this risk. Also supporting the development of alternative credit-scoring models based on seasonal income can improve repayment flexibility. Impact investors can also help PAYG providers integrate safety nets such as temporary payment suspensions or emergency credit extensions to reduce service interruptions.

Infrastructure Limitations: In areas with weak mobile network coverage and unreliable digital payment systems, PAYG services face significant operational challenges. Limited connectivity prevents seamless transactions, while inadequate financial infrastructure hampers digital payment adoption. Impact investors can finance infrastructure development projects that expand mobile network coverage and improve digital payment reliability in rural areas. Investments in last-mile connectivity solutions, satellite-based communication networks, and fintech partnerships can enhance the effectiveness of PAYG models, ensuring uninterrupted service access.

6.5 Conclusion

It is established that impact investing can be a meaningful driver of financial inclusion, playing a transformative role in reshaping financial systems to be more equitable, resilient, and accessible. Beyond serving as a mechanism for capital allocation, impact investing is instrumental in addressing the structural inefficiencies that perpetuate financial exclusion, particularly for underserved populations. Impact investors contribute to the creation of sustainable economic opportunities, enabling individuals and communities to build financial resilience and long-term prosperity, by directing investment capital towards inclusive financial services. A central insight from this discussion is that financial inclusion is not merely about providing access to financial products but about integrating marginalised populations into the broader economy in ways that enhance economic participation and agency. Impact investing, when effectively deployed, advances multiple developmental objectives, including poverty alleviation, gender equity, and inclusive economic growth. These objectives align directly with the Sustainable Development Goals (SDGs), reinforcing the role of financial inclusion as a cornerstone of sustainable development.

The persistence of financial exclusion among rural populations, informal workers, and women underscores the need for targeted strategies that address infrastructure gaps, regulatory constraints, and disparities in financial literacy. Addressing these challenges requires a coordinated, multi-stakeholder approach that brings together impact investors, governments, financial institutions, and technology providers. By fostering collaboration and aligning investments with local economic realities, impact investing can create more inclusive financial ecosystems that serve the needs of diverse populations. Looking forward, the effectiveness of impact investing in financial inclusion will depend on its ability to scale while maintaining a strong commitment to social impact. This will require greater adoption of blended finance models, improved risk-sharing mechanisms, and more rigorous impact measurement frameworks to ensure accountability and effectiveness. The long-term success of financial inclusion efforts depends on strategic alignment with local socio-economic conditions and continuous adaptation to emerging challenges and opportunities. Financial inclusion must be embedded within broader economic policies and development strategies to create lasting, transformative impact. By leveraging capital for inclusive growth, impact investors can drive meaningful change, ensuring that financial services are not a privilege for the few but a fundamental right for all. The path forward demands financial commitment and a renewed emphasis on equity, sustainability, and long-term value creation to foster a more just and inclusive global financial system.

Notes

1. See empirical studies such as Adegbite and Machethe (2022), Akande et al. (2023), and Mahato and Jha (2023).
2. See Donovan (2012), also GSMA (2010).
3. Suri and Jack (2016), Okello and Ntayi (2020), and Abiona and Koppensteiner (2022).
4. Mothobi et al. (2018), Koo and Kim (2023), Safari et al. (2023), Dorfleitner and Nguyen (2024).

5. Several published studies, industry reports, and case analyses have documented the case of M-Pesa, for example Eijkman et al. (2010), Mbiti and Weil (2015), Rosengard (2016), Winn (2016), and Kingiri and Fu (2020).
6. Hughes and Lonie (2007) and Jack and Suri (2011) documented the rapid adoption of M-Pesa due to its ability to fill this financial gap.
7. Porteous (2006) analysed the various regulatory conditions that would enable mobile money services.
8. See https://tala.co/; https://branch.co/.
9. See https://www.kiva.org/; https://www.upstart.com/.
10. See https://lenddoefl.com/; https://www.credolab.com/.
11. Also see case study in Tinsley and Agapitova (2018).
12. See Yunus (2007).
13. This is extensively documented in development finance literature for example see Mukendi and Manda (2022), Ingale (2024), and ALIAS (2024).
14. See Kumar (2006) for detail discussion.

References

Abiona, O., & Koppensteiner, M. F. (2022). Financial inclusion, shocks, and poverty: Evidence from the expansion of mobile money in Tanzania. *Journal of Human Resources, 57*(2), 435–464.

Adegbite, O., & Machethe, C. L. (2022). The impact of financial inclusion on the livelihoods of rural smallholder farmers in Nigeria. *African Finance Journal, 24*(2), 38–50.

Alliance for Financial Inclusion. (2024). *2023 Global Policy Forum report*. Alliance for Financial Inclusion.

Akande, J., Hosu, Y., Kabiti, H., Ndhleve, S., & Garidzirai, R. (2023). Financial literacy and inclusion for rural agrarian change and sustainable livelihood in the Eastern Cape, South Africa. *Heliyon, 9*(6), e16330.

Alias, N. A. L. (2024). *The contribution of microfinance institutions to personal financial management*. Available at SSRN 4845831.

Babalola, F. I., Mhlongo, N. Z., Obinyeluaku, M. I., Oladayo, G. O., & Darao-jimba, C. (2023). Microfinance and economic empowerment in Nigeria: A critical review of impact and sustainability. *Finance & Accounting Research Journal, 5*(12), 381–404.

Babatunde, R. (2018). *Impact of remittances on food security and nutrition of migrant s household: Evidence from Nigeria.*

Bauchet, J., Marshall, C., Starita, L., Thomas, J., & Yalouris, A. (2011). *Latest findings from randomized evaluations of microfinance, consultative group to assist the poor*. Washington D.C.: World Bank.

Banerjee, S. B., & Jackson, L. (2017). Microfinance and the business of poverty reduction: Critical perspectives from rural Bangladesh. *Human Relations, 70*(1), 63–91. https://doi.org/10.1177/0018726716640865

Baptista, P., & Heitmann, S. (2010). *Unleashing the power of convergence to advance mobile money ecosystems*. IFC and the Harvard Kennedy School.

Borges, P. (2007). *Women empowered: Inspiring change in the emerging world.* Rizzoli International Publications.

Bylander, M., Jacoby, L., Bradley, P., & Pérez, A. B. (2019). Over-indebtedness and microcredit in Cambodia: Moving beyond borrower-centric frames. *Development Policy Review, 37*, O140–O160.

CGAP. (2020). *CGAP launches global microfinance survey to inform COVID-19 response.*

CGAP. (2023). *Building value in microfinance through digitization: Lessons from loan automation pilots*. Consultative Group to Assist the Poor. https://www.cgap.org/research/publication/building-value-in-microfinance-through-digitization-lessons-loan-automation

Christen, R. P., Lauer, K., Lyman, T., & Rosenberg, R. (2012). *A guide to regulation and supervision of microfinance* (p. 128). Consultative Group to Assist the Poor (CGAP).

Croxson, K., Frost, J., Gambacorta, L., & Valletti, T. (2022). Platform-based business models and financial inclusion: Policy trade-offs and approaches1. *Journal of Competition Law & Economics, 19*(1), 75–102. https://doi.org/10.1093/joclec/nhac010

Cull, R., Asli Demirgüç-Kunt, A., & Morduch, J. (2009). Microfinance Meets the Market. *Journal of Economic Perspectives, 23*(1), 167–192. https://doi.org/10.1257/jep.23.1.167

Cull, R., Demirguc-Kunt, A., & Morduch, J. (2021). *Banking the world: Empirical foundations of financial inclusion.* MIT Press.

Demirgüç-Kunt, A., Klapper, L., Singer, D., Ansar, S., & Hess, J. (2018). The Global Findex Database 2017: Measuring Financial Inclusion and the Fintech Revolution. *The World Bank.* https://globalfindex.worldbank.org/

Deloitte. (2021). *Accelerating toward greater financial inclusion.* https://www2.deloitte.com/us/en/insights/industry/financial-services/alternative-data-innovation-financial-inclusion.html

Deloitte. (2023). *Tech trends 2023.* Deloitte Insights.

Donovan, K. (2012). Mobile money for financial inclusion [Article|Book Chapter]. *Information and communications for development 2012: Maximizing mobile, 61*(1), 61–73.

Dorfleitner, G., & Nguyen, Q. A. (2024, July 01). Mobile money for women's economic empowerment: The mediating role of financial management practices. *Review of Managerial Science, 18*(7), 1807–1836. https://doi.org/10.1007/s11846-022-00564-2

Duvendack, M., Palmer-Jones, R., Copestake, J., Hooper, L., Loke, Y., & Rao, N. (2011). *What is the evidence of the impact of microfinance on the well-being of poor people?* . EPPI-Centre, Social Science Research Unit, Institute of Education, University of London. https://www.givedirectly.org/pdf/DFID_microfinance_evidence_review.pdf

Eijkman, F., Kendall, J., & Mas, I. (2010). Bridges to cash: The retail end of M-PESA. *Savings and Development, 34*(2), 219–252.

Equity Group Holdings PLC. (2023). *Integrated financial report 2023.* https://equitygroupholdings.com/investor-relations/annual-reports

Fareed, F., Gabriel, M., Lenain, P., & Reynaud, J. (2017). *Financial inclusion and women entrepreneurship: Evidence from Mexico.* OECD.

Fortune Business Insights. (2025). *Crowdfunding market size, share & industry analysis, by type (equity-based, debt-based, blockchain-based, and others), by end-user (startups, ngos, and individuals), and regional forecast, 2024–2032.* https://www.fortunebusinessinsights.com/crowdfunding-market-107129

Fourmy, R. (2020). The impact of mobile money on remittance recipients' household welfare and education: Evidence from Kenya. In *Migration, remittances, and sustainable development in Africa* (pp. 157–176). Routledge.

Gas, S. (2017, July 15). Mobile money, cashless society and financial inclusion: case study on somalia and kenya. *Cashless Society and Financial Inclusion: Case Study on Somalia and Kenya.*

Ghosh, I. (2018). *Challenging the dominant narratives of a digital financial inclusion.* University of California, Berkeley.

GSMA. (2020). *The Mobile Economy 2020.* GSMA. https://www.gsma.com/mobileeconomy/wp-content/uploads/2020/03/GSMA_MobileEconomy2020_Global.pdf

GSMA. (2022). *The Mobile Economy 2022.* GSMA. https://www.gsma.com/mobileeconomy/wp-content/uploads/2022/02/280222-The-Mobile-Economy-2022.pdf

GSMA. (2023). *The Mobile Economy 2023.* GSMA. https://www.gsma.com/mobileeconomy/wp-content/uploads/2023/03/270223-The-Mobile-Economy-2023.pdf

GSMA. (2024). *State of the industry report on mobile money 2024.* https://www.gsma.com/sotir/

Guérin, I., & Venkatasubramanian, G. (2022). The socio-economy of debt. Revisiting debt bondage in times of financialization. *Geoforum, 137*, 174–184. https://doi.org/https://doi.org/10.1016/j.geoforum.2020.05.020

Helwig, K., Hill-O'Connor, C., Mikulewicz, M., Mugiraneza, P., & Christensen, E. (2020). *The role of microfinance in climate change adaptation: Evidence from rural Rwanda*. Glasgow Caledonian University.

Henry, C., Coleman, S., Orser, B., & Foss, L. (2022). Women's entrepreneurship policy and access to financial capital in different countries: An institutional perspective. *Entrepreneurship Research Journal, 12*(3), 227–262.

Hussain, R., Ahmad, R., & Fatima, Q. (2025). Impact of microfinance on agriculture and livestock production: Insights from Southern Punjab, Pakistan. *The Critical Review of Social Sciences Studies, 3*(1), 2126–2151.

Hughes, N., & Lonie, S. (2007). M-PESA: Mobile money for the "unbanked" turning cellphones into 24-hour tellers in Kenya. *Innovations: Technology, Governance, Globalization, 2*(1–2), 63–81. https://doi.org/10.1162/itgg.2007.2.1-2.63

International Finance Corporation. (2023). *IFC Annual Report 2023: Strategy in action–Building a better future*. World Bank Group. https://www.ifc.org/en/annual-report/2023

IFC. (2024a). *Microfinance*. https://www.ifc.org/en/what-we-do/sector-expertise/financial-institutions/archive-do-not-publish/msme-finance/microfinance

Ingale, A. (2024). *The role of microfinance in empowering women entrepreneurs*. Available at SSRN 5024302.

Jack, W., & Suri, T. (2011). *Mobile money: The economics of M-PESA*. NBER

Kikulwe, E. M., Fischer, E., & Qaim, M. (2014). Mobile money, smallholder farmers, and household welfare in Kenya. *PLoS ONE, 9*(10), Article e109804.

Kochar, A., Nagabhushana, C., Sarkar, R., Shah, R., & Singh, G. (2022). Financial access and women's role in household decisions: Empirical evidence from India's national rural livelihoods project. *Journal of Development Economics, 155*, 102821.

Konte, M., & Tetteh, G. K. (2023). Mobile money, traditional financial services and firm productivity in Africa. *Small Business Economics, 60*(2), 745–769.

Koo, Minji , & Kim, C.-S. (2023). The impact of digital transformation on financial inclusion: Through the case of mobile money account in senegal. *World Regional Studies Journal, 41*(3), 259–286. https://scholar.kyobobook.co.kr/article/detail/4010067732839

Koomson, I., Villano, R. A., & Hadley, D. (2020, June 01). Effect of financial inclusion on poverty and vulnerability to poverty: Evidence using a multidimensional measure of financial inclusion. *Social Indicators Research, 149*(2), 613–639. https://doi.org/10.1007/s11205-019-02263-0

Kumar, A. (2006). *Expanding bank outreach through retail partnerships: correspondent banking in Brazil*. World Bank Publications.

Kumar, S. S., & Jie, Q. (2023, December 01). Exploring the role of financial inclusion in poverty reduction: An empirical study. *World Development Sustainability, 3*, 100103. https://doi.org/10.1016/j.wds.2023.100103

Lal, T. (2021). Impact of financial inclusion on economic development of marginalized communities through the mediation of social and economic empowerment. *International Journal of Social Economics, 48*(12), 1768–1793. https://doi.org/10.1108/IJSE-12-2020-0830

Liu, J., Chen, Y., Chen, X., & Chen, B. (2024). Digital financial inclusion and household financial vulnerability: An empirical analysis of rural and urban disparities in China. *Heliyon, 10*(15). https://doi.org/10.1016/j.heliyon.2024.e35540

Mahato, J., & Jha, M. K. (2023). Does financial inclusion promote sustainable livelihood development? Mediating effect of microentrepreneurship. *Journal of Financial Economic Policy, 15*(4/5), 485–499.

Manford, M. (2022). *Crisis in microfinance institutions in Ghana a case study of selected microfinance institutions in the Ga East Municipality.* University of Education, Winneba.

Market.US. (2023). *Neobanking market.* https://market.us/report/neobanking-market/#utm_source=scoop.market.us&utm_medium=END_FLOATING_A2

Mbiti, I., & Weil, D. N. (2015). Mobile banking: The impact of M-Pesa in Kenya. In *African successes, Volume III: Modernization and development* (pp. 247–293). University of Chicago Press.

McKinsey. (2019). *Digital Identification: A key to inclusive growth.* https://www.mckinsey.com/capabilities/mckinsey-digital/our-insights/digital-identification-a-key-to-inclusive-growth

Moniruzzaman, M. (2022). The Impact of remittances on household food security: Evidence from a survey in Bangladesh. *Migration and Development, 11*(3), 352–371. https://doi.org/10.1080/21632324.2020.1787097

Morawczynski, O., & Pickens, M. (2012). *Poor people using mobile financial services: Observations on customer usage and impact from M-PESA.* World Bank.

Mothobi, O., Moshi, G., & Deen-Swarray, M. (2018). *Mobile money and financial inclusion in Sub-Saharan Africa countries.* CPR South.

Mukendi, S., & Manda, S. (2022). Micro-financial institutions and processes of women empowerment in Zambia. *World Development Perspectives, 28*, Article 100466.

Ndungu, N. (2021). *A digital financial services revolution in Kenya: The M-Pesa case study.* African Economic Research Consortium, Nairobi.

Okello, C. B. G., & Ntayi, J. M. (2020). Mobile money adoption and usage and financial inclusion: Mediating effect of digital consumer protection. *Digital*

Policy, Regulation and Governance, 22(3), 157–176. https://doi.org/10.1108/DPRG-01-2019-0005

Pattnaik, D., Ray, S., & Hassan, M. K. (2024, May 30). Microfinance: A bibliometric exploration of the knowledge landscape. *Heliyon, 10*(10), e31216. https://doi.org/10.1016/j.heliyon.2024.e31216

Porteous, D. (2006). *The enabling environment for mobile banking in Africa.*

Rosengard, J. K. (2016). *A quantum leap over high hurdles to financial inclusion: The mobile banking revolution in Kenya.* Harvard Kennedy School.

Sa-Dhan. (2022). *Promoting women-led enterprises: Microfinance leads the way conference on financial inclusion.*

Safari, L. I., Kilindo, A., & Nyoni, F. (2023). The effect of mobile money services usage on financial inclusion in nyamigamba village. *International Journal for Multidisciplinary Research, 5*(6). https://www.ijfmr.com/research-paper.php?id=9778

Suri, T., & Jack, W. (2016). The long-run poverty and gender impacts of mobile money. *Science, 354*, 1288–1292. https://doi.org/10.1126/science.aah5309

Thomas, D. (1990). Intra-household resource allocation: An inferential approach. *Journal of Human Resources, 25*(4), 635–664.

Tinsley, E., & Agapitova, N. (2018). *Private sector solutions to helping smallholders succeed: Social enterprise business models in the agriculture sector.* World Bank.

Tiwasing, P., Addae, J. A., Naab, F. Z., & Naab, G. Z. (2024). Do mobile money services enhance business performance? An empirical analysis of Kenyan businesses. *The International Journal of Entrepreneurship and Innovation*, 14657503241230946.

Uzma, S. H., & Pratihari, S. K. (2019). Financial modelling for business sustainability: A study of business correspondent model of financial inclusion in India. *Vikalpa, 44*(4), 211–231.

White, O., Madgavkar, A., Manyika, J., Mahajan, D., Bughin, J., McCarthy, M., & Spe, O. (2019). *Digital identification: A key to inclusive growth.* McKinsey.

White, O., Madgavkar, A., Townsend, Z., Manyika, J., Olanrewaju, T., Sibanda, T., & Kaufman, S. (2021). *Financial data unbound: The value of open data for individuals and institutions.* McKinsey Global Institute.

Winn, J. K. (2016). Mobile payments and financial inclusion: Kenya, Brazil, and India as case studies. In *Research handbook on electronic commerce law* (pp. 62–88). Edward Elgar Publishing.

World Bank. (2014). *Financial inclusion* (Global Financial Development Report). T. W. B. Group. https://openknowledge.worldbank.org/handle/10986/16238

World Bank. (2019). *Disruptive technologies in the credit information sharing industry: Developments and implications.* World Bank.

World Bank. (2023). *Digital progress and trends report 2023.* World Bank Group.

World Economic Forum. (2023, November 21). *Here's why Africa is the world leader in digital and mobile banking*. https://www.weforum.org/agenda/2023/11/africa-digital-mobile-banking-leader/

Yunus, M. (2007). *Banker to the poor*. Penguin Books India.

Zogning, F. (2023, January 02). Financial inclusion, inclusive entrepreneurship, and alternative financing options. *Journal of Small Business & Entrepreneurship, 35*(1), 8–13. https://doi.org/10.1080/08276331.2022.2120345

CHAPTER 7

Impact Capital for Inclusive Finance

7.1 Introduction

Significant segments of the global population, particularly in Emerging Markets and Developing Economies (EMDEs), continue to face systemic barriers to financial access. These barriers disproportionately affect marginalised groups, including Micro, Small, and Medium Enterprises (MSMEs), low-income households, women, and rural communities, exacerbating cycles of poverty and economic stagnation. In response to these challenges, targeted investment strategies have gained prominence to address financial disparities while generating measurable social and economic returns. Investors are actively reshaping the financial ecosystem and unlocking opportunities for those traditionally excluded from formal economic participation by directing capital towards underserved markets and pioneering new financial models. This chapter explores the transformative potential of impact investing in financial access as it makes a case for it. A key aspect of the discussion is the role of policy in shaping the impact investment landscape. Effective regulatory frameworks, Public–Private Partnerships, and targeted financial inclusion strategies ensure that impact investments translate into sustainable development outcomes. The chapter provides a comprehensive understanding of how capital can be deployed effectively to empower underserved communities by situating impact investing within the broader discourse on financial inclusion.

N. Udohaya, *Impact Investing and Financial Inclusion*, Sustainable Development Goals Series,
https://doi.org/10.1007/978-3-031-87841-1_7

Through a multidimensional analysis, it sheds light on both the opportunities and challenges of leveraging impact finance to reshape the global financial landscape. The discussion advocates for a concerted effort from investors, policymakers, and financial institutions to create a more equitable and resilient financial system that serves the needs of all, particularly those historically excluded from mainstream financial markets.

7.2 The Case for Impact Investing for Financial Access

The rationale for impact investing in financial access is twofold. First, it provides much-needed capital to underserved communities, enabling financial inclusion through innovative financial products, microfinance initiatives, and digital banking solutions. Second, it addresses systemic inefficiencies by fostering financial ecosystems that prioritise equitable growth, thereby enhancing economic resilience and reducing poverty. Impact investing has demonstrated its potential to close critical financing gaps, particularly for Micro, Small, and Medium Enterprises (MSMEs), low-income households, and women entrepreneurs—sectors that often struggle to secure affordable credit from traditional financial institutions. Moreover, integrating fintech solutions and alternative credit assessment models has expanded financial accessibility, reducing dependency on collateral-based lending and mitigating risk perceptions associated with marginalised groups. Impact investing is poised to be pivotal in reshaping financial inclusion strategies. Impact investing can create a more inclusive and equitable financial landscape by leveraging innovative investment frameworks, multi-stakeholder partnerships, and policy interventions. This section explores the key challenges, opportunities, and mechanisms through which impact investing can drive financial access and empower underserved communities.

7.2.1 Unmet Financial Needs: Highlighting the Gap

The global financial inclusion gap remains a critical obstacle to inclusive growth and development, with 1.4 billion adults globally unbanked. This persistent exclusion affects nearly a quarter of the world's adult population. It highlights the shortcomings of traditional financial systems in addressing the needs of underserved groups such as women, microenterprises, and rural communities. Despite recent advancements in financial inclusion, systemic barriers continue to perpetuate cycles of financial exclusion. International institutions such as the OECD and the World Bank underscore the urgency of this issue, revealing that access to financial services remains uneven and fragmented (Maravalle & Pandiella, 2022). The unbanked population is concentrated in low-income countries and among specific vulnerable groups. For instance, rural populations may face geographic isolation, micro-enterprises may struggle with limited access to affordable credit due to perceived risks and inadequate financial histories and women often face systemic discrimination from gender-biased lending practices to legal barriers in property ownership exacerbating their financial exclusion (Andersson Djurfeldt, 2020; Khan et al., 2024). These overlapping challenges demand targeted interventions to address both the structural and cultural barriers that prevent equitable access to financial resources.

Micro, Small, and Medium Enterprises (MSMEs)

Micro, Small, and Medium Enterprises (MSMEs) represent a cornerstone of economic development globally, serving as a critical engine for job creation, innovation, and income generation. The segment constitutes approximately 90 per cent of businesses worldwide and are responsible for over 50 per cent of global employment, underscoring their indispensable role in both developed and developing economies (IFC, 2024b). Despite their significance, MSMEs grapple with a profound and persistent financing gap. The IFC estimates the financing gap for formal MSMEs at $5.2 trillion and for informal MSMEs at $2.9 trillion. This gap substantially constrains their ability to sustain and expand operations, innovate, and contribute more robustly to economic development. The financing challenges MSMEs face are particularly acute in regions such as Sub-Saharan Africa and Latin America. In these areas, less than 20 per cent of MSMEs have access to formal credit, a stark contrast to their counterparts in more developed economies (IFC, 2022). This limited access is often attributed to several structural and systemic barriers, including the lack of collateral, high transaction costs associated with small loans, and inadequate financial infrastructure. These constraints are further exacerbated by the perceived riskiness of lending to small businesses, especially those operating in volatile or informal sectors. The reliance of MSMEs on informal lending networks and personal savings highlights a critical vulnerability. Informal financing mechanisms, while providing an immediate recourse, are often insufficient and can impose exorbitant interest rates, further straining the financial viability of these enterprises.[1] Personal savings, on the other hand, are typically inadequate to support significant capital investments or operational scaling. As a result, many MSMEs are unable to invest in technology, enhance productivity, or expand market reach, limiting their capacity to compete effectively in the global economy (Simba et al., 2024).

This financing deficit has far-reaching implications for economic growth and resilience. The inability of MSMEs to access affordable and adequate financing stifles innovation and productivity, key drivers of economic progress. Additionally, it undermines their ability to withstand economic shocks, such as fluctuations in demand, supply chain disruptions, or macroeconomic instability. For instance, during the COVID-19 pandemic, MSMEs with limited financial buffers were disproportionately affected, with many forced to downsize operations or shut down entirely. Addressing the MSME financing gap requires a multifaceted approach that integrates policy, financial innovation, and capacity-building. Governments and Development Finance Institutions must prioritise the creation of enabling environments that facilitate MSME access to credit, such as credit guarantee schemes, financial inclusion strategies, and regulatory frameworks that reduce barriers to entry for alternative financing providers. At the same time, the private sector can play a pivotal role by leveraging financial technology (fintech) to lower transaction costs, improve credit assessments through alternative data, and expand the reach of financial services to underserved areas. Moreover, tailored capacity-building programmes are essential to empower MSMEs with the skills and knowledge required to navigate formal financial systems and enhance their creditworthiness. These programmes should focus on financial literacy, business management, and strategic planning, enabling MSMEs to present compelling cases for investment and effectively utilise the resources they secure. While MSMEs are undoubtedly the backbone of global economies, their potential remains constrained by significant financing challenges. Bridging the $5.2 trillion financing gap in developing countries necessitates concerted efforts from policymakers, financial institutions, and the international community. By addressing these challenges through innovative solutions and supportive ecosystems, the global economy can unlock the transformative potential of MSMEs to drive inclusive and sustainable growth.

Low-Income Households

Lack of access to basic financial services, such as savings accounts, credit, and insurance, has profound implications for low-income households. It exacerbates their vulnerability to economic shocks, limits opportunities for upward mobility, and entrenches cycles of poverty (Dabla-Norris, 2015). The intersection of financial exclusion and socio-economic deprivation necessitates a rigorous examination of the underlying challenges and potential solutions to foster financial inclusion. The absence of formal financial access for low-income households often forces them to rely on informal lending networks, which carry significant risks and costs. Informal loans typically come with exorbitant interest rates, lack regulatory oversight, and can perpetuate exploitative practices. These mechanisms, while sometimes accessible in emergencies, fail to provide a sustainable pathway for economic stability or growth. The inability to access formal savings accounts and insurance products leaves low-income households particularly exposed to economic shocks. Unexpected events, such as illness, natural disasters, or job loss, can swiftly erode limited resources, pushing families further into poverty. The absence of a financial safety net constrains their capacity to invest in essential areas such as healthcare, education, and income-generating activities, which are critical for breaking intergenerational poverty cycles. Studies reveal that financial exclusion undermines broader economic growth by limiting aggregate demand and reducing the potential for wealth accumulation in underserved communities (Kumar & Jie, 2023). When a significant portion of the population lacks access to financial services, the overall economy suffers from reduced productivity and a diminished ability to mobilise domestic resources for investment. The global challenge of financial exclusion among low-income households demands a comprehensive understanding of the systemic barriers and their far-reaching consequences. Addressing this gap is crucial to fostering resilience and promoting sustainable development. Bridging the financial inclusion gap is not merely an economic imperative but a vital step towards achieving social equity and poverty alleviation on a global scale.

Women: Gender Disparities in Financial Access

Gender disparities in financial access remain deeply entrenched worldwide, perpetuating inequalities that hinder the full economic participation of women. According to the latest Global Findex report women are 4 per cent less likely than men to have a formal bank account globally.

In low-income countries, this gap widens to 6 per cent (Demirgüç-Kunt et al., 2022). This reflects a structural and systemic barrier that disproportionately impacts women. These disparities are exacerbated by discriminatory practices, lack of collateral, limited financial literacy, and socio-cultural norms that restrict women's economic agency. A significant driver of the gender gap in financial access is the limited asset ownership among women. In many regions, inheritance laws, cultural traditions, and gender-biased property rights prevent women from owning land or other significant assets (Perrin & Hyland, 2023). Without collateral, women face substantial barriers in accessing credit, which limits their ability to invest in income-generating activities or entrepreneurial ventures. Perrin and Hyland (2023) showed that women-led businesses, particularly in Sub-Saharan Africa and South Asia, are disproportionately small-scale and underfunded due to these constraints, perpetuating a cycle of economic marginalisation. Another critical factor contributing to financial exclusion is the limited participation of women in formal employment. Women are overrepresented in informal economies, where wages are irregular, and employment lacks social protections. This informality not only restricts their financial stability but also limits access to financial products and services that require proof of steady income, such as loans or mortgages. In Sub-Saharan Africa, for example, women constitute a significant proportion of the agricultural workforce, yet their contributions are often undervalued and remain outside formal financial systems, further marginalising their economic potential.

The implications of gender disparities in financial access are far-reaching. Economically disempowered women are less able to invest in education, healthcare, and other critical needs for their families, perpetuating cycles of intergenerational poverty. Increasing women's financial inclusion can transform household welfare and community development. Women's control over financial resources often translates into higher spending on children's education, nutrition, and health, creating a multiplier effect that benefits society as a whole (Hendriks, 2019; Siddik & Kabiraj, 2018). Furthermore, financial exclusion limits the broader economic contributions that women can make. Studies show that bridging the gender gap in financial access could significantly boost GDP growth in many developing regions. McKinsey estimates that advancing gender equality in economic participation, including financial access, could add \$12 trillion to global GDP by 2025 (Woetzel, 2015). This underscores the urgency of addressing gender disparities as an economic imperative, not just a social one. In regions like South Asia and Sub-Saharan Africa, where socio-cultural norms play a significant role in shaping women's financial behaviours, targeted interventions are essential. These should include increasing women's financial literacy, reforming discriminatory legal frameworks, and expanding access to digital financial services that offer flexible and inclusive solutions. Digital platforms, in particular can bridge gender gaps by providing women with secure and convenient ways to save, borrow, and transact, bypassing traditional barriers such as mobility constraints or lack of physical banking infrastructure. Addressing gender disparities in financial access requires a comprehensive approach that tackles the structural and systemic barriers faced by women. By ensuring equitable access to financial resources, stakeholders can unlock the untapped economic potential of women, fostering inclusive growth and sustainable development. Reducing the financial gender gap is a matter of equity and a strategic imperative to enhance global economic resilience and prosperity.

Rural Populations

Financial exclusion is a pervasive issue in rural areas, where limited infrastructure, geographic isolation, and low levels of digital connectivity create formidable barriers to accessing financial services (van Zanden, 2023). Rural communities, particularly in regions like Sub-Saharan Africa and South Asia, face compounded challenges that restrict their integration into formal financial systems. These obstacles hinder their economic potential and heighten their vulnerability to economic and environmental shocks, perpetuating cycles of poverty and marginalisation. One of the primary barriers to financial inclusion in rural areas is the lack of adequate infrastructure. Financial institutions are often concentrated in urban centres, leaving rural populations without access to physical banking facilities. This absence necessitates long and costly travel to access basic services such as opening a savings account or obtaining credit. The scarcity of ATMs and point-of-sale terminals further exacerbates this issue, limiting the ability of rural residents to engage in cashless transactions or save securely. Geographic isolation also poses a significant challenge. Many rural areas are characterised by dispersed populations and poor transportation networks, making it economically unviable for financial institutions to establish branches. This isolation is compounded by low levels of digital connectivity, which impede the adoption of mobile banking and other digital financial services. In regions with limited internet penetration, the potential for leveraging technology to bridge the financial inclusion gap remains largely unrealised. Additionally, the economic structures of rural communities often exacerbate financial exclusion. A significant proportion of rural households rely on seasonal incomes derived from agriculture or informal labour. These irregular income patterns make it difficult for individuals to meet the eligibility criteria for formal financial products, such as loans or insurance. The absence of consistent cash flow also reduces the ability of rural households to save or invest, constraining their economic opportunities. The lack of market access further limits the financial integration of rural populations. In many cases, rural producers face challenges in selling their goods at fair prices due to the absence of organised markets or transportation infrastructure. This situation restricts their income-generating potential and reinforces their dependence on informal credit sources, which often come with exploitative terms. Financial exclusion in rural areas has far-reaching implications for economic development. The inability to access credit, savings, and insurance undermines the resilience of rural households to economic shocks, such as price volatility or crop failures. Environmental shocks, including floods and droughts, further compound these vulnerabilities, often pushing families deeper into poverty. Without access to formal financial services, rural populations lack the resources needed to invest in adaptive measures or recover from such crises. The broader economy also suffers from the exclusion of rural populations from formal financial systems. When a substantial segment of the population cannot participate in financial markets, aggregate savings and investments decline, limiting the capital available for economic growth.

Furthermore, the underutilisation of rural economic potential perpetuates regional disparities and slows overall development. Financial exclusion in rural areas is a multifaceted challenge that requires targeted interventions. Addressing the systemic barriers faced by rural populations is essential for unlocking their economic potential and fostering inclusive growth. By recognising the unique needs and constraints of rural communities, stakeholders can develop innovative solutions to bridge the financial inclusion gap, promote resilience, and reduce inequality on a global scale.

Insurance Gap: Global Protection Deficits

The global insurance protection gap, estimated in the trillions of dollars, reflects a critical unmet need for risk coverage across households, businesses, and communities. This gap, defined as the difference between the amount of insurance coverage required and what is currently in place, is most pronounced in developing economies. The disparity in coverage has profound implications for economic resilience and societal equity, leaving vulnerable populations exposed to a range of risks without adequate financial safeguards. In low-income countries, insurance penetration often remains below 1 per cent of GDP, a stark contrast to developed economies where penetration rates are substantially higher (OECD, 2024). This disparity is underpinned by structural and systemic barriers, including limited access to insurance markets, lack of financial literacy, and high costs relative to household incomes. Furthermore, in many developing regions, cultural factors and mistrust of formal financial institutions deter individuals and businesses from seeking insurance products. The consequences of this protection gap are particularly severe in the face of natural disasters, health crises, and economic shocks. For instance, uninsured households affected by climate-related events, such as floods or droughts, often bear the full financial burden of recovery. This lack of coverage exacerbates poverty cycles, as families deplete savings, incur debt, or forego essential expenditures to cope with the aftermath of such events. Similarly, businesses without adequate insurance face prolonged disruptions, reduced productivity, and potential closure, undermining local economies and employment.

Health crises further highlight the critical nature of the insurance gap. In many low- and middle-income countries, the absence of health insurance exposes individuals to catastrophic out-of-pocket expenses. Studies indicate that medical costs are a leading cause of impoverishment in these regions, forcing families to choose between accessing healthcare and meeting other basic needs. The COVID-19 pandemic underscored the vulnerability of uninsured populations, as governments and aid organisations struggled to fill the void left by insufficient private coverage. The protection gap also extends to agriculture, a sector that is particularly vulnerable to environmental and market risks. In regions where agriculture constitutes a significant share of the economy, the lack of crop and livestock insurance leaves farmers exposed to yield losses, price fluctuations, and extreme weather events. This insecurity not only reduces agricultural productivity but also discourages investment in sustainable practices, further entrenching economic vulnerability.

Marginalised populations bear a disproportionate share of the financial burden associated with the insurance gap. Low-income households, women, and rural communities are often excluded from formal insurance markets due to affordability constraints, geographic isolation, and discriminatory practices. As a result, these groups are left with limited recourse in the event of crises, deepening existing inequalities and reducing social mobility (Beech et al., 2021; Taconet et al., 2020). Addressing the global insurance protection gap requires coordinated efforts from governments, insurers, and international organisations. Innovative approaches, such as parametric insurance products and micro-insurance schemes, have demonstrated potential to extend coverage to underserved populations. By leveraging technology, such as mobile platforms and blockchain, insurers can reduce administrative costs, enhance accessibility, and build consumer trust. The global insurance protection gap represents a significant barrier to economic resilience and social equity. Bridging this gap is essential to mitigating the impacts of risks on vulnerable populations and fostering inclusive development. Through targeted interventions and collaborative efforts, stakeholders can ensure that the benefits of insurance extend to all, promoting a more equitable and sustainable global economy.

Multifaceted Marginalisation

Marginalised groups often experience overlapping and mutually reinforcing challenges that exacerbate their exclusion from economic, social, and political opportunities. These intersectional barriers are particularly pronounced for individuals and communities that sit at the nexus of multiple forms of marginalisation. For example, rural women face compounded disadvantages due to both gender-based and geographic exclusions. Similarly, Indigenous communities often contend with systemic discrimination, historical injustices, and geographic isolation that collectively undermine their financial inclusion and socio-economic participation. Understanding and addressing these multifaceted barriers require context-specific and inclusive strategies that recognise the diverse needs of these populations. Rural women represent a case study of how intersectional barriers operate. These women often lack access to formal financial services due to the dual impact of gender discrimination and rural isolation. Limited physical banking infrastructure in rural areas is compounded by socio-cultural norms that restrict women's mobility and economic agency. As a result, rural women are disproportionately excluded from credit, savings, and insurance markets, which hinders their ability to invest in education, health, and income-generating activities. This exclusion not only perpetuates poverty at the household level but also limits broader economic development in rural communities. Indigenous communities, on the other hand, grapple with systemic challenges rooted in historical and institutional discrimination. In many cases, these communities are geographically isolated, with limited access to education, healthcare, and financial institutions. Discriminatory policies and practices have historically excluded Indigenous peoples from participating in mainstream economic systems, leading to entrenched cycles of poverty. Furthermore, the lack of culturally appropriate financial products and services often discourages Indigenous communities from engaging with formal financial systems, further marginalising them from opportunities for economic growth and resilience.

The implications of intersectional barriers extend beyond financial exclusion. These overlapping challenges limit access to education, healthcare, and political representation, creating a multidimensional cycle of marginalisation. For instance, the lack of financial inclusion often restricts access to resources that could be used to invest in children's education or healthcare, perpetuating intergenerational poverty. Similarly, limited political representation among marginalised groups reduces their ability to advocate for policies and programmes that address their unique needs. Addressing intersectional barriers requires a nuanced and comprehensive approach that goes beyond one-size-fits-all solutions. Policies and programmes must be tailored to the specific contexts and experiences of marginalised groups, taking into account the unique challenges they face. For example, efforts to improve financial inclusion among rural women could involve the implementation of community-based cooperative savings and lending groups for rural women. These groups provide an alternative to traditional banking by allowing members to pool their savings and access small loans without requiring formal credit history or collateral. Additionally, financial literacy workshops tailored to their specific socio-economic and cultural contexts can empower women to make informed financial decisions, helping to bridge the gap in financial inclusion. Similarly, initiatives targeting Indigenous communities should prioritise culturally appropriate financial products, community-driven development programmes, and the preservation of indigenous knowledge and practices. Collaboration among stakeholders is essential for tackling intersectional barriers effectively. Governments, non-governmental organisations, and the private sector must work together to create enabling environments that support marginalised populations. This includes reforming discriminatory policies, investing in infrastructure, and fostering inclusive economic systems. Furthermore, involving marginalised communities in the design and implementation of policies and programmes ensures that their voices are heard and their needs are met. Intersectional barriers represent a significant challenge to achieving equity and inclusion on a global scale. By recognising and addressing the overlapping challenges faced by marginalised groups, stakeholders can promote sustainable development, social justice, and economic resilience. Context-specific and inclusive strategies are both a moral imperative and a practical necessity for building a more equitable and prosperous world.

7.2.2 Market Failure and Systemic Barriers

Existing financial systems often fail to serve underserved populations, perpetuating cycles of exclusion and economic inequality. These failures stem from structural inefficiencies and entrenched systemic barriers that limit equitable access to financial services. This section explores key reasons why financial systems fall short of addressing the needs of marginalised groups such as women, micro-enterprises, and rural communities.

Biases in Credit Allocation

Biases in credit allocation reflect a systemic market failure that undermines financial inclusion, perpetuating economic inequities for women, minority groups, and low-income populations. This issue is deeply rooted in discriminatory lending practices, institutional risk aversion, and the reliance on collateral-based lending models, all of which serve to entrench the exclusion of marginalised groups from formal financial systems. These biases are symptomatic of structural inefficiencies and entrenched social and economic disparities within the financial sector.

Discriminatory Lending Practices: Discriminatory lending practices are a critical manifestation of biases in credit allocation, reflecting cultural and institutional prejudices that disproportionately affect certain demographics. Financial institutions frequently perpetuate systemic discrimination through credit approval processes, where loan officers may unconsciously or consciously favour male applicants, influenced by ingrained assumptions about men's financial capabilities compared to women's. Empirical evidence from the World Bank (2021) underscores this disparity, revealing that women-owned businesses in developing countries are 30 per cent less likely to secure formal loans than male-owned businesses, even after accounting for factors like business performance and creditworthiness. The exclusion is not limited to gender but extends to minority groups, who are often deemed "high-risk" borrowers based on discriminatory stereotypes rather than objective financial metrics. Additionally, societal expectations around gender roles exacerbate this issue. Women in many cultures are primarily viewed as caretakers rather than economic contributors, limiting their participation in formal financial systems. The OECD (2022) reports that 60 per cent women in low-income countries face cultural barriers discouraging them from seeking formal financial services. Minority groups such as indigenous populations face systemic exclusion driven by historical inequities and socio-political marginalisation, further perpetuating cycles of financial exclusion and economic disempowerment. Moreover, discriminatory lending practices are compounded by the lack of tailored financial products that address the specific needs of underserved groups. Traditional financial services often cater to urban, middle-income men as the default customer profile, neglecting the unique realities of women, rural populations, and minority groups. Microloans tailored to rural women have demonstrated exceptional repayment rates exceeding 95 per cent (GIIN, 2022), challenging prevailing stereotypes about the financial risk associated with marginalised borrowers. However, such products remain underdeveloped due to institutional blind spots and the prioritisation of more conventional customer bases.

Institutional Risk Aversion: Risk aversion among financial institutions further exacerbates biases in credit allocation. Banks and other lenders often categorise low-income individuals and micro-enterprises as inherently high-risk borrowers, despite empirical evidence to the contrary. The International Finance Corporation (2024) found that microfinance borrowers in South Asia had a default rate of less than 3 per cent, comparable to or better than traditional corporate clients. Despite this, financial institutions frequently operate under the assumption that low-income borrowers are more likely to default, leading to restricted credit flows to these groups. Risk mitigation strategies heavily favour large-scale corporate clients with established credit histories, concentrating resources among larger firms and sidelining smaller borrowers. The World Economic Forum (2021) reveals that 75 per cent of total bank lending in developing countries is directed towards established corporations, forcing micro-enterprises to rely on informal or predatory lenders. Furthermore, many financial institutions lack diversified credit portfolios, focusing predominantly on low-risk, high-return segments. This approach perpetuates systemic exclusion and limits the potential for financial inclusion to drive broader economic growth. Diversifying credit portfolios to include micro-enterprises can improve risk management and enhance financial stability.

Collateral-Based Lending Models: Collateral-based lending models further entrench systemic barriers to financial access. Traditional financial systems heavily rely on tangible assets, such as land or property, as collateral to mitigate lending risks. However, this approach disproportionately disadvantages marginalised populations, particularly those in low-income and rural areas, who often lack the necessary assets. The reliance on collateral as a risk mitigation tool disproportionately excludes these groups. Only 10 per cent of women in Sub-Saharan Africa own land, a common form of collateral (Behr et al., 2023), this effectively barring them from accessing formal credit channels. Structural barriers such as weak property rights frameworks exacerbate this issue. In regions where land titles are not formally recognised, even individuals who own land are unable to leverage it for financial access. The World Bank (2022) estimates that 70 per cent of land in developing countries is untitled or improperly documented, further limiting the ability of marginalised groups to utilise their assets as collateral. Additionally, the heavy reliance on collateral restricts entrepreneurship among underserved groups, particularly micro-enterprises that form the backbone of local economies. The Global Partnership for Financial Inclusion highlights that 45 per cent of micro-enterprises cite lack of collateral as the primary barrier to accessing formal financial services. By prioritising asset-based lending models, financial institutions systematically exclude significant segments of the population, leaving their economic potential untapped and perpetuating cycles of poverty and inequality (GPFI, 2024).

Information Asymmetry

Information asymmetry constitutes a systemic market failure and a formidable barrier to financial inclusion, particularly for underserved populations. It arises when financial institutions and potential customers operate with unequal access to critical information, impeding efficient credit allocation and perpetuating economic inequities. The roots of this market failure lie in limited credit histories, inadequate financial education, and data fragmentation, which collectively reinforce the exclusion of marginalised groups from formal financial systems.

Limited Credit Histories: One of the most evident manifestations of information asymmetry is the absence of formal credit histories among underserved populations. Many individuals and micro-enterprises in low-income communities lack the financial documentation required by traditional credit assessment models. This limitation stems from their reliance on informal economies, where financial transactions often go unrecorded. Without formal credit histories, lenders face challenges in objectively evaluating repayment capacity, leading to the exclusion of these borrowers. Research by the Consultative Group to Assist the Poor (CGAP, 2020) emphasises this issue, noting that the absence of credit histories entrenches financial exclusion by rendering borrowers invisible to formal financial systems. For example, rural entrepreneurs in developing countries frequently find themselves trapped in a cycle where their inability to secure formal loans prevents them from building the financial records necessary to access credit in the future.

Inadequate Financial Education: Inadequate financial education further exacerbates information asymmetry, widening the gap between financial institutions and potential customers. Low levels of financial literacy prevent individuals and small businesses from understanding available financial products and services, limiting their ability to make informed decisions. Many underserved populations remain unaware of microfinance opportunities, savings instruments, or insurance schemes that could enhance their financial security. This lack of knowledge diminishes demand for financial products and increases susceptibility to financial pitfalls, such as overborrowing or exploitation by predatory lenders (de la Cuesta-Gonzalez et al., 2021). The deficit creates a perpetual disconnect between financial service providers and the very populations they aim to serve, further entrenching cycles of economic vulnerability and exclusion.

Data Fragmentation: Data fragmentation within financial institutions presents another critical dimension of information asymmetry. Many banks and lenders lack integrated systems capable of analysing the diverse needs of underserved populations. This often results in inefficiencies in both product design and service delivery. For instance, siloed data systems may fail to recognise repayment patterns among informal workers, thereby missing opportunities to extend credit to viable but overlooked market segments. Financial institutions in developing economies grapple with challenges related to data integration, Mohsen et al. (2024) highlight how this remains a significant barrier in the region. This inefficiency perpetuates the systemic exclusion of marginalised groups, as financial institutions are unable to adapt their offerings to the unique needs of these populations. Furthermore, the fragmentation creates additional barriers by increasing operational costs, often passed on to customers through higher fees, further discouraging the uptake of formal financial services.

The cumulative effects of limited credit histories, inadequate financial education, and data fragmentation reinforce the systemic nature of information asymmetry. These interconnected barriers exacerbate financial exclusion, trapping underserved populations in cycles of poverty and preventing them from leveraging financial services to improve their economic prospects. The persistence of these market failures underscores the need for structural changes within the financial ecosystem to address the root causes of information asymmetry and its detrimental impact on inclusive economic growth.

High Transaction Costs

High transaction costs represent a systemic market failure and a significant barrier to achieving financial inclusion. These costs, rooted in operational inefficiencies, underdeveloped infrastructure, and regulatory complexities, create an uneven playing field that excludes low-income and rural populations from accessing essential financial services. This persistent barrier highlights the structural flaws in financial ecosystems that disproportionately affect the most vulnerable.

Operational Inefficiencies: The high costs associated with serving dispersed and remote populations are emblematic of operational inefficiencies in financial service delivery. Rural areas, often characterised by low population density and geographic remoteness, require financial institutions to expend significant resources to establish and maintain a presence. The need to deploy field agents, set up physical branches, or transport cash to underserved regions significantly increases operational overheads. Unfortunately, these costs are often passed on to the end-users in the form of inflated interest rates and service fees. Rural borrowers in developing countries frequently face interest rates that are 20–30 per cent higher than those in urban centres (Mori, 2023), reflecting the additional logistical burdens borne by financial institutions. This creates a cycle of exclusion, as unaffordable financial products effectively lock out those who need them most, reinforcing systemic inequities within the financial system.

Underdeveloped Infrastructure: The issue is compounded by the chronic underdevelopment of infrastructure in many regions, further inflating transaction costs. Poor road networks, unreliable electricity, and limited digital connectivity obstruct the seamless delivery of financial services, making outreach to underserved populations prohibitively expensive. Sub-Saharan Africa, for instance, provides a stark example, where only 43 per cent of the population has access to stable electricity (Blimpo & Cosgrove-Davies, 2019). Financial institutions operating in these contexts face significant challenges in adopting cost-saving digital innovations, such as mobile banking. An African Development Bank study highlights that financial services in regions with inadequate infrastructure can cost up to 50 per cent more than in better-equipped areas (Ondiege et al., 2013). In addition to raising cost, this systemic neglect of discourages financial institutions from venturing into unprofitable markets, perpetuating geographic inequalities in access to finance.

Regulatory Complexity: Regulatory complexity further exacerbates high transaction costs, particularly for smaller financial institutions and microfinance organisations that often prioritise serving marginalised groups. Navigating intricate and often fragmented regulatory environments imposes considerable compliance costs, straining already limited resources. These costs are disproportionately borne by institutions operating in low-income markets, where profit margins are thin and scalability is constrained.[2] Smaller institutions, lacking economies of scale, are forced to pass these costs onto their customers or withdraw services altogether, thereby exacerbating the financial exclusion of underserved populations.

High transaction costs, as a systemic and structural barrier, underscore the need for transformative changes within the financial ecosystem.

Addressing operational inefficiencies, bridging infrastructure gaps, and reforming regulatory frameworks are essential steps towards dismantling the barriers that perpetuate financial exclusion. Such efforts are critical for empowering underserved populations and fostering more inclusive and sustainable economic development.

Entrenched Socio-Economic Inequalities
Systemic barriers such as wealth and income inequality, cultural and social norms, and geographic disparities exacerbate this challenge, creating structural inefficiencies that perpetuate economic marginalisation and exclusion. These barriers are deeply rooted in societal and institutional frameworks, necessitating a comprehensive analysis to understand their pervasive impact on financial access and inclusion.

Wealth and Income Inequality: Wealth and income inequality represent significant systemic barriers to financial inclusion, as economic disparities create an uneven playing field in accessing financial resources. The concentration of wealth among higher-income groups often skews the allocation of financial resources, leaving lower-income populations underserved. Financial institutions, driven by profit motives, prioritise servicing wealthier clients due to their higher profitability and lower perceived risk. This results in a financial landscape where low-income individuals and households struggle to access credit, savings, and insurance products that could help them escape poverty. The perpetuation of poverty cycles is closely tied to these economic disparities. Without access to affordable financial services, low-income populations are forced to rely on informal or predatory lending systems, further exacerbating their economic vulnerability. Households in the bottom income quintile are three times more likely to remain unbanked than those in the top quintile (Demirgüç-Kunt & Klapper, 2012) The systemic exclusion of low-income groups limits their economic opportunities and reinforces broader societal inequalities, making financial inclusion a critical lever for poverty alleviation and equitable growth.

Cultural and Social Norms: Cultural and social norms significantly influence financial behaviour and access, often serving as barriers to inclusion. Gender roles and societal expectations, for instance, frequently limit women's financial independence and decision-making power. In many cultures, women are viewed primarily as caregivers rather than economic agents, restricting their ability to engage with formal financial systems.

This gender bias manifests in lower account ownership, reduced access to credit, and limited participation in entrepreneurial activities among women. Marginalised communities, such as ethnic minorities and indigenous populations, face additional systemic discrimination that restricts their access to formal financial services. Historical and socio-political marginalisation often translates into lower levels of financial literacy and trust in financial institutions. This dynamic is particularly evident in regions with significant indigenous populations, where exclusion from land ownership and economic resources further isolates these communities from formal financial systems. Addressing these cultural and social barriers requires a nuanced approach that considers the intersectionality of identity, economic status, and historical context.

Geographic Disparities: Geographic disparities present another systemic barrier to financial inclusion, with urban-centric financial systems often neglecting rural populations. The urban–rural divide in financial access is stark; urban areas typically benefit from dense banking networks and digital infrastructure, while rural regions are frequently underserved or entirely excluded. Remote areas often face a lack of basic banking infrastructure, such as branches, ATMs, and reliable internet connectivity, further isolating these communities from formal financial systems. For example, in Sub-Saharan Africa, rural populations must often travel significant distances to access banking services, incurring additional costs and time burdens. This geographic exclusion not only limits individual financial empowerment but also hampers broader economic development by constraining the flow of capital and investment in rural economies.

Wealth and income inequality, cultural and social norms, and geographic disparities collectively represent systemic barriers that inhibit financial inclusion. These challenges are deeply interwoven into societal and institutional structures, perpetuating cycles of poverty, exclusion, and economic stagnation. Addressing these barriers requires a multidimensional strategy that includes policy reforms, technological innovations, and targeted interventions aimed at empowering marginalised populations. Dismantling these systemic barriers allows stakeholders to create a more inclusive financial ecosystem that promotes equitable economic participation and sustainable development.

7.3 Impact Capital for Financial Access in the Emerging Markets and Developing Economies (EMDEs)

Financial exclusion remains a fundamental constraint to economic growth and social development in EMDEs. With almost 27 per cent of the adults globally lacking access to basic financial services, 80 per cent of these individuals reside in Asia and Sub-Saharan Africa (World Bank, 2024). Approximately 70 per cent of the 162 million formal MSMEs in developing economies face severe financing constraints, limiting their ability to expand, innovate, and create employment (IFC, 2024b). Access to finance is essential for economic mobility and resilience. Without reliable credit and savings mechanisms, individuals and businesses in EMDEs struggle to make long-term investments, mitigate financial risks, and respond effectively to economic shocks. The informal financial systems that often fill this void—such as rotating savings groups and informal lenders—lack stability, regulation, and consumer protections, leaving borrowers vulnerable to predatory practices. By channelling impact capital into financial services, investors can directly address these gaps by supporting the expansion of Microfinance Institutions, digital banking platforms, and fintech solutions that increase access to credit, savings, and insurance. Financial inclusion fosters entrepreneurship, empowers individuals to manage economic shocks, and drives broader economic participation, making it a key enabler of sustainable development. Moreover, impact investing in financial services aligns with broader macroeconomic goals such as reducing income inequality, increasing workforce participation, and stabilising financial systems in fragile economies.

EMDEs represent some of the most promising yet underfunded regions for impact investment. These economies account for most of the world's population and house a significant proportion of the global unbanked population. Despite their high growth potential, EMDEs receive only a fraction of global impact investment flows, with over 70 per cent of impact assets under management concentrated in North America and Western Europe. This imbalance presents a missed opportunity for both investors and policymakers, as impact investments in EMDEs can drive financial inclusion, catalyse economic growth, and enhance resilience in vulnerable communities. Beyond economic potential, the demographic and technological shifts in EMDEs reinforce the urgency for increased impact investment. With rapidly growing populations and rising urbanisation, the demand for accessible financial services is surging. Moreover, fintech innovations have lowered entry barriers, enabling new business models that extend banking and credit services to previously underserved populations. Impact investing can thus play a crucial role in leveraging these trends to create sustainable, long-term economic benefits in the region.

7.3.1 *EMDEs as Good Investment for Impact Capital*

EMDEs present a compelling opportunity for impact investors due to their rapid population growth, urbanisation, and increasing technological adoption. Africa and South Asia, in particular, have burgeoning middle classes with rising demand for financial services. The widespread use of mobile phones and internet connectivity has paved the way for digital financial innovations such as mobile money, peer-to-peer lending, and blockchain-based financial solutions. Unlike saturated financial markets in developed economies, EMDEs offer untapped growth potential. The financial service penetration rate remains low in many regions, presenting ample opportunities for first-movers to establish market leadership. Fintech startups in EMDEs have capitalised on these trends, providing cost-effective, scalable solutions that cater to underserved populations. Investment in such enterprises generates financial returns and contributes to systemic economic transformation. Investments in these areas yield financial returns and contribute to systemic economic transformation. Microfinance, digital lending platforms, and innovative credit-scoring mechanisms can bridge financial gaps while offering investors exposure to high-growth markets with relatively untapped potential. Fintech-enabled financial services, when aligned with inclusive policies, can generate sustainable returns while improving financial resilience for millions of people.

Growth Opportunities

The potential for impact investing in Emerging Markets and Developing Economies (EMDEs) is underscored by the vast growth opportunities presented by these regions. Unlike saturated markets in developed economies, EMDEs offer a dynamic environment characterised by rapid economic expansion, demographic shifts, and evolving consumer demands.

Untapped Markets: The sheer size of the underserved population in EMDEs presents a unique opportunity for investors to provide essential services. Many individuals and businesses lack access to basic financial, healthcare, and infrastructure services, making these markets ripe for investment. The high level of financial exclusion, for instance, underscores the need for scalable financial solutions such as microfinance, digital banking, and mobile payment platforms. Additionally, sectors such as education, healthcare, and clean energy remain underdeveloped in many EMDEs, creating a compelling case for targeted impact investments that address these gaps while generating sustainable returns.

Rapid Urbanisation: Urban centres in EMDEs are expanding at an unprecedented rate, creating both challenges and investment opportunities. The demand for sustainable infrastructure, affordable housing, clean energy, and efficient transportation systems is growing rapidly. Failure to address these needs could exacerbate socio-economic inequalities and environmental degradation. Impact investments directed towards urban development can facilitate the construction of resilient cities through smart infrastructure, efficient public transportation, and sustainable waste management systems. These investments not only generate financial returns but also contribute to long-term social stability and environmental sustainability.

Youthful Demographics: Many EMDEs have young and fast-growing populations, creating high demand for education, employment, and innovative financial products. The rising middle class and increasing digital literacy further enhance the attractiveness of these markets. By investing in education technology (EdTech), vocational training, and entrepreneurship programmes, impact investors can empower the next generation with skills essential for economic productivity. This youthful demographic also presents an opportunity for investors in consumer services, fintech, and e-commerce, where digital-native consumers are increasingly driving economic activities.

Adoption of Technology

Leapfrogging Technologies: Unlike developed markets that are constrained by legacy systems, many EMDEs have demonstrated an ability to leapfrog traditional technological progressions. The widespread adoption of mobile banking, for instance, has bypassed the need for brick-and-mortar banking infrastructure, enabling millions to access financial services remotely. The success of Kenya's M-Pesa has paved the way for similar innovations in other EMDEs, demonstrating how digital financial services can enhance economic participation, improve transaction efficiency, and reduce costs. The rapid adoption of blockchain technology, artificial intelligence, and decentralised finance solutions further highlights how emerging economies are positioned to benefit from disruptive innovations.

Innovation Hubs: Emerging markets are increasingly becoming centres for fintech and social innovation, attracting significant global investment. Cities in Africa, Latin America, and South Asia are fostering a culture of technological entrepreneurship, driving solutions tailored to local challenges. Impact investors have the opportunity to support and scale disruptive innovations that can generate both financial and social returns. With access to venture capital, accelerator programmes, and policy support, these innovation hubs can drive technological progress that aligns with both commercial viability and developmental goals.

Policy Reforms and Economic Liberalisation

Regulatory Improvements: Many EMDE governments have implemented reforms to improve the business environment, making these markets more attractive for investment. Streamlined regulations, investment incentives, and financial sector modernisation are encouraging foreign direct investment (FDI) and fostering private-sector-led development. Governments have also recognised the need to create enabling environments for financial inclusion, green energy transition, and digital transformation. Policies that promote fintech adoption, enhance financial literacy, and protect investor rights are strengthening the foundation for long-term impact investments in these regions.

Trade Agreements: Participation in regional and international trade agreements enhances market accessibility and facilitates investment flows. Agreements such as the African Continental Free Trade Area (AfCFTA) create unified markets that allow investors to scale their operations more effectively across multiple countries. Open trade policies provide impact investors with broader market reach, reducing investment risks associated with regulatory inconsistencies across jurisdictions.

Local Entrepreneurship and Talent

Rise of Social Enterprises: Entrepreneurs in emerging markets are increasingly developing solutions tailored to local challenges, creating investment opportunities that blend financial returns with social impact. The growth of Microfinance Institutions, renewable energy cooperatives, and agritech solutions illustrates how local entrepreneurship can drive inclusive economic development. Impact investors can play a crucial role in providing capital, mentorship, and market access to these enterprises, ensuring their scalability and long-term sustainability.

Capacity-Building: Investment in education and skill development enhances workforce quality, supporting long-term economic growth. By funding initiatives in vocational training, STEM education, and digital literacy, impact investors contribute to a more skilled and adaptable workforce, increasing productivity and driving innovation. Programmes that support women entrepreneurs, youth employment, and financial literacy initiatives create ripple effects that enhance economic resilience and equitable growth. Addressing structural barriers such as gender inequality and access to capital through targeted investments further enhances social and economic inclusivity.

Impact on Global Development

Driving Sustainable Development: Progress in EMDEs is essential for global economic stability and long-term sustainability. Investments in financial inclusion, resilient infrastructure, education, and climate adaptation contribute to economic and social transformation. Expanding access to credit and financial services ensures that more individuals and businesses can participate in the economy, fostering inclusive growth. Similarly, investment in sustainable energy, waste management, and climate-smart agriculture strengthens local economies while addressing pressing environmental concerns. As EMDEs continue to develop, impact investors must align their strategies with sustainable development priorities to maximise long-term benefits.

Poverty Reduction: Investments that promote inclusive economic growth have a significant impact on poverty alleviation. By supporting access to capital for MSMEs, improving agricultural productivity through agritech solutions, and expanding healthcare infrastructure, impact investors can drive large-scale poverty reduction. Sustainable business models that integrate economic empowerment with financial returns ensure that growth benefits are widely shared, creating more resilient and self-sustaining economies. Additionally, targeted investments in social infrastructure, such as affordable housing and public health, help reduce disparities that have historically limited upward mobility in many EMDEs.

Impact investing in financial services for EMDEs represents both a high-return opportunity for investors and a critical strategy for inclusive and sustainable economic development. Policymakers have a vital role to play in facilitating the growth of impact investing in EMDEs. Governments can create an enabling environment that attracts private capital to high-impact sectors through regulatory reforms, Public–Private Partnerships, and targeted financial incentives. Meanwhile, investors must adopt a long-term perspective, integrating financial returns with measurable social and environmental outcomes. Scaling impact investments in financial services for EMDEs is a pathway to economic growth and an essential component of global financial and climate stability. The time to act is now, with impact capital positioned to drive the next wave of innovation and inclusivity in the world's fastest-growing markets. Investors who recognise and seize this opportunity will generate financial returns and contribute to reshaping economic landscapes in ways that foster shared prosperity and resilience.

7.4 Strategies for Engagement in Impact Investing for Inclusive Finance

Impact investing for inclusive finance requires well-structured engagement strategies to ensure sustainable and transformative financial inclusion. Investors and stakeholders must navigate diverse socio-economic landscapes, mitigate risks, and build long-term capacity to generate financial and social returns. An effective engagement strategy not only improves the reach and efficiency of financial inclusion efforts but also fosters a more stable and resilient financial ecosystem in emerging markets and developing economies (EMDEs). This section explores key strategies to optimise engagement in impact investing, particularly in EMDEs, through deep integration with local actors, comprehensive risk mitigation, sustainable capacity-building initiatives, technological innovation, and active policy engagement.

7.4.1 Local Partnerships for Market Entry and Sustainable Impact

Collaborating with local organisations and stakeholders is fundamental to ensuring successful market entry and long-term sustainability. Establishing strong partnerships with local actors provides access to market intelligence, regulatory insights, and socio-cultural understanding, which are critical for designing and implementing financial solutions that effectively meet the needs of underserved populations.

Leveraging Local Knowledge: Local financial institutions, NGOs, cooperatives, and microfinance organisations have extensive knowledge of market dynamics, consumer behaviour, and regulatory frameworks. Partnering with these institutions allows impact investors to navigate complex economic environments and tailor financial services to fit specific regional needs. Such collaboration ensures that investment strategies are contextually relevant and responsive to community-specific challenges.

Enhancing Cultural Relevance: Financial inclusion efforts must align with local traditions, norms, and economic structures. Many underserved populations rely on informal savings groups, community lending schemes, and alternative credit models. By integrating these existing practices into formal financial systems, impact investors can increase adoption rates and reduce resistance to new financial products and services.

Facilitating Efficient Market Penetration: Entering new markets is resource-intensive and requires significant groundwork. Engaging with local fintech startups, cooperatives, and established financial service providers allows investors to reach previously unbanked populations efficiently. These partnerships enable access to distribution networks and customer trust, reducing operational costs and accelerating adoption.

Strengthening Stakeholder Buy-In: Involving community leaders, regulatory bodies, and government institutions ensures that financial inclusion initiatives gain legitimacy and public support. Local partnerships facilitate greater transparency, ethical compliance, and long-term sustainability by aligning investment activities with local economic development priorities.

7.4.2 Risk Mitigation to Ensure Investment Stability

Investing in financial inclusion in EMDEs presents several risks, including political instability, currency fluctuations, regulatory uncertainties, and market inefficiencies. A robust risk mitigation framework is essential to safeguard investments and ensure long-term impact.

Political Risk Insurance (PRI): Political instability, abrupt regulatory changes, and government intervention pose significant risks to financial investments. Investors can mitigate these risks by securing political risk insurance from multilateral organisations such as the World Bank's Multilateral Investment Guarantee Agency (MIGA). PRI provides coverage against expropriation, currency inconvertibility, and breach of contract, enhancing investor confidence.

Hedging Strategies: Currency risks remain a significant challenge in impact investing, as fluctuations in exchange rates can erode returns. Investors should implement hedging mechanisms such as forward contracts, currency swaps, and local currency lending structures to mitigate exposure to foreign exchange volatility.

Diversification of Investment Portfolios: Spreading investments across different sectors, regions, and financial instruments reduces dependency on single markets and minimises systemic risks. Diversification enhances resilience against localised economic downturns and regulatory changes.

Regulatory Risk Assessment: Conducting thorough due diligence on policy frameworks, financial regulations, and compliance requirements allows investors to anticipate and adapt to regulatory changes. Engaging legal experts and maintaining proactive communication with policymakers ensures alignment with national financial inclusion strategies.

Flexible Investment Structures: Designing financial products with adaptive repayment terms, impact-linked returns, and blended finance models enhances financial resilience. By incorporating concessional financing and revenue-sharing agreements, investors can balance financial sustainability with impact objectives.

7.4.3 Capacity-Building for Long-Term Success

Investing in local capacity development is crucial for ensuring that financial inclusion initiatives are sustainable and yield long-term benefits. Strengthening the capabilities of financial institutions, entrepreneurs, and communities fosters a resilient financial ecosystem that continues to function effectively beyond the initial investment period.

Training Financial Institutions and Local Entrepreneurs: Capacity-building programmes targeted at banks, Microfinance Institutions, and fintech startups help enhance governance, risk management, and financial product development. Training in sustainable business practices, financial planning, and technology integration improves institutional efficiency.

Developing Human Capital: Impact investments should include support for vocational training, digital skills development, and business incubation programmes. A well-trained workforce is essential for sustaining financial inclusion efforts and ensuring communities fully leverage available financial products.

Encouraging Knowledge Transfer: Facilitating mentorship programmes and technical assistance between international investors, financial experts, and local financial institutions enhances the sharing of best practices. Cross-border collaboration fosters innovation and strengthens financial service delivery.

Investing in Digital and Financial Infrastructure: Strengthening the digital banking ecosystem, expanding mobile payment networks, and supporting blockchain-based financial solutions ensure broader and more efficient access to financial services. These investments create infrastructure that facilitates long-term financial inclusion.

Promoting Financial Education: Ensuring that individuals and small businesses understand financial products, risk management, and investment opportunities maximises the benefits of inclusive finance. Comprehensive financial education campaigns empower individuals to make informed financial decisions.

7.4.4 Leveraging Technology and Data-Driven Approaches

Digital financial solutions offer a pathway to reaching underserved populations, but investors must actively shape and deploy these innovations in ways that enhance market integration, sustainability, and accessibility. Effective engagement involves fostering collaboration with fintech firms, supporting digital infrastructure expansion, and ensuring that technological innovations align with the realities of low-income and excluded communities. Investors engaging in inclusive finance must strategically deploy technology to maximise impact and efficiency.

Building Strategic Fintech Collaborations: Investors should actively partner with fintech companies that specialise in digital financial services, such as mobile banking, peer-to-peer lending, and automated savings platforms. These partnerships facilitate the expansion of financial access to unbanked populations while leveraging innovative business models to improve financial literacy and customer engagement.

Developing Digital Infrastructure for Scalability: Impact investors should support the expansion of digital payment systems, interoperable financial platforms, and mobile banking networks. By funding or co-developing essential infrastructure, investors help establish a solid foundation for inclusive finance while enabling cost-efficient transactions for low-income users.

Utilising AI and Machine Learning for Better Risk Assessment: AI-driven credit-scoring models allow financial institutions to assess creditworthiness based on alternative data, such as transaction history, social network analysis, and utility payments. This enables broader access to credit and microloans for individuals and small businesses with limited formal financial records.

Ensuring Data Protection and Cybersecurity Measures: The growing reliance on digital finance requires strong data security frameworks to prevent fraud and protect users' financial information. Investors must engage with financial technology providers to establish robust cybersecurity standards that safeguard consumer trust and regulatory compliance.

Implementing Digital Identity Solutions for Seamless Access: Biometric identification and blockchain-based identity management enable individuals without formal documentation to access financial services securely. Investors can play a role in funding and scaling these innovations to facilitate more inclusive banking models.

7.4.5 Engaging with Policymakers to Foster an Enabling Environment

Policymakers play a critical role in shaping the regulatory landscape that determines the success of impact investments in inclusive finance. A stable, transparent, and well-regulated financial ecosystem fosters investor confidence, encourages innovation, and ensures financial products align with national economic goals.

Advocating for Pro-Inclusion Regulations: Financial inclusion requires policy frameworks that support responsible lending, fintech innovation, and consumer protection. Investors must engage with regulators to advocate for balanced policies that facilitate financial access while maintaining systemic stability. Clear guidelines on interest rate caps, digital lending, and data privacy can enhance the effectiveness of impact investments.

Influencing National Financial Inclusion Strategies: Many governments have developed financial inclusion strategies that outline national priorities for expanding banking access, increasing microfinance penetration, and enhancing digital payment systems. Aligning impact investment strategies with these national priorities ensures better integration of private-sector solutions into public policy frameworks.

Promoting Public-Private Partnerships (PPPs): Governments and Development Finance Institutions (DFIs) can collaborate with impact investors to develop blended finance models that leverage public funds to attract private capital. Risk-sharing mechanisms, such as government-backed loan guarantees, can help mitigate perceived risks in emerging markets and increase private-sector participation in inclusive finance.

Encouraging Transparent Regulatory Frameworks: Unclear or inconsistent financial regulations can hinder investment flows. Policymakers must provide predictable and transparent regulatory environments that support financial innovation while safeguarding financial stability. Initiatives such as regulatory sandboxes allow financial service providers to test innovative products under controlled conditions before full market deployment.

Participating in Policy Dialogues: Active participation in multi-stakeholder policy discussions ensures that impact investors' perspectives contribute to shaping inclusive financial systems. Engagement with central banks, financial regulators, and international development organisations can help establish best practices and align global financial inclusion initiatives with local market conditions

7.5 Policy Implications for Impact Investing in Inclusive Finance

The effectiveness of impact investments hinges on the presence of an enabling policy environment that ensures regulatory clarity, financial stability, and accessibility for all stakeholders. Policymakers play a pivotal role in shaping the legal and institutional frameworks that promote responsible investment, mitigate risks, and facilitate financial services reaching underserved populations. This section provides a comprehensive analysis of the policy implications that can enhance the effectiveness and scalability of impact investing in inclusive finance.

7.5.1 Strengthening Regulatory Frameworks for Financial Inclusion

Robust and adaptive regulatory frameworks are fundamental to fostering financial inclusion while ensuring financial system stability. Policymakers must develop clear, well-defined, and forward-looking regulations that balance investor confidence, consumer protection, and financial sector integrity.

Developing Clear and Predictable Regulations: Uncertain or fragmented regulatory environments deter investment. Governments must establish coherent legal frameworks that reduce bureaucratic inefficiencies, harmonise financial sector policies across institutions, and provide investors with a transparent roadmap for compliance. This involves ensuring that financial regulations do not create bottlenecks for financial service providers while maintaining sufficient safeguards against financial instability. Governments must engage financial stakeholders, including banks, Microfinance Institutions, and impact investors, in the policy-making process to ensure that regulations are both practical and effective.

Enhancing Consumer Protection Mechanisms: While financial inclusion expands access, it also exposes vulnerable populations to risks such as predatory lending, over-indebtedness, and fraud. Effective policies must include strong consumer protection laws, transparent pricing structures, and grievance redress mechanisms. Financial regulators should establish dedicated oversight bodies that monitor compliance and ensure financial institutions uphold ethical lending practices. Additionally, financial literacy programmes should be introduced to equip consumers with knowledge of their rights, the risks of borrowing, and the implications of financial decisions.

Improving Financial Transparency and Accountability: Standardised reporting requirements on social and financial performance should be enforced to improve investor confidence. Mandatory impact reporting frameworks should require financial institutions to disclose lending patterns, service accessibility, and impact assessment metrics. This ensures that financial inclusion initiatives do not merely increase access but also deliver measurable improvements in socio-economic conditions. Furthermore, independent audits and public reporting of financial institutions' lending and investment practices can enhance accountability.

Encouraging Regulatory Sandboxes and Innovation Hubs: Policymakers should provide controlled environments where fintech firms and financial institutions can test new models and technologies under regulatory oversight. This approach reduces barriers to entry for innovative financial service providers while allowing regulators to evaluate risks before full-scale market adoption. Regulatory sandboxes enable experimentation in a controlled manner, reducing the potential systemic risks that new financial models might pose. Innovation hubs facilitate collaboration between regulators and financial technology providers to develop frameworks that accommodate emerging financial solutions while ensuring regulatory compliance.

7.5.2 Incentivising Private-Sector Participation

The private sector is a crucial driver of financial inclusion, yet many investors remain hesitant due to perceived risks, uncertain returns, and the lack of enabling conditions. Policymakers must implement strategic interventions that incentivise the private sector to deploy capital towards financial inclusion initiatives.

Tax Incentives and Targeted Subsidies: Governments should offer tax breaks, investment credits, and targeted subsidies to impact investors and financial institutions that focus on underserved markets. These incentives lower the cost of capital for investors while expanding financial services to marginalised groups. Additionally, tax deductions for financial institutions that engage in microfinance and digital inclusion initiatives can catalyse investment. Such incentives can be structured to reward long-term investments rather than short-term speculative activities.

Public-Private Partnerships (PPPs) for De-Risking Investments: By blending public funds with private capital, PPPs create de-risking structures that enhance the viability of financial inclusion projects. Governments can co-finance projects, provide first-loss guarantees, or establish investment funds that offer concessional capital to high-risk, high-impact ventures. These initiatives encourage the private sector to invest in high-risk markets by reducing their exposure to potential losses, thus making inclusive finance more attractive.

Credit Guarantees and Risk-Sharing Instruments: Many impact investors hesitate to lend in emerging markets due to high default risks. Governments and multilateral institutions should establish credit guarantee schemes that absorb a portion of investment losses, thereby mitigating investor risk and enabling more flexible lending models for SMEs and informal-sector entrepreneurs. Such mechanisms can be structured to cover different levels of risk exposure, ensuring that financial institutions remain engaged in lending to underserved sectors.

Concessional Finance for Social Enterprises: Policymakers can facilitate concessional loans, grants, and blended finance structures to support social enterprises working in financial inclusion. By lowering interest rates for Microfinance Institutions and digital lenders, concessional finance ensures affordability for end-users while maintaining investor returns. These instruments can be particularly useful in scaling innovative financial products tailored to low-income and underserved populations.

7.5.3 Leveraging Digital Financial Services for Broader Reach

Digital financial services are revolutionising financial inclusion by providing scalable, low-cost solutions that reach previously unbanked populations. However, realising the full potential of digital finance requires policymakers to establish regulatory and infrastructure frameworks that ensure sustainability, security, and accessibility.

Developing Interoperable Payment Systems: Fragmented financial infrastructures create inefficiencies and limit the scalability of digital financial services. Governments should facilitate the interoperability of mobile payment systems, banks, and fintech platforms to create seamless transaction experiences for users. Ensuring interoperability reduces transaction costs, enhances efficiency, and fosters a more inclusive financial ecosystem by allowing different service providers to connect and serve a larger market.

Enhancing Cybersecurity and Consumer Protection: The rise of digital financial platforms exposes users to risks such as data breaches, fraudulent transactions, and identity theft. Policymakers must implement cybersecurity regulations that mandate strong encryption protocols, secure authentication mechanisms, and user data protection frameworks. Additionally, public awareness campaigns on digital financial security should be rolled out to educate consumers on safe financial practices and how to detect fraud.

Expanding Digital Identity Systems: The absence of formal identification remains a major barrier to financial inclusion. Digital identity verification solutions, such as biometric authentication and blockchain-based digital IDs, can streamline financial onboarding processes and enhance security. Governments must prioritise the integration of digital identity infrastructure with financial services to improve access for marginalised populations. This will facilitate Know Your Customer processes and make it easier for underserved individuals to access banking and financial services.

7.5.4 Supporting Financial Literacy and Inclusion Programmes

Financial inclusion goes beyond providing access to financial services—it requires ensuring that individuals, businesses, and communities have the necessary knowledge and skills to effectively utilise financial products. Without adequate financial literacy, the benefits of inclusive finance may not be fully realised, leading to poor financial decision-making, over-indebtedness, and economic instability. Policymakers must integrate financial literacy initiatives into broader national financial inclusion strategies to ensure that financial services translate into meaningful economic empowerment.

Embedding Financial Education in School Curricula: Financial literacy should be institutionalised from an early age through formal education. Governments should incorporate financial education into national school curricula, covering essential topics such as budgeting, savings, investment planning, insurance, and responsible borrowing. Early exposure to financial principles fosters responsible financial behaviour and prepares young people to engage with financial institutions in a productive and informed manner.

Developing Nationwide Financial Literacy Campaigns: Public awareness campaigns should be developed to reach diverse populations, including rural communities, informal sector workers, and marginalised groups. These campaigns should leverage various communication platforms, including radio, television, social media, and community workshops, to disseminate information on financial management, fraud prevention, and responsible borrowing. Governments should collaborate with financial service providers and civil society organisations to ensure maximum outreach and impact.

Collaborating with Financial Institutions to Provide Tailored Training: Financial institutions, including banks, microfinance organisations, and fintech companies, should be incentivised to provide financial literacy training as part of their customer engagement strategies. These institutions should offer workshops, interactive online courses, and personalised financial coaching to equip individuals and small businesses with the skills to manage their finances effectively. Additionally, financial literacy programmes should be customised to cater to different target groups, such as women entrepreneurs, smallholder farmers, and first-time borrowers.

Implementing Behavioural Finance Techniques: Traditional financial education programmes often face challenges related to engagement and retention. Governments should integrate behavioural finance techniques, such as gamification, nudging strategies, and personalised financial advisory services, to make financial literacy programmes more engaging and effective. Digital financial literacy tools, including AI-powered financial coaching apps, can also help individuals develop better financial habits through interactive learning and real-time financial advice.

7.5.5 Addressing Gender and Socio-Economic Barriers to Finance

Women, rural populations, and low-income communities have historically faced systemic financial exclusion, limiting their ability to participate in economic growth and social mobility. Addressing these barriers is critical to ensuring that impact investing in inclusive finance leads to equitable development outcomes. Policymakers must implement targeted interventions that tackle gender biases, geographic disparities, and socio-economic constraints that impede access to financial services.

Mandating Gender-Responsive Financial Policies: Women entrepreneurs and female-led businesses remain disproportionately underserved in financial markets due to structural biases and traditional collateral-based lending models. Financial institutions must be incentivised or mandated to design financial products tailored to women's needs, including unsecured credit lines, lower collateral requirements, and business development support. Governments can require financial service providers to set gender-based lending quotas or offer incentives to institutions that exceed inclusion targets.

Expanding Financial Access in Rural and Marginalised Communities: Many financial institutions remain concentrated in urban centres, leaving rural populations with limited access to formal banking services. Policymakers must prioritise the expansion of financial infrastructure, including mobile banking agents, community cooperatives, and rural banking initiatives. Regulatory frameworks should encourage innovative delivery models such as digital lending platforms, village savings groups, and partnerships with agricultural cooperatives to ensure that rural populations have tailored and accessible financial solutions.

Providing Alternative Credit Mechanisms: Traditional credit-scoring systems exclude vast segments of the population who lack formal banking history or collateral. Governments and regulators must promote the adoption of alternative credit assessment tools that consider non-traditional data, such as mobile money transactions, utility bill payments, and local trade records. Artificial intelligence (AI) and machine learning can help develop fairer credit-scoring models that reduce systemic discrimination against financially excluded populations.

Ensuring Equitable Distribution of Financial Resources: Policymakers must ensure that financial inclusion initiatives do not inadvertently reinforce existing inequalities. National financial inclusion strategies should include mechanisms for monitoring and reporting on the distribution of financial resources across gender, income groups, and geographic locations. This can be achieved through gender-disaggregated data collection, financial access scorecards, and periodic impact assessments.

7.5.6 Anticipating Future Trends and Innovations

The financial landscape is evolving rapidly, with emerging technologies and global economic shifts reshaping the way financial services are accessed and delivered. Policymakers must adopt a forward-looking approach that anticipates these changes and ensures that financial inclusion strategies remain relevant, adaptive, and resilient in the face of technological advancements and market disruptions.

Regulating Artificial Intelligence and Big Data in Finance: The increasing reliance on AI and big data analytics in financial services presents opportunities and risks. AI-driven credit scoring, predictive lending, and automated investment advisory services have the potential to expand financial inclusion, but they also pose risks related to algorithmic bias and data privacy. Policymakers must establish ethical guidelines and governance structures that ensure transparency, accountability, and fairness in AI-driven financial services. Regulatory frameworks should mandate explainable AI models and establish mechanisms to detect and mitigate bias in automated decision-making.

Harnessing Blockchain for Financial Transparency: Distributed ledger technologies (DLT) have the potential to revolutionise financial transactions by improving security, efficiency, and transparency. Blockchain-based financial services, such as decentralised lending platforms and smart contract-enabled transactions, can reduce costs and expand financial access. Governments must establish regulatory clarity on the use of blockchain in financial services, ensuring that such innovations enhance, rather than disrupt, financial stability. Additionally, policymakers should explore the use of blockchain for remittance transfers and digital identity verification to facilitate secure and low-cost financial transactions.

Aligning Green Finance with Inclusive Finance: Climate change disproportionately affects financially vulnerable populations, particularly those in developing economies. Policymakers should integrate financial inclusion strategies with climate resilience initiatives, such as microfinance for climate-adaptive agriculture, green bonds for smallholder farmers, and insurance products that mitigate climate-related risks. Financial regulators should encourage financial institutions to develop sustainability-linked financial products that cater to low-income and climate-vulnerable communities.

Encouraging Decentralised Finance (DeFi) for Financial Inclusion: The emergence of decentralised finance (DeFi) presents new opportunities for financial inclusion by removing intermediaries and lowering transaction costs. DeFi platforms can enable peer-to-peer lending, cross-border remittances, and decentralised savings mechanisms that are accessible to unbanked populations. However, DeFi also poses regulatory challenges, including fraud risks, volatility, and lack of consumer protection. Policymakers must create regulatory sandboxes that allow for controlled experimentation with DeFi solutions while ensuring compliance with Anti-Money Laundering (AML) and consumer protection regulations.

7.6 Conclusion

The persistent financing gaps experienced by Micro, Small, and Medium Enterprises (MSMEs), low-income households, women, and rural populations illustrate the extent to which conventional financial models continue to overlook the needs of marginalised groups. Market failures, systemic inefficiencies, and entrenched socio-economic barriers further reinforce these disparities, necessitating innovative approaches that go beyond traditional lending models. While impact capital can provide a means to direct financial resources towards underserved communities while maintaining financial sustainability, the effectiveness of the intervention depends significantly on the presence of enabling environments, supportive regulatory frameworks, and multi-stakeholder collaborations that align financial inclusion objectives with broader development goals. Policy plays a crucial role in shaping the trajectory of financial inclusion through impact investing. Governments, Development Finance Institutions (DFIs), and regulatory bodies must work in tandem with private-sector actors to create conditions conducive to sustainable investment. This involves reducing systemic barriers such as information asymmetries, high transaction costs, and institutional biases in credit allocation. Furthermore, the integration of digital financial services and fintech solutions presents a promising avenue for expanding financial reach, provided that concerns around cybersecurity, data privacy, and consumer protection are adequately addressed. By deepening collaboration across sectors, refining financial instruments to cater to diverse needs, and continuously innovating in response to market failures, impact investing can serve as a cornerstone in the global effort to close financial access gaps and drive inclusive growth. Impact investing is not a panacea, but it is an indispensable tool in the broader effort to create equitable financial systems that serve all members of society.

Notes

1. See discussion on how informal lending exacerbate financial strain on small enterprises in Allen et al. (2019) and Yang et al. (2023).
2. Akang (2023) discusses the burden of compliance-related expenses.

References

Akang, A. U. (2023). Regulatory compliance and access to finance: Implications for business growth in developing economies. *Sciental Journal of Education Humanities and Social Sciences, 1*(2), 8–23.

Allen, F., Qian, M., & Xie, J. (2019, July 01). Understanding informal financing. *Journal of Financial Intermediation, 39*, 19–33. https://doi.org/10.1016/j.jfi.2018.06.004

Andersson Djurfeldt, A. (2020, January 01). Gendered land rights, legal reform and social norms in the context of land fragmentation—A review of the literature for Kenya, Rwanda and Uganda. *Land Use Policy, 90*, 104305. https://doi.org/10.1016/j.landusepol.2019.104305

Beech, B. M., Ford, C., Thorpe, R. J., Jr., Bruce, M. A., & Norris, K. C. (2021). Poverty, racism, and the public health crisis in America. *Frontiers in Public Health, 9*, Article 699049. https://doi.org/10.3389/fpubh.2021.699049

Behr, D., Groussard, H., Khaitina, V., & Shen, L. (2023). *Women's land rights in sub-Saharan Africa: Where do we stand in practice.* World Bank Group Global Indicators Brief No. 23.

Blimpo, M. P., & Cosgrove-Davies, M. (2019). *Electricity access in Sub-Saharan Africa: Uptake, reliability, and complementary factors for economic impact.* World Bank Publications.

CGAP. (2020). *CGAP launches global microfinance survey to inform COVID-19 response.*

Dabla-Norris, E. (2015). *Causes and consequences of income inequality: A global perspective*. International Monetary Fund.
de la Cuesta-Gonzalez, M., Paredes-Gazquez, J., Ruza, C., & Fernandez-Olit, B. (2021, February 01). The relationship between vulnerable financial consumers and banking institutions. A qualitative study in Spain. *Geoforum, 119*, 163–176. https://doi.org/10.1016/j.geoforum.2021.01.006
Demirgüç-Kunt, A., & Klapper, L. F. (2012). *Measuring financial inclusion: The global findex database* (Working Paper 6025). World Bank Policy Research. https://papers.ssrn.com/sol3/papers.cfm?abstract_id=2043012
Demirgüç-Kunt, A., Klapper, L., Singer, D., & Ansar, S. (2022). *The global findex database 2021: Financial inclusion, digital payments, and resilience in the age of COVID-19* (The Global Findex Database). W. B. Publications. https://www.worldbank.org/en/publication/globalfindex/Report
GPFI. (2024). *G20 global partnership for financial inclusion action plan for micro, small, and medium enterprise financing*.
Global Impact Investing Network. (2022). *GIINSIGHT: Enhancing gender equality through financial inclusion*. GIIN. https://thegiin.org/research/publication/giinsight-gender-equality-financial-inclusion
Hendriks, S. (2019). The role of financial inclusion in driving women's economic empowerment. *Development in Practice, 29*(8), 1029–1038.
IFC. (2022). *Banking on SMEs: Driving growth, creating jobs global SME finance facility progress report*.
IFC. (2024b). *MSME Finance*. https://www.ifc.org/en/what-we-do/sector-expertise/financial-institutions/msme-finance
International Finance Corporation. (2024, March). *IFC portfolio default rate analysis*. World Bank Group. https://www.ifc.org
Khan, F. U., Nouman, M., Negrut, L., Abban, J., Cismas, L. M., & Siddiqi, M. F. (2024, December 31). Constraints to agricultural finance in underdeveloped and developing countries: a systematic literature review. *International Journal of Agricultural Sustainability, 22*(1), 2329388. https://doi.org/10.1080/14735903.2024.2329388
Kumar, S. S., & Jie, Q. (2023, December 01). Exploring the role of financial inclusion in poverty reduction: An empirical study. *World Development Sustainability, 3*, 100103. https://doi.org/10.1016/j.wds.2023.100103
Maravalle, A., & Pandiella, A. G. (2022). *Expanding access to finance to boost growth and reduce inequalities in Mexico*. OECD.
Mohsen, S. E., Hamdan, A., & Shoaib, H. M. (2024). Digital transformation and integration of artificial intelligence in financial institutions. *Journal of Financial Reporting and Accounting, 23*(2), 680–699.
Mori, M. (2023). Urban-rural polarization: Financial implications, challenges, and opportunities. *Journal of Innovation in Business and Economics, 7*(02), 159–166.

OECD. (2022). *Policy brief on access to finance for inclusive and social entrepreneurship: What role can fintech and financial literacy play?*

OECD. (2024). *Global insurance market trends 2024*. O. Publishing. https://doi.org/10.1787/5b740371-en

Ondiege, P., Moyo, J. M., & Verdier-Chouchane, A. (2013). Developing Africa's infrastructure for enhanced competitiveness. *The Africa Competitiveness Report, 2013*, 69–91.

Perrin, C., & Hyland, M. (2023). *Gendered laws and women's financial inclusion*. World Bank.

Siddik, M., & Kabiraj, S. (2018). Does financial inclusion induce financial stability? Evidence from cross-country analysis [Article]. *Australasian Accounting Business and Finance Journal, 12*(1), 33–46. https://doi.org/10.14453/aabfj.v12i1.3

Simba, A., Tajeddin, M., Dana, L.-P., & Ribeiro Soriano, D. E. (2024, January 01). Deconstructing involuntary financial exclusion: A focus on African SMEs. *Small Business Economics, 62*(1), 285–305. https://doi.org/10.1007/s11187-023-00767-1

Taconet, N., Mejean, A., & Guivarch, C. (2020). Influence of climate change impacts and mitigation costs on inequality between countries. *Climatic Change, 160*(1), 15–34.

van Zanden, J. L. (2023). Examining the relationship of information and communication technology and financial access in Africa. *Journal of Business and Economic Options, 6*(3), 26–36.

Woetzel, J. (2015). *The power of parity: How advancing women's equality can add $12 trillion to global growth*. McKinsey Global Institute.

World Economic Forum. (2021, November). *Future readiness of SMEs: Mobilizing the SME sector to drive widespread sustainability and prosperity*. https://www.weforum.org/reports/future-readiness-of-smes/

World Bank. (2022). *World development report 2022: Finance for an equitable recovery*. Washington, DC: World Bank. https://doi.org/10.1596/978-1-4648-1843-1

World Bank. (2024). *Financial Inclusion : A key enabler to reducing poverty and boosting prosperity*. The World Bank. https://www.worldbank.org/en/topic/financialinclusion/overview

Yang, F., Ye, X., Huang, W., & Zhao, X. (2023, December 01). The impacts on informal financing strategy of small and micro enterprises by interest rate risks and public health emergencies. *International Entrepreneurship and Management Journal, 19*(4), 1673–1705. https://doi.org/10.1007/s11365-023-00872-3

PART IV

Looking Ahead

CHAPTER 8

Long-Term Sustainability of Financial Inclusion Strategies

8.1 Introduction

Ensuring the longevity and relevance of financial inclusion initiatives requires a thoughtful approach to sustainability. This entails designing systems and solutions that not only address immediate challenges but also remain effective in the face of evolving socio-economic conditions. True sustainability ensures that individuals and communities are equipped to continue participating in the financial ecosystem without external dependency. Sustainable financial inclusion initiatives aim to tackle the underlying causes of exclusion, fostering long-term resilience and adaptability. Central to this is building local capacity through financial literacy programmes that empower individuals to make informed decisions. Solutions must align with the specific needs of communities, leveraging local knowledge to enhance relevance and acceptance. Additionally, integrating technology ensures that initiatives are scalable and cost-effective, enabling widespread impact. This chapter delves into key factors critical for achieving sustainability. These include the role of partnerships in pooling resources and expertise, the importance of aligning with supportive policies, fostering community engagement for ownership and feedback, designing scalable and adaptable models, and incorporating environmental and social considerations. By addressing these foundational

N. Udohaya, *Impact Investing and Financial Inclusion*, Sustainable Development Goals Series,
https://doi.org/10.1007/978-3-031-87841-1_8

elements, financial inclusion efforts can create enduring impact and pave the way for inclusive growth and prosperity.

8.2 The Role of Partnerships

Expanding Reach and Effectiveness: Partnerships are the backbone of successful financial inclusion initiatives, enabling diverse stakeholders to combine their resources, expertise, and networks. Collaborative efforts between governments, private-sector entities, non-governmental organisations (NGOs), and international development agencies create synergies that amplify the reach and impact of these programmes. Governments play a critical role in creating enabling environments by providing regulatory frameworks, infrastructure, and funding. National policies that prioritise financial inclusion, such as simplifying account-opening processes or offering incentives for financial literacy programmes, create fertile ground for private and non-profit collaboration. NGOs and development agencies often bring localised expertise and a deep understanding of community needs, ensuring that initiatives are relevant and culturally sensitive. Meanwhile, private-sector entities, including banks and fintech companies, contribute technological innovation, operational efficiency, and capital, which are essential for scaling financial services effectively. For example, the mobile money platform, developed through a partnership between Safaricom and the Kenyan government, has become a global benchmark for digital financial inclusion. By leveraging Safaricom's technological expertise and the government's policy support, M-Pesa transformed access to financial services for millions of Kenyans, particularly those in rural areas. The initiative's success underscores the transformative potential of multi-stakeholder collaborations when aligned with a shared mission.

Aligning Public Policy with Private Innovation: Public–Private Partnerships (PPPs) represent a powerful model for advancing financial inclusion. These partnerships align the social objectives of public policy with the innovation, efficiency, and agility of the private sector. By fostering collaboration between governments and private entities, PPPs help bridge gaps in infrastructure, funding, and expertise, accelerating the delivery of financial services to underserved populations (Carbonara & Pellegrino, 2020). One key advantage of PPPs is their ability to combine the public sector's focus on equity and access with the private sector's ability to innovate and scale. Governments provide the regulatory frameworks and public trust needed for such initiatives to succeed, while private companies supply the technological tools and operational capacity. India's Pradhan Mantri Jan Dhan Yojana (PMJDY) exemplifies this. The government-led financial inclusion programme partnered with financial institutions to open over 400 million bank accounts for previously unbanked individuals. The programme's success was driven by the integration of government subsidies, private banking services, and technological platforms such as the Aadhaar biometric identification system. This collaboration not only increased access to formal financial services but also enhanced the delivery of social welfare programmes through direct benefit transfers.

Cross-Sector Knowledge Exchange: Cross-sector knowledge exchange forms a cornerstone of innovation in financial inclusion. It leverages the unique expertise of each stakeholder, fostering solutions that are both innovative and contextually relevant. NGOs, for example, provide an on-the-ground understanding of social and cultural norms, enabling financial institutions to tailor products to community needs effectively. This is particularly critical in rural areas, where trust and cultural alignment significantly impact adoption rates. Private-sector organisations, on the other hand, bring advanced technological solutions and operational efficiency, enhancing the scalability and accessibility of financial services. Governments benefit from both these inputs by adopting successful models into national policy frameworks, ensuring broad implementation. The Bangladesh's microfinance sector, cross-sector partnerships have led to innovative lending products that address rural needs while maintaining regulatory compliance, significantly increasing outreach and impact.

Global and Regional Alliances: Global and regional alliances are pivotal in overcoming systemic barriers to financial inclusion, such as fragmented infrastructure and regulatory disparities. International organisations like the Better Than Cash Alliance provide platforms for stakeholders worldwide to collaborate on promoting digital financial systems, focusing on transitioning from cash to digital payments, and have facilitated large-scale government payment digitalisation in numerous countries. These alliances create global standards, pool resources, and share best practices, allowing member nations to accelerate financial inclusion initiatives. At a regional level, collaborations address specific geographic challenges. For example, the African Development Bank (AfDB) has played a critical role in Sub-Saharan Africa by working with governments and telecom operators to extend mobile money services to rural areas. These efforts have transformed access to financial services for millions, overcoming the constraints of traditional banking infrastructure. Similarly, in Southeast Asia, regional partnerships focus on improving cross-border payment systems, which are crucial for migrant workers sending remittances home.

Technology Partnerships: Technology partnerships revolutionise financial inclusion by enabling financial services to reach underserved populations in an efficient and cost-effective manner. Fintech companies, mobile network operators, and tech giants provide the technological backbone for initiatives that extend financial services into remote and underserved areas. For instance, mobile money services in Africa have benefited from collaborations between telecom companies and banks, delivering banking capabilities to communities without physical branches. Emerging technologies such as blockchain have been instrumental in developing secure and transparent cross-border payment systems, catering to the needs of migrant workers and their families. Cloud computing, another game-changer, empowers financial institutions to store and analyse vast amounts of customer data, enabling more informed decision-making and personalised service delivery. A prime example is the partnership between financial institutions and mobile operators in Nigeria, which has enabled the proliferation of mobile money services. This partnership model demonstrates how technology can lower barriers and expand access to underserved demographics.

Community-Based Partnerships: Community-based partnerships anchor financial inclusion efforts by bridging the gap between formal financial systems and local populations. Collaborations with local leaders, cooperatives, and community organisations enhance outreach efforts by leveraging the trust and credibility these entities hold within their communities. This trust is particularly important in areas where formal institutions are viewed with scepticism or where literacy levels are low. Microfinance initiatives have successfully used these partnerships to deliver credit and financial literacy programmes to rural populations. For example, in India, self-help groups (SHGs) facilitated by NGOs and banks have enabled women in rural areas to access small loans for entrepreneurship. These partnerships not only improve access to financial services but also create pathways for economic empowerment. Furthermore, such collaborations often include financial education components, ensuring that individuals use financial products effectively, thereby reinforcing the long-term impact of financial inclusion initiatives.

Partnerships must be designed with a focus on equity, accountability, and shared goals. Clear communication and mutual understanding among stakeholders are critical for maintaining alignment and achieving desired outcomes. Moreover, partnerships should include mechanisms for regular monitoring and evaluation, allowing for adaptive strategies that respond to changing needs and challenges. By fostering inclusive and well-coordinated partnerships, financial inclusion initiatives can achieve greater scale and effectiveness. These collaborations serve as a vital catalyst for innovation, resource mobilisation, and knowledge-sharing, ultimately driving sustainable progress towards a more inclusive financial ecosystem.

8.3 Policy Alignment

Supportive Regulatory Frameworks: A well-structured regulatory framework serves as the foundation for advancing financial inclusion, ensuring that financial services remain accessible, affordable, and secure for underserved populations. One of the most effective strategies in achieving this is through simplified account-opening processes. Traditional banking procedures, often laden with stringent Know Your Customer (KYC) requirements, pose a significant barrier to low-income individuals who may lack formal identification. To enhance financial inclusion in a more innovative and timely manner, governments and regulatory bodies should explore the adoption of AI-driven identity verification and digital KYC processes. These technologies streamline account opening by leveraging biometric authentication, real-time data analysis, and blockchain verification, reducing reliance on traditional documentation while ensuring robust security measures against financial crimes. The rise of mobile banking and digital financial services has further underscored the importance of regulatory adaptation. Policies that encourage the growth of mobile money platforms and branchless banking create new avenues for financial inclusion. Clear and well-defined guidelines on licensing, interoperability, and consumer protection within the digital financial ecosystem foster competition, enhance service delivery, and broaden access to financial products. Additionally, regulations supporting the growth of Microfinance Institutions (MFIs) and

Small and Medium-sized Enterprise (SME) lenders can provide much-needed financial support to excluded groups. By ensuring that MFIs have clear operational guidelines and sustainable funding sources to maintain liquidity and expand their outreach, policymakers can strengthen their role in providing essential financial services to low-income communities. Balancing financial stability with inclusive growth remains a delicate act for regulators. While policies should promote access, they must also mitigate risks such as fraud and money laundering. An effective regulatory framework ensures that financial institutions operate within ethical and sustainable parameters while serving the broader goal of financial inclusion.

Incentives for Private Investment: Governments can introduce a range of incentives designed to attract investment into underserved financial markets. Tax benefits, for instance, provide financial institutions with the necessary motivation to expand their services to marginalised groups or operate in rural areas. These tax incentives can be structured to reward banks and fintech companies that actively engage in financial inclusion projects. Beyond tax relief, subsidies and matching funds can further stimulate private-sector investment. Public–Private Partnerships (PPPs) become particularly effective when the government provides subsidies or matches funds for institutions that invest in inclusive financial services. This approach not only encourages innovation but also mitigates the risks that private investors might face when entering underserved markets. Interest rate subsidies serve as a powerful incentive for private investment in financial inclusion. By lowering the cost of capital for financial institutions that lend to underserved populations, interest rate subsidies reduce the perceived risk and improve the commercial viability of lending to low-income individuals and small businesses. Additionally, direct grants for capacity-building and technology development encourage financial service providers to innovate and expand their reach to marginalised groups. Additionally, credit guarantee schemes offer another layer of security for financial institutions willing to lend to higher-risk segments, such as small businesses, women entrepreneurs, and rural borrowers. These schemes reduce lenders' exposure to default risk, making financial inclusion a more viable and sustainable investment. When aligned with market-driven incentives, policies designed to encourage private-sector engagement create a dynamic ecosystem where financial institutions can contribute meaningfully to inclusive finance without sacrificing their commercial sustainability.

Consumer Protection: Trust is a cornerstone of financial inclusion, and ensuring robust consumer protection mechanisms is vital in fostering confidence among underserved communities. Without proper safeguards, financial exclusion can deepen as individuals become wary of engaging with financial institutions. One of the fundamental elements of consumer protection is ensuring transparency in financial services. Financial institutions must provide clear, standardised information on fees, interest rates, and terms of service. Complex financial jargon often deters individuals from making informed decisions, and as such, disclosures should be simplified and made available in local languages and culturally relevant formats. In addition to transparency, fair pricing regulations help prevent predatory lending practices. Without adequate oversight, some financial institutions may exploit vulnerable customers through excessive interest rates and hidden fees. To prevent such exploitation, governments can introduce interest rate caps on microloans and establish regulations against aggressive debt collection practices. These measures ensure that financial services remain accessible and fair, rather than contributing to cycles of unsustainable debt. Another pressing concern in digital financial services is data privacy and security. As more individuals engage with mobile banking and digital payment platforms, the risk of data breaches and unauthorised access to personal financial information increases. Strong data protection regulations must be in place to safeguard consumers from potential harm. Clear guidelines on data storage, usage, and consent empower individuals with greater control over their financial data, fostering a sense of security in digital transactions. Beyond safeguarding financial transactions, accessible and independent dispute resolution mechanisms play a critical role in upholding consumer rights. Ombudsman services, customer protection agencies, and alternative dispute resolution frameworks provide recourse for individuals facing fraud, service failures, or unfair treatment. When consumers trust that they have avenues to seek justice and redress, they are more likely to engage with financial products and services, thereby enhancing the sustainability of financial inclusion.

Policy alignment is a fundamental strategy in ensuring the long-term sustainability of financial inclusion. A supportive regulatory framework lays the groundwork for financial accessibility, while incentives for private-sector engagement foster sustainable investments in inclusive finance. At the same time, robust consumer protection measures instil trust, ensuring that financial inclusion efforts remain both ethical and impactful. By aligning policies in these three key areas, governments and stakeholders can create resilient financial systems that serve the needs of marginalised populations while contributing to broader economic and social development objectives. The interplay between regulatory support, private-sector incentives, and consumer protection forms a comprehensive strategy that enhances financial inclusion's sustainability, making it not just an initiative, but a long-term driver of economic empowerment.

8.4 Community Engagement

Local Ownership: The sustainability of financial inclusion initiatives is greatly enhanced when communities actively participate in their design and implementation. When financial inclusion efforts are imposed externally without considering local contexts, they often fail to gain traction. However, by engaging communities from the outset, financial service providers and policymakers can foster a sense of ownership among local populations, ensuring that solutions are relevant and embraced by the people they aim to serve. Local ownership can take various forms, such as community-based financial cooperatives, participatory governance models, and grassroots financial education programmes. These structures empower community members to participate actively in decision-making, ensuring that financial services align with their cultural, economic, and social realities. Furthermore, leveraging local networks, such as religious institutions, community leaders, and informal savings groups, can improve trust in financial products and encourage widespread adoption. Embedding financial inclusion initiatives within existing community structures fosters long-term sustainability, as communities develop a personal stake in maintaining and improving the systems established.

Financial Literacy Programmes: Financial literacy serves as the foundation for meaningful financial inclusion. Without adequate knowledge and skills, individuals may lack the confidence to engage with financial products effectively. Education initiatives aimed at enhancing financial literacy can empower individuals to make informed financial decisions, reduce the risk of over-indebtedness, and maximise the benefits of financial inclusion. To be effective, financial literacy programmes must be tailored to the specific needs and contexts of the communities they serve. Traditional classroom-based approaches may not always be the most effective, especially in rural or underserved areas with low literacy levels. Alternative approaches, such as interactive workshops, peer-to-peer learning, gamification, and digital tools, can enhance engagement and improve knowledge retention. Additionally, integrating financial literacy into school curricula can instil positive financial habits from an early age, creating a more financially aware future generation. Moreover, partnerships with financial institutions, fintech firms, and non-governmental organisations (NGOs) can enhance the reach and effectiveness of financial literacy initiatives. Leveraging technology, such as mobile apps and social media campaigns, allows financial education to be disseminated widely, overcoming geographical and logistical barriers. Ensuring that financial literacy programmes are accessible in local languages and culturally sensitive formats further enhances their impact, making financial education an inclusive and empowering tool for long-term sustainability.

Feedback Mechanisms: A critical yet often overlooked aspect of community engagement in financial inclusion is the establishment of effective feedback mechanisms. Financial service providers and policymakers must create channels through which communities can voice their experiences, challenges, and suggestions regarding financial products and services. Without these mechanisms, financial inclusion initiatives risk becoming static, failing to adapt to evolving community needs and technological advancements. Feedback mechanisms can take various forms, including community forums, surveys, digital reporting platforms, and participatory policy reviews. Leveraging technology, such as mobile-based feedback tools and AI-driven sentiment analysis, can enable real-time monitoring of customer experiences, allowing for prompt corrective actions. Additionally, engaging local community representatives as financial inclusion ambassadors can facilitate two-way communication between service providers and end-users, ensuring that voices from the grassroots level are heard and considered in decision-making processes. Beyond gathering feedback, it is crucial to establish systems for implementing the insights gained. Policymakers and financial institutions should demonstrate responsiveness by adjusting policies, refining financial products, and improving service delivery based on community feedback. This iterative approach not only enhances trust but also ensures that financial inclusion remains dynamic and attuned to the realities of the communities it seeks to serve.

Community engagement is a fundamental pillar in ensuring the long-term sustainability of financial inclusion. Fostering local ownership, enhancing financial literacy, and establishing effective feedback mechanisms enable financial inclusion initiatives to become deeply embedded within communities, promoting resilience and adaptability. When communities actively participate in shaping financial services, they are more likely to embrace and sustain these systems, creating a more inclusive and equitable financial landscape. A people-centric approach to financial inclusion ensures that interventions are impactful and enduring, contributing to broader socio-economic development goals.

8.5 Scalability and Adaptability

For financial inclusion to achieve sustained impact, it must be designed with scalability in mind. A financial inclusion model that succeeds in a small pilot phase but fails to expand effectively across larger populations risks becoming a fragmented or isolated initiative. Scalable solutions leverage technology, institutional partnerships, and innovative distribution channels to ensure broader outreach and long-term sustainability. Product standardisation and regulatory harmonisation are key factors in achieving scalability. Ensuring that financial products adhere to consistent standards across different jurisdictions allows for seamless expansion, reducing the complexity and cost of entering new markets. Financial literacy and awareness campaigns should also be designed with scalability in mind, using mass communication platforms like radio, social media, and community outreach programmes to educate large populations about financial products and services.

While scalability ensures that financial inclusion initiatives reach a broader audience, adaptability guarantees their continued relevance in an ever-changing socio-economic landscape. Financial inclusion strategies must remain dynamic and flexible in responding to shifts in technology, economic conditions, and consumer behaviour. One of the most pressing reasons for adaptability is the rapid advancement of financial technology (fintech). Innovations such as blockchain-based financial services, artificial intelligence-driven credit scoring, and decentralised finance (DeFi) solutions are continuously transforming the financial landscape. Financial inclusion strategies must integrate these technological advancements to ensure they remain relevant and effective. Regulators and policymakers must also adopt flexible regulatory approaches that allow for the safe experimentation and integration of new financial technologies while maintaining consumer protection and financial stability. Socio-economic shifts also necessitate adaptability in financial inclusion efforts. Crises such as pandemics, economic recessions, and climate-related disasters can dramatically alter the financial needs of communities. For instance, the COVID-19 pandemic accelerated the demand for contactless payments and digital lending platforms, highlighting the need for financial inclusion initiatives to adapt quickly to emerging realities. Governments and financial institutions must implement adaptive measures such as emergency financial relief programmes, digital disbursements, and targeted financial literacy campaigns that address the evolving needs of affected populations.

Additionally, consumer preferences and behaviours evolve over time, requiring financial service providers to assess and refine their offerings continually. Conducting regular consumer research and employing agile product development approaches can ensure that financial products remain user-centric and inclusive of diverse demographic groups. Adaptive financial inclusion models also require strong feedback loops that allow for continuous learning and improvement. Mechanisms such as community consultations, user surveys, and real-time data analytics can help identify emerging challenges and opportunities, enabling financial service providers to adjust their strategies accordingly. Maintaining a flexible and responsive approach ensures that financial inclusion efforts remain effective and resilient, even in the face of uncertainty.

Scalability and adaptability are indispensable strategies for ensuring the long-term sustainability of financial inclusion. Scalable models enable financial services to reach a larger population efficiently, leveraging digital technologies, partnerships, and standardised frameworks to facilitate expansion. Meanwhile, adaptability ensures that financial inclusion initiatives remain relevant in response to changing technological, economic, and consumer dynamics. A dual focus on both scalability and adaptability allows financial inclusion strategies to achieve widespread and enduring impact, fostering inclusive financial ecosystems that are resilient, innovative, and responsive to the needs of diverse communities.

8.6 Environmental and Social Considerations

Integrating Environmental Sustainability: Financial inclusion cannot be considered sustainable unless it incorporates environmental sustainability. Climate change, environmental degradation, and resource scarcity pose significant risks to economic stability, particularly for vulnerable communities. Integrating green finance initiatives within financial inclusion efforts ensures that economic development does not come at the cost of environmental harm (Tran & Oanh, 2023). One effective approach is the promotion of green financing options that support climate resilience while advancing financial inclusion. Green microfinance, for instance, enables small-scale entrepreneurs and farmers to access funding for environmentally sustainable projects, such as solar energy solutions, water conservation systems, and climate-smart agriculture. Additionally, financial institutions can introduce preferential lending rates for businesses that adopt sustainable practices, thereby incentivising environmentally responsible economic activities. The use of fintech also plays a crucial role in driving environmentally sustainable financial inclusion. Digital banking and paperless transactions reduce carbon footprints, while blockchain-based tracking systems enhance transparency in green financing. Governments and financial institutions can collaborate to establish regulatory frameworks that encourage sustainable investment practices and hold financial actors accountable for environmental impacts.

Promoting Social Equity: In addition to economic participation, financial inclusion should also address social equity to ensure that all individuals, regardless of gender, socio-economic status, or geographical location, have equal access to financial services. Inclusive finance can serve as a powerful tool for reducing inequality and empowering marginalised communities, particularly women, youth, and indigenous populations. A key strategy for promoting social equity in financial inclusion is the development of tailored financial products that cater to the specific needs of underserved groups. For example, Microfinance Institutions can design loan products that consider the financial constraints of low-income women entrepreneurs while offering flexible repayment structures. Similarly, youth-friendly financial products can help young individuals build credit histories and access capital for education or business ventures. Another crucial aspect is ensuring accessibility in financial services. Digital financial inclusion has expanded opportunities for many, yet technological barriers still exclude some populations. Financial institutions must invest in literacy programmes and user-friendly interfaces that accommodate those with limited digital skills. Moreover, regulatory frameworks should enforce anti-discriminatory practices in lending and banking, preventing systemic biases that exclude marginalised communities from accessing financial services. Financial inclusion initiatives should also incorporate social impact metrics to measure their effectiveness in improving social equity. Tracking data on gender balance in loan approvals, economic upliftment of disadvantaged groups, and access to financial resources in rural areas enables policymakers and financial institutions to make data-driven decisions that enhance inclusivity.

The Intersection of Environmental and Social Sustainability: Environmental sustainability and social equity are interdependent elements of sustainable financial inclusion. Many underserved communities are disproportionately affected by environmental challenges, making it imperative to integrate both aspects into financial inclusion strategies. For example, climate-resilient financial products that provide insurance and emergency funds for communities prone to natural disasters can mitigate climate-induced economic shocks while promoting financial stability. Additionally, sustainable infrastructure investment can drive both environmental and social benefits. Initiatives that finance renewable energy projects in off-grid rural areas not only reduce carbon emissions but also expand financial access to previously unbanked populations. Governments and financial institutions should prioritise such dual-impact investments that generate long-term economic, environmental, and social benefits. Collaboration between policymakers, financial institutions, and community organisations is key to ensuring the successful implementation of environmentally and socially inclusive financial models. Fostering multi-stakeholder partnerships allows financial inclusion initiatives to align with global sustainability goals and contribute to holistic development that benefits both people and the planet.

Integrating environmental sustainability and social equity into financial inclusion strategies ensures holistic and long-lasting development. Green financing mechanisms, inclusive financial products, and regulatory frameworks that promote environmental responsibility and social inclusion are essential for the sustainability of financial inclusion efforts. Adopting a comprehensive approach that considers both environmental and social factors enables financial inclusion to drive not just economic growth, but also a more equitable and sustainable future for all.

8.7 Conclusion

Sustaining financial inclusion requires more than providing access. It demands resilience, adaptability, and integration within broader socio-economic systems. Financial inclusion must be embedded in local realities, responsive to evolving needs, and designed to generate lasting empowerment rather than dependency. A key insight is that collaboration is indispensable. Effective partnerships across public, private, and civil society sectors allow for shared expertise, innovative models, and sustainable resource mobilisation. Without aligned and coordinated efforts, financial inclusion risks fragmentation and inefficiency. Financial inclusion policies must balance accessibility with consumer protection. It is also crucial that regulatory frameworks simplify entry while safeguarding against exploitation are essential. Trust remains the foundation of financial participation, and clear, ethical regulations ensure long-term engagement from individuals and institutions alike. Community-driven approaches is a linchpin of sustainability. When communities are active stakeholders, they shape solutions that reflect their needs and realities. Financial literacy, local ownership, and feedback mechanisms are not peripheral considerations but central to ensuring financial inclusion translates into economic empowerment. The longevity of financial inclusion strategies will be determined by how scalable and adaptable they are. The rapid evolution of technology and economic conditions necessitates models that can expand without losing efficiency while remaining flexible enough to adjust to shifting landscapes. A rigid approach will quickly become obsolete in the face of technological disruption and socio-economic changes. Finally, sustainability must be viewed holistically, integrating environmental and social considerations. Financial inclusion should not only serve economic objectives but also align with green finance principles and social equity. Inclusive and responsible financial systems contribute to resilience of individuals but of entire communities and economies. In sum, financial inclusion must transcend a narrow focus on access and evolve into an adaptive, community-driven, and ethically governed system. Embedding sustainability at its core enables it to serve as a powerful driver of equitable development and long-term prosperity.

References

Carbonara, N., & Pellegrino, R. (2020, February 01). The role of public private partnerships in fostering innovation. *Construction Management and Economics, 38*(2), 140–156. https://doi.org/10.1080/01446193.2019.1610184

Tran, O., & Oanh, K. (2023, November 09). Sustainable development: Driving force from the relationship between finance inclusion, green finance and green growth. *Sustainable Development, 32*. https://doi.org/10.1002/sd.2808

CHAPTER 9

Conclusion

From its ethical and religious roots into a sophisticated financial movement that integrates financial returns with tangible social and environmental impact, impact investing has evolved tremendously. The shift from SRI to proactive impact investing marks an ideological change and a redefinition of financial priorities. The historical milestones that shaped this transformation—from the Pax World Fund and the Sullivan Principles to the rise of shareholder activism—highlight the growing demand for investments beyond profit maximisation. Movements against apartheid, environmental degradation, and poor corporate governance further reinforced the necessity of embedding ethical considerations into financial strategies. The introduction of ESG factors and the UN-backed PRI signalled a turning point, legitimising the integration of social responsibility within mainstream finance. Impact investing is distinguished from traditional finance because it emphasises intentionality, where investors actively seek measurable social and environmental benefits alongside financial returns. With key principles such as intentionality, financial performance expectations, impact measurement and management (IMM), and contributions beyond capital, impact investing aligns capital with broader societal and environmental objectives. These principles ensure that impact investing generates financial returns and fosters

N. Udohaya, *Impact Investing and Financial Inclusion*, Sustainable Development Goals Series,
https://doi.org/10.1007/978-3-031-87841-1_9

long-term, sustainable development by integrating accountability, measurable impact, and a commitment to responsible stewardship of capital. Investment structures such as green bonds, social impact bonds, and blended finance serve as mechanisms to align capital with purpose, demonstrating that investments can be structured to yield both financial viability and long-term sustainability. However, for impact investing to achieve scale, regulatory frameworks must evolve, and financial incentives must be aligned to support a sustained commitment to these principles.

At the same time, financialisation has altered growth paradigms. While it has contributed to economic expansion and capital accumulation, the relentless prioritisation of shareholder value has often led to the neglect of broader social and environmental considerations. This singular focus on short-term financial gains has, in many cases, undermined sustainability efforts and deepened economic inequalities, leaving marginalised communities without adequate access to financial resources or development opportunities. A central debate within impact investing is whether financialisation is inherently exploitative or whether it can be leveraged to drive positive social change. The tendency of capital to concentrate in profitable markets often marginalises vulnerable communities, exacerbating structural inequalities. Sustainable finance must integrate mechanisms that align financial incentives with long-term environmental and social well-being, moving beyond the narrow focus on short-term gains. Technology is proving to be a crucial enabler in this transition. Innovations such as blockchain, artificial intelligence (AI), and regenerative metrics allow for more accurate measurement of sustainability outcomes. These tools enhance transparency, accountability, and efficiency in the deployment of impact capital. Real-time financial decision-making powered by emerging technologies ensures greater transparency and responsiveness, strengthening the credibility of impact investing.

As more and more investors participate in impact investing, North America and Europe currently dominate capital allocation, and emerging markets in Sub-Saharan Africa, Southeast Asia, and Latin America are gaining traction. Challenges remain, including fragmented impact measurement frameworks, regulatory uncertainties, and macroeconomic shifts. A promising development within the field is the rise of Gender Lens Investing (GLI), which explicitly targets financial inclusion and economic empowerment for women and women-led businesses. The approach recognises the economic advantages of gender equity and the importance of directing capital towards underrepresented groups. Financial inclusion remains among the most potent economic empowerment and poverty reduction levers. Traditional financial markets have long excluded low-income and marginalised communities due to perceived risks, perpetuating cycles of financial exclusion. Impact investing can promote inclusive financial systems by providing access to savings, credit, insurance, and payment solutions. Technology, particularly fintech and mobile banking, has become a key enabler in reaching unbanked populations. Many emerging markets and developing economies (EMDEs) still struggle with limited financial access, highlighting the need for comprehensive policy frameworks, Public–Private Partnerships, and regulatory interventions to facilitate inclusive finance. To ensure that financial inclusion efforts are sustainable, they must be designed as long-term solutions rather than short-term fixes. A multi-stakeholder approach is crucial, with investors, policymakers, financial institutions, and civil society collaborating to dismantle systemic barriers to financial access. Education and financial literacy are pivotal in equipping individuals with the knowledge necessary for informed financial decision-making. While technology offers scalable and cost-effective solutions, community engagement remains essential to ensure that financial services are contextually relevant and accessible. Partnerships between governments, NGOs, and the private sector amplify impact, fostering economic resilience and disrupting poverty cycles.

The true transformative potential of impact investing lies in its ability to bridge the financing gap that hinders sustainable development. Directing capital towards underfunded sectors, underserved communities, and climate resilience initiatives necessitates a concerted effort from governments, private investors, and international organisations. Aligning capital with purpose-driven objectives provides a framework for impact investing to address economic inequality, financial exclusion, and environmental challenges. Integrating sustainable financial mechanisms, ethical investment instruments, and innovative technologies will ensure that impact investing continues to drive meaningful, systemic change. As the field matures, it possesses the potential to redefine global finance, demonstrating that investments can serve both economic prosperity and the broader well-being of society and the planet. Only through collective action can the full potential of impact investing be unlocked, ensuring that it becomes a cornerstone of a more inclusive and sustainable global economy.

Bibliography

Aalbers, M., & Haila, A.-K. E. (2018). The financialization of housing: A political economy approach. *Urban Studies, 55*(15), 1821–1835.
Abiona, O., & Koppensteiner, M. F. (2022). Financial inclusion, shocks, and poverty: Evidence from the expansion of mobile money in Tanzania. *Journal of Human Resources, 57*(2), 435–464.
Acumen. (2024). *Acumen announces $300M agricultural adaptation commitment*. Retrieved November 28, 2024, from https://acumen.org/news/acumen-announces-300m-agricultural-adaptation-commitment/
Adegbite, O., & Machethe, C. L. (2022). The impact of financial inclusion on the livelihoods of rural smallholder farmers in Nigeria. *African Finance Journal, 24*(2), 38–50.
Adeoye, O. B., Chigozie, A. E., Nwakamma, N.-E., Danny, J. M., Usman, F. O., & Olu-Lawal, K. A. (2024). A conceptual framework for data-driven sustainable finance in green energy transition. *World Journal of Advanced Research and Reviews, 21*(2), 1791–1801.
AgDevCo. (2023). *Annual Report 2023.*
Aggarwal, D., & Elembilassery, V. (2018). Sustainable finance in emerging markets: A venture capital investment decision dilemma. *South Asian Journal of Business and Management Cases, 7*(2), 131–143. https://doi.org/10.1177/2277977918774651
Agrawal, A. (2018). Effectiveness of impact-investing at the base of the pyramid: An empirical study from India. In *Social entrepreneurship and sustainable business models: The case of India* (pp. 207–246). Palgrave Macmillan.

N. Udohaya, *Impact Investing and Financial Inclusion*, Sustainable Development Goals Series,
https://doi.org/10.1007/978-3-031-87841-1

Agrawal, A., & Hockerts, K. (2021, April 03). Impact investing: Review and research agenda. *Journal of Small Business & Entrepreneurship, 33*(2), 153–181. https://doi.org/10.1080/08276331.2018.1551457

Akande, J., Hosu, Y., Kabiti, H., Ndhleve, S., & Garidzirai, R. (2023). Financial literacy and inclusion for rural agrarian change and sustainable livelihood in the Eastern Cape, South Africa. *Heliyon, 9*(6), e16330.

Akang, A. U. (2023). Regulatory compliance and access to finance: Implications for business growth in developing economies. *Sciental Journal of Education Humanities and Social Sciences, 1*(2), 8–23.

Alias, N. A. L. (2024). *The contribution of microfinance institutions to personal financial management*. Available at SSRN 4845831.

Allen, F., Qian, M., & Xie, J. (2019, July 01). Understanding informal financing. *Journal of Financial Intermediation, 39*, 19–33. https://doi.org/10.1016/j.jfi.2018.06.004

Andersson Djurfeldt, A. (2020, January 01). Gendered land rights, legal reform and social norms in the context of land fragmentation—A review of the literature for Kenya, Rwanda and Uganda. *Land Use Policy, 90*, 104305. https://doi.org/10.1016/j.landusepol.2019.104305

Angelides, P. (2000). *The double bottom line: Investing in California's emerging markets.* California State Treasurer's Office.

Antoninis, M., Alcott, B., Al Hadheri, S., April, D., Fouad Barakat, B., Barrios Rivera, M., Baskakova, Y., Barry, M., Bekkouche, Y., & Caro Vasquez, D. (2023). *Global education monitoring report 2023: Technology in education: A tool on whose terms?* UNESCO.

Ashta, A., Khan, S., & Otto, P. (2011, 03/03). Does microfinance cause or reduce suicides? Policy recommendations for reducing borrower stress. *Strategic Change, 24.* https://doi.org/10.2139/ssrn.1715442

Avery, C. (2001). Business and human rights in a time of change. In *Liability of multinational corporations under international law* (pp. 17–73). Brill Nijhoff.

Babalola, F. I., Mhlongo, N. Z., Obinyeluaku, M. I., Oladayo, G. O., & Daraojimba, C. (2023). Microfinance and economic empowerment in Nigeria: A critical review of impact and sustainability. *Finance & Accounting Research Journal, 5*(12), 381–404.

Babatunde, R. (2018). *Impact of remittances on food security and nutrition of migrant s household: Evidence from Nigeria.*

Bajulaiye, O., Fenwick, M., Skultetyova, I., & Vermeulen, E. P. (2020). *Digital transformation in the hedge fund and private equity industry* (Working Paper 2020–1). Lex Research Topics in Corporate Law & Economics.

Banerjee, S. B., & Jackson, L. (2017). Microfinance and the business of poverty reduction: Critical perspectives from rural Bangladesh. *Human Relations, 70*(1), 63–91. https://doi.org/10.1177/0018726716640865

Baptista, P., & Heitmann, S. (2010). *Unleashing the power of convergence to advance mobile money ecosystems*. IFC and the Harvard Kennedy School.

Beech, B. M., Ford, C., Thorpe, R. J., Jr., Bruce, M. A., & Norris, K. C. (2021). Poverty, racism, and the public health crisis in America. *Frontiers in Public Health, 9*, Article 699049. https://doi.org/10.3389/fpubh.2021.699049

Behr, D., Groussard, H., Khaitina, V., & Shen, L. (2023). *Women's land rights in sub-Saharan Africa: Where do we stand in practice*. World Bank Group Global Indicators Brief No. 23.

Bell, M. (2021). *Why ESG performance is growing in importance for investors*. Ernst & Young.

Bengo, I., Borrello, A., & Chiodo, V. (2021, March 06). Preserving the integrity of social impact investing: Towards a distinctive implementation strategy. *Sustainability, 13*, 2852. https://doi.org/10.3390/su13052852

Bernards, N. (2024, January 02). Where is finance in the financialization of development? *Globalizations, 21*(1), 88–102. https://doi.org/10.1080/14747731.2023.2222481

Bhandari, M. P. (2024, 2024, July 15). What is next for the sustainable development goals, what are the challenges concerning SDG 10—Reduced inequalities? *Sustainable Earth Reviews, 7*(1), 23. https://doi.org/10.1186/s42055-024-00093-8

Blimpo, M. P., & Cosgrove-Davies, M. (2019). *Electricity access in Sub-Saharan Africa: Uptake, reliability, and complementary factors for economic impact*. World Bank Publications.

Borges, P. (2007). *Women empowered: Inspiring change in the emerging world*. Rizzoli International Publications.

Bouri, A., Mudaliar, A., Schiff, H., Bass, R., & Dithrich, H. (2018). *Roadmap for the future of impact investing: Reshaping financial markets*. GIIN.

Brest, P. (2015). *Impact investing in developing countries: Legal institutions and work-arounds*. Stanford Law and Policy Lab.

Brest, P., & Born, K. (2013). Unpacking the impact in impact investing. *Stanford Social Innovation Review, 11*(4), 22–31.

Bruyn, S. (1991). *The field of social investment*. Cambridge University Press.

Bugg-Levine, A., & Emerson, J. (2011). *Impact investing: Transforming how we make money while making a difference*. Wiley. https://books.google.co.uk/books?id=LhxIwPYYhfEC

Bylander, M., Jacoby, L., Bradley, P., & Pérez, A. B. (2019). Over-indebtedness and microcredit in Cambodia: Moving beyond borrower-centric frames. *Development Policy Review, 37*, O140–O160.

Carbonara, N., & Pellegrino, R. (2020, February 01). The role of public private partnerships in fostering innovation. *Construction Management and Economics, 38*(2), 140–156. https://doi.org/10.1080/01446193.2019.1610184

Casasnovas, G., & Jones, J. (2022, September 01). Who has a seat at the table in impact investing? Addressing inequality by giving voice. *Journal of Business Ethics, 179*(4), 951–969. https://doi.org/10.1007/s10551-022-05154-6

Castellas, E.I.-P., Ormiston, J., & Findlay, S. (2018). Financing social entrepreneurship: The role of impact investment in shaping social enterprise in Australia. *Social Enterprise Journal, 14*(2), 130–155.

CGAP. (2020). *CGAP launches global microfinance survey to inform COVID-19 response.*

Chen, S., Song, Y., & Gao, P. (2023, November 01). Environmental, social, and governance (ESG) performance and financial outcomes: Analyzing the impact of ESG on financial performance. *Journal of Environmental Management, 345*, 118829. https://doi.org/10.1016/j.jenvman.2023.118829

Chiapello, E. (2020). Financialization as a socio-technical process. In *The Routledge international Handbook of financialization* (pp. 81–91). Routledge.

Christen, R. P., Lauer, K., Lyman, T., & Rosenberg, R. (2012). *A guide to regulation and supervision of microfinance* (p. 128). Consultative Group to Assist the Poor (CGAP).

Clark, C., Emerson, J., & Thornley, B. (2014). *The impact investor: Lessons in leadership and strategy for collaborative capitalism.* John Wiley & Sons.

Convergence. (2022). *Blended finance for health and education* (Data Brief, Issue).

Craven, M., Liu, L., Mysore, M., & Wilson, M. (2020). *COVID-19: Implications for business* (p. 8). McKinsey & Company.

Croxson, K., Frost, J., Gambacorta, L., & Valletti, T. (2022). Platform-based business models and financial inclusion: Policy trade-offs and approaches1. *Journal of Competition Law & Economics, 19*(1), 75–102. https://doi.org/10.1093/joclec/nhac010

Cuccia, A. (2023). Venture philanthropy. In *Encyclopedia of sustainable management* (pp. 3925–3930). Springer.

Dabla-Norris, E. (2015). *Causes and consequences of income inequality: A global perspective.* International Monetary Fund.

Dadush, S. (2012). Impact investment indicators: A critical assessment. In Davis et al. (Eds.), *Governance by indicators: Global power through quantification and rankings.* Oxford University Press.

Daggers, J. (2019). *'Solving' social problems with markets and measurement: A critical study of social impact investing.* Birkbeck, University of London.

Dalmia, N., & Kanodia, S. (2020). *EdTech in India: A turning point.* Omidyar Network India. Retrieved July 12, 2024 from, https://www.omidyarnetwork.in/insights/edtech-in-india-redseer-report-2019-20

de la Cuesta-Gonzalez, M., Paredes-Gazquez, J., Ruza, C., & Fernandez-Olit, B. (2021, February 01). The relationship between vulnerable financial consumers and banking institutions. A qualitative study in Spain. *Geoforum, 119*, 163–176. https://doi.org/10.1016/j.geoforum.2021.01.006

Deloitte. (2021). *Accelerating toward greater financial inclusion*. https://www2.deloitte.com/us/en/insights/industry/financial-services/alternative-data-innovation-financial-inclusion.html

Demirgüç-Kunt, A., & Klapper, L. F. (2012). *Measuring financial inclusion: The global findex database* (Working Paper 6025). World Bank Policy Research. https://papers.ssrn.com/sol3/papers.cfm?abstract_id=2043012

Demirgüç-Kunt, A., Klapper, L., Singer, D., & Ansar, S. (2022). *The global findex database 2021: Financial inclusion, digital payments, and resilience in the age of COVID-19* (The Global Findex Database). W. B. Publications. https://www.worldbank.org/en/publication/globalfindex/Report

Donovan, K. (2012). Mobile money for financial inclusion [Article|Book Chapter]. *Information and communications for development 2012: Maximizing mobile, 61*(1), 61–73.

Dorfleitner, G., & Nguyen, Q. A. (2024, July 01). Mobile money for women's economic empowerment: The mediating role of financial management practices. *Review of Managerial Science, 18*(7), 1807–1836. https://doi.org/10.1007/s11846-022-00564-2

Drucker, P. F. (1986). *Innovation and entre-preneurship: Practice and principles*. Harper & Row.

Duvendack, M., Palmer-Jones, R., Copestake, J., Hooper, L., Loke, Y., & Rao, N. (2011). *What is the evidence of the impact of microfinance on the well-being of poor people?* . EPPI-Centre, Social Science Research Unit, Institute of Education, University of London. https://www.givedirectly.org/pdf/DFID_microfinance_evidence_review.pdf

Ebrahim, A., & Rangan, V. K. (2014). What impact? A framework for measuring the scale and scope of social performance. *California Management Review, 56*(3), 118–141.

Eccles, R. G., Serafeim, G., Seth, D., & Ming, C. C. Y. (2013). The performance frontier: Innovating for a sustainable strategy: Interaction. *Harvard Business Review, 91*(7), 17–18.

Eijkman, F., Kendall, J., & Mas, I. (2010). Bridges to cash: The retail end of M-PESA. *Savings and Development, 34*(2), 219–252.

Ely, A., & Hearn, D. (2021). Impact investors need to share power, not just capital. *Stanford Social Innovation Review*. https://ssir.org/articles/entry/impact_investors_need_to_share_power_not_just_capital

Emerson, J., & Twersky, F. (1996). *New social entrepreneurs: The success, challenge and lessons of non-profit enterprise creation*. Roberts Foundation Homeless Economic Development Fund.

Enbridge. (2021). *Enbridge Publishes Sustainability-Linked Bond Framework*. Enbridge. Retrieved November 11, from, https://www.enbridge.com/media-center/news/details?id=123680&lang=en

Engel, S., & Pedersen, D. (2019). Microfinance as poverty-shame debt. *Emotions and Society, 1*(2), 181–196.

[Record #3513 is using a reference type undefined in this output style.]

Ermann, M. D., & Clements, W. H. (1984). The interfaith center on corporate responsibility and its campaign against marketing infant formula in the third world. *Social Problems, 32*(2), 185–196. https://doi.org/10.2307/800787

Escobar, A. (2011). *Encountering development: The making and unmaking of the third world* (Vol. 1). Princeton University Press.

Fanconi, P. A., & Scheurle, P. (2017). *Small money big impact: Fighting poverty with microfinance*. John Wiley & Sons.

Fareed, F., Gabriel, M., Lenain, P., & Reynaud, J. (2017). *Financial inclusion and women entrepreneurship: Evidence from Mexico*. OECD.

Fehling, M., Nelson, B. D., & Venkatapuram, S. (2013). Limitations of the millennium development goals: A literature review. *Global Public Health, 8*(10), 1109–1122.

Feyen, E., Frost, J., Gambacorta, L., Natarajan, H., & Saal, M. (2021). *Fintech and the digital transformation of financial services: Implications for market structure and public policy*. BIS papers.

Fielding, M., Davis, M., Weitz, N., Cummings-John, I., Hickey, A., Johnson, F., Senyagwa, J., Martinez, L., & Sun, M. (2015). *Agricultural investment and rural transformation: A case study of the Makeni bioenergy project in Sierra Leone*. JSTOR.

Financial Conduct Authority. (2023). *Sustainability disclosure requirements and investment labels*.

Findlay, S., & Moran, M. (2019). Purpose-washing of impact investing funds: Motivations, occurrence and prevention. *Social Responsibility Journal, 15*(7), 853–873. https://doi.org/10.1108/SRJ-11-2017-0260

Foreign Commonwealth & Development Affairs. (2023). *International development in a contested world: Ending extreme poverty and tackling climate change*.

Fortune Business Insights. (2025). *Crowdfunding market size, share & industry analysis, by type (equity-based, debt-based, blockchain-based, and others), by end-user (startups, ngos, and individuals), and regional forecast, 2024–2032*. https://www.fortunebusinessinsights.com/crowdfunding-market-107129

Fourmy, R. (2020). The impact of mobile money on remittance recipients' household welfare and education: Evidence from Kenya. In *Migration, remittances, and sustainable development in Africa* (pp. 157–176). Routledge.

Frede, J. (2021). *Impact value added (IVA)*. Available at SSRN 3868250.

Freshfields Bruckhaus Deringer. (2005). *A legal framework for the integration of environmental, social and governance issues into institutional investment.* European Company Law.

Friedman, M. (1970, September 13). A Friedman Doctrine: The social responsability of business. *The New York Times.* https://www.nytimes.com/1970/09/13/archives/a-friedman-doctrine-the-social-responsibility-of-business-is-to.html

Gabor, D. (2019). Securitization for sustainability. *Does it help achieve the sustainability development goals.* Heinrich Boll Foundation.

Gas, S. (2017, July 15). Mobile money, cashless society and financial inclusion: Case study on somalia and kenya. *Cashless Society and Financial Inclusion: Case Study on Somalia and Kenya.*

Gasper, D. (2019, May 04). The road to the Sustainable Development Goals: Building global alliances and norms. *Journal of Global Ethics, 15*(2), 118–137. https://doi.org/10.1080/17449626.2019.1639532

Geczy, C., Jeffers, J. S., Musto, D. K., & Tucker, A. M. (2021, November 01). Contracts with (Social) benefits: The implementation of impact investing. *Journal of Financial Economics, 142*(2), 697–718. https://doi.org/10.1016/j.jfineco.2021.01.006

Gianoncelli, A., & Gaggiotti, G. (2021). Managing for impact: The role of investors' networks in enhancing appropriate impact measurement and management practices. In A. Richards & J. Nicholls (Eds.), *Generation impact* (pp. 219–233). Emerald Publishing Limited. https://doi.org/10.1108/978-1-78973-929-920200022

GIIN. (2019). *The core characteristics of impact investing.* https://thegiin.org/publication/post/core-characteristics-of-impact-investing/

GIIN. (2021). *The state of impact measurement and management practice.*

GIIN. (2023). *Impact investing: A guide to this dynamic market.*

GIIN. (2024). *About GIIN.* Retrieved May 12, 2024 from, https://thegiin.org/about/#:~:text=Since%202009%2C%20the%20Global%20Impact,impact%20investing%20around%20the%20world.

Gittell, R., Magnusson, M., & Merenda, M. (2012). *The sustainable business case book.* Saylor Foundation.

Gond, J.-P., & Piani, V. (2012, March 01). Enabling Institutional investors' collective action: The role of the principles for responsible investment initiative. *Business & Society, 52*(1), 64–104. https://doi.org/10.1177/0007650312460012

GPFI. (2024). *G20 global partnership for financial inclusion action plan for micro, small, and medium enterprise financing.*

[Record #3361 is using a reference type undefined in this output style.]

GSMA. (2024). *State of the industry report on mobile money 2024.* https://www.gsma.com/sotir/

Guerin, I., Labie, M., & Morvant-Roux, S. (2018). Inadequate growth, over-indebtedness, and crises in microcredit: What have we learned? *Enterprise Development & Microfinance, 29*(2)118–132.

Gutterman, A. (2024). *Impact investment tools, structures and instruments.*

Gutterman, A. S. (2020). *What is impact investing*. Available at SSRN 3823837.

Hahn, T., McDermott, C., Ituarte-Lima, C., Schultz, M., Green, T., & Tuvendal, M. (2015, December 01). Purposes and degrees of commodification: Economic instruments for biodiversity and ecosystem services need not rely on markets or monetary valuation. *Ecosystem Services, 16*, 74–82. https://doi.org/10.1016/j.ecoser.2015.10.012

Hand, D., Sunderji, S., & Pardo, N. (2023). *Impact investing allocations, activity & performance*. Market GIINSight.

Hand, D., Sunderji, S., Ulanow, M., Remsberg, R., & Xiao, K. (2024). *State of the market 2024: Trends, performance and allocations*. https://thegiin.org/publication/research/state-of-the-market-2024-trends-performance-and-allocations/

Hart, S. L., & Ahuja, G. (1996). Does it pay to be green? An empirical examination of the relationship between emission reduction and firm performance. *Business Strategy and the Environment, 5*(1), 30–37.

Helwig, K., Hill-O'Connor, C., Mikulewicz, M., Mugiraneza, P., & Christensen, E. (2020). *The role of microfinance in climate change adaptation: Evidence from rural Rwanda*. Glasgow Caledonian University.

Hendriks, S. (2019). The role of financial inclusion in driving women's economic empowerment. *Development in Practice, 29*(8), 1029–1038.

Henry, C., Coleman, S., Orser, B., & Foss, L. (2022). Women's entrepreneurship policy and access to financial capital in different countries: An institutional perspective. *Entrepreneurship Research Journal, 12*(3), 227–262.

Hockerts, K., Hehenberger, L., Schaltegger, S., & Farber, V. (2022, September 01). Defining and conceptualizing impact investing: Attractive nuisance or catalyst. *Journal of Business Ethics, 179*(4), 937–950. https://doi.org/10.1007/s10551-022-05157-3

Holtslag, M., Chevrollier, N., & Nijhof, A. (2021). Impact investing and sustainable market transformations: The role of venture capital funds. *Business Ethics, the Environment & Responsibility, 30*(4), 522–537.

Hughes, N., & Lonie, S. (2007). M-PESA: Mobile money for the "unbanked" turning cellphones into 24-hour tellers in Kenya. *Innovations: Technology, Governance, Globalization, 2*(1–2), 63–81. https://doi.org/10.1162/itgg.2007.2.1-2.63

Hummels, H., & de Leede, M. (2014). The emergence of impact investments: The case of microfinance. In *socially responsible investment in the 21st century: Does it make a difference for society?* (Vol. 7, pp. 91–115). Emerald

Group Publishing Limited. https://doi.org/10.1108/S2043-905920140000007004

IFC. (2022). *Banking on SMEs: Driving growth, creating jobs global SME finance facility progress report.*

IFC. (2024a). *Microfinance.* https://www.ifc.org/en/what-we-do/sector-expertise/financial-institutions/archive-do-not-publish/msme-finance/microfinance

IFC. (2024b). *MSME Finance.* https://www.ifc.org/en/what-we-do/sector-expertise/financial-institutions/msme-finance

Ingale, A. (2024). *The role of microfinance in empowering women entrepreneurs.* Available at SSRN 5024302.

International Energy Agency. (2024). *World energy investment report.* https://www.iea.org/reports/world-energy-investment-2024

Iqbal, A., Hassan, S., Mahmood, H., & Tanveer, M. (2022, November 01). Gender equality, education, economic growth and religious tensions nexus in developing countries: A spatial analysis approach. *Heliyon, 8*(11), e11394. https://doi.org/10.1016/j.heliyon.2022.e11394

IRENA. (2023). *Renewables readiness assessment honduras.*

Jack, W., & Suri, T. (2011). *Mobile money: The economics of M-PESA.* NBER

Jansen, A., Wang, R., Behrens, P., & Hoekstra, R. (2024, September 01). Beyond GDP: A review and conceptual framework for measuring sustainable and inclusive wellbeing. *The Lancet Planetary Health, 8*(9), e695-e705. https://doi.org/10.1016/S2542-5196(24)00147-5

Jenik, I., & Lauer, K. (2017). *Regulatory sandboxes and financial inclusion.* CGAP

John, R. (2006). *Venture philanthropy: The evolution of high engagement philanthropy in Europe.*

Jones, G. G. (2016). *Joan Bavaria and multi-dimensional capitalism.* Harvard Business Review Press (China Case Studies).

Jordan-Kirwan, S., & Tengtio, K. (2020). *How to measure the gender impact of investments: Using the 2X challenge indicators in alignment with IRIS+.*

Kasyanenko, S., Kenworthy, P. G., Ruch, F. U., Vashakmadze, E. T., Vorisek, D. L., & Wheeler, C. M. (2023). *Investment in emerging market and developing economy regions: Trends, prospects, and policy options.* World Bank Group.

Khan, F. U., Nouman, M., Negrut, L., Abban, J., Cismas, L. M., & Siddiqi, M. F. (2024, December 31). Constraints to agricultural finance in underdeveloped and developing countries: A systematic literature review. *International Journal of Agricultural Sustainability, 22*(1), 2329388. https://doi.org/10.1080/14735903.2024.2329388

Khera, P., Ogawa, M. S., & Sahay, M. R. (2021). *Is digital financial inclusion unlocking growth?* International Monetary Fund.

Kikulwe, E. M., Fischer, E., & Qaim, M. (2014). Mobile money, smallholder farmers, and household welfare in Kenya. *PLoS ONE, 9*(10), Article e109804.

Kingiri, A. N., & Fu, X. (2020, January 02). Understanding the diffusion and adoption of digital finance innovation in emerging economies: M-Pesa money mobile transfer service in Kenya. *Innovation and Development, 10*(1), 67–87. https://doi.org/10.1080/2157930X.2019.1570695

Klapper, L., & Lusardi, A. (2020). Financial literacy and financial resilience: Evidence from around the world. *Financial Management, 49*(3), 589–614. https://doi.org/10.1111/fima.12283

Knoll, L., & Fraser, A. (2024, January 02). Social impact bond assetization struggles: A comparative case study of the United Kingdom and Germany. *Economy and Society, 53*(1), 92–111. https://doi.org/10.1080/03085147.2024.2309802

Knoll, M. S. (2002). Ethical screening in modern financial markets: The conflicting claims underlying socially responsible investment. *The Business Lawyer, 57*(2), 681–726.

Knox-Hayes, J. (2013, 2013/12/01/). The spatial and temporal dynamics of value in financialization: Analysis of the infrastructure of carbon markets. *Geoforum, 50*, 117–128. https://doi.org/10.1016/j.geoforum.2013.08.012

Konte, M., & Tetteh, G. K. (2023). Mobile money, traditional financial services and firm productivity in Africa. *Small Business Economics, 60*(2), 745–769.

Koo, Minji , & Kim, C.-S. (2023). The impact of digital transformation on financial inclusion: Through the case of mobile money account in senegal. *World Regional Studies Journal, 41*(3), 259–286. https://scholar.kyobobook.co.kr/article/detail/4010067732839

Koomson, I., Villano, R. A., & Hadley, D. (2020, June 01). Effect of financial inclusion on poverty and vulnerability to poverty: Evidence using a multidimensional measure of financial inclusion. *Social Indicators Research, 149*(2), 613–639. https://doi.org/10.1007/s11205-019-02263-0

Kumar, A. (2006). *Expanding bank outreach through retail partnerships: Correspondent banking in Brazil*. World Bank Publications.

Kumar, S. S., & Jie, Q. (2023, December 01). Exploring the role of financial inclusion in poverty reduction: An empirical study. *World Development Sustainability, 3*, 100103. https://doi.org/10.1016/j.wds.2023.100103

Kuokkanen, N. (2024, March 01). A problematizing review of the financialization of living beings. *Critical Perspectives on Accounting, 99*, 102739. https://doi.org/10.1016/j.cpa.2024.102739

Lal, T. (2021). Impact of financial inclusion on economic development of marginalized communities through the mediation of social and economic empowerment. *International Journal of Social Economics, 48*(12), 1768–1793. https://doi.org/10.1108/IJSE-12-2020-0830

Lapierre, D., & Moro, J. (2009). *Five past midnight in Bhopal: The epic story of the world's deadliest industrial disaster*. Hachette UK.

Larson, Z. (2020). The Sullivan principles: South Africa, apartheid, and globalization. *Diplomatic History, 44*(3), 479–503. https://doi.org/10.1093/dh/dhaa002

Liu, J., Chen, Y., Chen, X., & Chen, B. (2024). Digital financial inclusion and household financial vulnerability: An empirical analysis of rural and urban disparities in China. *Heliyon, 10*(15). https://doi.org/10.1016/j.heliyon.2024.e35540

Lu, Y., Naik, N. Y., & Teo, M. (2023). Diverse hedge funds. *The Review of Financial Studies, 37*(2), 639–683. https://doi.org/10.1093/rfs/hhad064

[Record #3438 is using a reference type undefined in this output style.]

Lyons, A. C., Kass-Hanna, J., Liu, F., Greenlee, A. J., & Zeng, L. (2020). *Building financial resilience through financial and digital literacy in South Asia and Sub-Saharan Africa.*

Ma, W., & Rahut, D. B. (2024, April 29). Climate-smart agriculture: Adoption, impacts, and implications for sustainable development. *Mitigation and Adaptation Strategies for Global Change, 29*(5), 44. https://doi.org/10.1007/s11027-024-10139-z

Mader, P. (2013). Rise and fall of microfinance in India: The Andhra Pradesh crisis in perspective. *Strategic Change, 22*(1–2), 47–66.

Mader, P., Mertens, D., & Van Der Zwan, N. (2020). Financialization: An introduction. In *The Routledge international Handbook of financialization* (pp. 1–16). Routledge.

Mahato, J., & Jha, M. K. (2023). Does financial inclusion promote sustainable livelihood development? Mediating effect of microentrepreneurship. *Journal of Financial Economic Policy, 15*(4/5), 485–499.

Manford, M. (2022). *Crisis in microfinance institutions in Ghana a case study of selected microfinance institutions in the Ga East Municipality.* University of Education, Winneba.

Maravalle, A., & Pandiella, A. G. (2022). *Expanding access to finance to boost growth and reduce inequalities in Mexico.* OECD.

Margolis, J. D., & Walsh, J. P. (2003). Misery loves companies: Rethinking social initiatives by business. *Administrative Science Quarterly, 48*(2), 268–305.

Marime-Ball, P., & Shaber, R. (2022). *The XX Edge: Unlocking higher returns and lower risk.* Simon and Schuster.

Market.US. (2023). *Neobanking market.* https://market.us/report/neobanking-market/#utm_source=scoop.market.us&utm_medium=END_FLOATING_A2

Martin, M. (2013). *Making impact investible.*

Mastercard. (2022). *Young women in africa: Agents of economic growth and transformation by 2030.*

Maula, M. V., & Lukkarinen, A. (2022). Attention across borders: Investor attention as a driver of cross-border equity crowdfunding investments. *Strategic Entrepreneurship Journal, 16*(4), 699–734.

Mbiti, I., & Weil, D. N. (2015). Mobile banking: The impact of M-Pesa in Kenya. In *African successes, Volume III: Modernization and development* (pp. 247–293). University of Chicago Press.

McKinsey. (2019). *Digital Identification: A key to inclusive growth.* https://www.mckinsey.com/capabilities/mckinsey-digital/our-insights/digital-identification-a-key-to-inclusive-growth

Mehra, R., Patel, P., Shetty, A., Golla, A., Kanesathasan, A., Jacobs, K., & Young, M. (2012). *Financial services for low-income women: Opportunities for economic empowerment*. International Center for Research on Women.

Mjuza, N., & Ntsalaze, L. (2019). Over-indebtedness of microfinance borrowers in South Africa. *African Review of Economics and Finance, 11*(2), 98–128.

Mohsen, S. E., Hamdan, A., & Shoaib, H. M. (2024). Digital transformation and integration of artificial intelligence in financial institutions. *Journal of Financial Reporting and Accounting, 23*(2), 680–699.

Moniruzzaman, M. (2022). The Impact of remittances on household food security: Evidence from a survey in Bangladesh. *Migration and Development, 11*(3), 352–371. https://doi.org/10.1080/21632324.2020.1787097

Morawczynski, O., & Pickens, M. (2012). *Poor people using mobile financial services: Observations on customer usage and impact from M-PESA*. World Bank.

Moretto, L., & Scola, B. (2017). *Development finance institutions and financial inclusion: From institution-building to market development.* CGAP Focus Note. https://www.cgap.org/sites/default/files/Focus-Note-DFIs-and-Financial-Inclusion-Mar-2017-rev_0.pdf

Mori, M. (2023). Urban-rural polarization: Financial implications, challenges, and opportunities. *Journal of Innovation in Business and Economics, 7*(02), 159–166.

Mothobi, O., Moshi, G., & Deen-Swarray, M. (2018). *Mobile money and financial inclusion in Sub-Saharan Africa countries*. CPR South.

Mudaliar, A., & Bass, R. (2017). GIIN perspectives: Evidence on the financial performance of impact investments. *Global impact investing network.*. Available at: https://thegiin.org/research/publication/financial-performance

Mukendi, S., & Manda, S. (2022). Micro-financial institutions and processes of women empowerment in Zambia. *World Development Perspectives, 28*, Article 100466.

Mulili, B. M. (2022). Digital financial inclusion: M-PESA in Kenya. In *Digital business in Africa: Social media and related technologies* (pp. 171–191). Springer.

Nachyła, P., & Justo, R. (2024, June 01). How do impact investors leverage non-financial strategies to create value? An impact-oriented value framework. *Journal of Business Venturing Insights, 21*, e00435. https://doi.org/10.1016/j.jbvi.2023.e00435

Nations, U. (2015). *The millennium development goals report 2015.*

Ndungu, N. (2021). *A digital financial services revolution in Kenya: The M-Pesa case study*. African Economic Research Consortium, Nairobi.

Nicholls, A. (2021). Impact measurement and management in sustainable finance. *Asian Development Outlook*, 1–54.

O'Brien, J. (2005). *Governing the corporation: Regulation and corporate governance in an age of scandal and global markets*. Wiley. https://books.google.co.uk/books?id=o5VEMbZQj94C

O'donohoe, N., Leijonhufvud, C., Saltuk, Y., Bugg-Levine, A., & Brandenburg, M. (2010). Impact investments: An emerging asset class. *JP Morgan, 6*, 3–91.

O'Reilly, C., Walsh, L., & Mottiar, Z. (2023, 05/17). Considerations for scaling a social enterprise: Key factors and elements. *The Irish Journal of Management, 42*. https://doi.org/10.2478/ijm-2023-0008

Odame, C., Appiah, K. O., & Gyimah, P. (2024). The financial inclusion–economic growth nexus: What is new now? *Journal of Money and Business, 4*(2), 121–137.

OECD. (2017). *Responsible business conduct for institutional investors: Key considerations for due diligence under the OECD Guidelines for Multinational Enterprises.*

OECD. (2019). Social impact investment 2019: The impact imperative for sustainable development. *O. Publishing*. https://doi.org/10.1787/9789264311299-en

OECD. (2022). *Policy brief on access to finance for inclusive and social entrepreneurship: What role can fintech and financial literacy play?*

OECD. (2024). *Global insurance market trends 2024*. O. Publishing. https://doi.org/10.1787/5b740371-en

OECD netFWD. (2014). *Venture philanthropy in development: Dynamics, challenges and lessons in the search for greater impact*. O. D. Centre.

Okello, C. B. G., & Ntayi, J. M. (2020). Mobile money adoption and usage and financial inclusion: Mediating effect of digital consumer protection. *Digital Policy, Regulation and Governance, 22*(3), 157–176. https://doi.org/10.1108/DPRG-01-2019-0005

Ondiege, P., Moyo, J. M., & Verdier-Chouchane, A. (2013). Developing Africa's infrastructure for enhanced competitiveness. *The Africa Competitiveness Report, 2013*, 69–91.

Orenstein, W., & Schneider, G. (2019). Enbridge's Line 3 project is becoming an issue in the 2020 presidential race. *MINNPOST*. https://www.minnpost.com/environment/2019/07/enbridges-line-3-project-is-becoming-an-issue-in-the-2020-presidential-race/

Orlitzky, M., Schmidt, F. L., & Rynes, S. L. (2003). Corporate social and financial performance: A meta-analysis. *Organization Studies, 24*(3), 403–441.

Ott, R. (2008). Not one drop: Betrayal and courage in the wake of the Exxon Valdez oil spill. Chelsea Green Publishing.

Pal, K. K., Piaget, K., Zahidi, S., & Baller, S. (2024). *Global gender gap report 2024.* World Economic Forum.

Pal, S., Vankila, S., & Fernandes, M. N. (2025, January 16). Interplay of financial inclusion and economic growth in emerging economies. *World Development Sustainability*, 100201. https://doi.org/10.1016/j.wds.2025.100201

Palandjian, T. (2017). *World's first social impact bond shows positive results.* https://socialfinance.org/insight/worlds-first-social-impact-bond-shows-positive-results/

Pattnaik, D., Ray, S., & Hassan, M. K. (2024, May 30). Microfinance: A bibliometric exploration of the knowledge landscape. *Heliyon, 10*(10), e31216. https://doi.org/10.1016/j.heliyon.2024.e31216

Pavicevic, O., Stepanovic, I., & Ilijic, L. (2021). Resilience in the times of Covid-19: Rethinking the neoliberal paradigm and creating new strategies for battling the global change. *Društvene i humanističke studije: časopis Filozofskog fakulteta u Tuzlil, 6*(3[16]), 165–182.

Perez, M. M., Patel, D. M., & Cuff, P. A. (2015). *Empowering women and strengthening health systems and services through investing in nursing and midwifery enterprise: Lessons from lower-income countries: Workshop summary.* National Academies Press.

Perrin, C., & Hyland, M. (2023). *Gendered laws and women's financial inclusion.* World Bank.

Pienitz, R., & Vincent, W. F. (2000). Effect of climate change relative to ozone depletion on UV exposure in subarctic lakes. *Nature, 404*(6777), 484–487.

Plokhy, S. (2018). *Chernobyl: The history of a nuclear catastrophe.* Basic Books.

Pomeroy, R., Arango, C., Lomboy, C. G., & Box, S. (2020). Financial inclusion to build economic resilience in small-scale fisheries. *Marine Policy, 118*, Article 103982.

Porteous, D. (2006). *The enabling environment for mobile banking in Africa.*

Preston, L. E., & O'Bannon, D. P. (1997). The corporate social-financial performance relationship: A typology and analysis. *Business & Society, 36*(4), 419–429.

Priyadarshee, A., & Ghalib, A. K. (2011). *The Andhra Pradesh microfinance crisis in India: Manifestation, causal analysis, and regulatory response* (Working Paper 157). Brooks World Poverty Institute.

Puaschunder, J. M. (2014). Ethical and Socially Responsible Investing. In *Investor Behavior* (pp. 513–532). https://doi.org/10.1002/9781118813454.ch28

Reisman, J., & Olazabal, V. (2021). Impact measurement and management techniques to achieve powerful results. In E. D. M. Sarmento & R. P. Herman (Eds.), *Global handbook of impact investing-solving global problems via smarter capital markets towards a more sustainable society* (pp. 667–696). Wiley.

Reuters. (2021). Enbridge raises funds from sustainability bonds in climate-goal push. *Reuters*. https://www.reuters.com/business/sustainable-business/enbridge-raises-funds-sustainability-bonds-climate-goal-push-2021-06-29/

Rizzi, F., Pellegrini, C., & Battaglia, M. (2018, January 01). The structuring of social finance: Emerging approaches for supporting environmentally and socially impactful projects. *Journal of Cleaner Production, 170*, 805–817. https://doi.org/10.1016/j.jclepro.2017.09.167

Robinson, J. A., & Acemoglu, D. (2012). *Why nations fail: The origins of power, prosperity and poverty*. Profile London.

Rogoff, K. S., & Reinhart, C. (2009). *This time is different*. Princeton University Press Princeton.

Root Capital. (2024). *Women in agriculture annual report October 2024*. Annual Reports. https://rootcapital.org/wai-report-2024/

Rosengard, J. K. (2016). *A quantum leap over high hurdles to financial inclusion: The mobile banking revolution in Kenya*. Harvard Kennedy School.

Roth, B. (2021). *Impact investing: A theory of financing social enterprises* (Working Paper 20–078). Harvard Business School Entrepreneurial Management.

Ruch, F. U. (2020). *Prospects, risks, and vulnerabilities in emerging and developing economies: Lessons from the past decade* (Working Paper 9181) World Bank Policy Research.

Sa-Dhan. (2022). *Promoting women-led enterprises: Microfinance leads the way conference on financial inclusion*.

Safari, L. I., Kilindo, A., & Nyoni, F. (2023). The effect of mobile money services usage on financial inclusion in nyamigamba village. *International Journal for Multidisciplinary Research, 5*(6). https://www.ijfmr.com/research-paper.php?id=9778

Sarmento, E. D. M., & Herman, R. P. (2020). Measuring and evaluating social impact in impact investing: An overview of the main available standards and methods. In E. D. M. Sarmento & R. P. Herman (Eds.), *Global handbook of impact investing: Solving global problems via smarter capital markets towards a more sustainable society* (pp. 555–666). John Wiley & Sons.

Schmidt, R. (2023, May 01). Are business ethics effective? A market failures approach to impact investing. *Journal of Business Ethics, 184*(2), 505–524. https://doi.org/10.1007/s10551-022-05133-x

Schueth, S. (2003). Socially responsible investing in the United States. *Journal of Business Ethics, 43*(3), 189–194. http://www.jstor.org/stable/25074988

Shome, S., Hassan, M. K., Verma, S., & Panigrahi, T. R. (2023, March 01). Impact investment for sustainable development: A bibliometric analysis. *International Review of Economics & Finance, 84*, 770–800. https://doi.org/10.1016/j.iref.2022.12.001

Siddik, M., & Kabiraj, S. (2018). Does financial inclusion induce financial stability? Evidence from cross-country analysis [Article]. *Australasian Accounting Business and Finance Journal, 12*(1), 33–46. https://doi.org/10.14453/aabfj.v12i1.3

Simba, A., Tajeddin, M., Dana, L.-P., & Ribeiro Soriano, D. E. (2024, January 01). Deconstructing involuntary financial exclusion: A focus on African SMEs. *Small Business Economics, 62*(1), 285–305. https://doi.org/10.1007/s11187-023-00767-1

Simon, J. G., Powers, C. W., & Gunnemann, J. P. (1972). *The ethical investor: Universities and corporate responsibility*. Yale University Press.

Simpson, N. (2023). *Line 3: An unthinkable pipeline powered by our cash*. bank.green. https://bank.green/blog/an-unthinkable-pipeline-powered-by-our-cash

Sinclair, S., McHugh, N., & Roy, M. J. (2021). Social innovation, financialisation and commodification: A critique of social impact bonds. *Journal of Economic Policy Reform, 24*(1), 11–27.

Sparkes, R. (2003). *Socially responsible investment: A global revolution*. John Wiley & Sons.

Sterner, T., Ewald, J., & Sterner, E. (2024, April 01). Economists and the climate. *Journal of Behavioral and Experimental Economics, 109*, 102158. https://doi.org/10.1016/j.socec.2023.102158

Subramanian, T., Muirow, A., & Anderson, J. (2022). Evolving the gender analysis in gender lens investing: Moving from counting women to valuing gendered experience. *Journal of Sustainable Finance & Investment, 12*(3), 684–703.

Sun, W. (2024). Unmasking the hidden risks in ESG reporting: How egoism distorts transparency and what can be done about it. *Research in Economics and Management Science, 1*(1).

Suri, T., & Jack, W. (2016, December 09). The long-run poverty and gender impacts of mobile money. *Science, 354*, 1288–1292. https://doi.org/10.1126/science.aah5309

Taconet, N., Mejean, A., & Guivarch, C. (2020). Influence of climate change impacts and mitigation costs on inequality between countries. *Climatic Change, 160*(1), 15–34.

The Rockerfeller Foundation. (2015). *City resilience framework*. https://www.rockefellerfoundation.org/wp-content/uploads/City-Resilience-Framework-2015.pdf

Thomson, E. (2023, July 16). Half-price energy bills if you live near a wind farm. Here's some incentives to accelerate the renewable energy transition. *Energy Transition—World Economic Forum.* https://www.weforum.org/stories/2023/05/renewable-energy-incentives-households-countries/

Thorarinsdottir, F. V., Óladóttir, Á. D., Darmstadt, G. L., & Benediktsdóttir, S. (2024). Gender lens investing: A scoping review from theoretical foundations to public equity applications. *Sustainability, 16*(20), 8916.

Tinsley, E., & Agapitova, N. (2018). *Private sector solutions to helping smallholders succeed: Social enterprise business models in the agriculture sector*. World Bank.

Tiwasing, P., Addae, J. A., Naab, F. Z., & Naab, G. Z. (2024). Do mobile money services enhance business performance? An empirical analysis of Kenyan businesses. *The International Journal of Entrepreneurship and Innovation*, 14657503241230946.

Tollison, R. D., & Wagner, R. E. (1988). *Social cost, rent seeking, and smoking: A public choice perspective*. Lexinton Books.

Tolliver, C., Keeley, A. R., & Managi, S. (2020, January 20). Drivers of green bond market growth: The importance of Nationally Determined Contributions to the Paris Agreement and implications for sustainability. *Journal of Cleaner Production, 244*, 118643. https://doi.org/10.1016/j.jclepro.2019.118643

Tran, O., & Oanh, K. (2023, November 09). Sustainable development: Driving force from the relationship between finance inclusion, green finance and green growth. *Sustainable Development, 32*. https://doi.org/10.1002/sd.2808

Trelstad, B. (2016). Impact investing: A brief history. *Capitalism & Society, 11*(2), 1–14.

UNEP FI & PRI. (2015). *Fiduciary duty in the 21st century.* Available at SSRN 2724866. https://www.unpri.org/fiduciary-duty/fiduciary-duty-in-the-21st-century-final-report/4998.article

UNEP Finance Initiative. (2005). *A legal framework for the integration of environmental, social and governance issues into institutional investment*.

United Nations. (2015). *Summit charts new era of sustainable development*. Retrieved September 18, from, https://www.un.org/sustainabledevelopment/blog/2015/09/summit-charts-new-era-of-sustainable-development-world-leaders-to-gavel-universal-agenda-to-transform-our-world-for-people-and-planet/

United Nations. (2024). *The sustainable development goals report 2024* (9210582594).

UNSCEB. (2017). *Leaving no one behind: Equality and non-discrimination at the heart of sustainable development*.

UNWOMEN. (2023). *Global gender equality in 2023: Urgent efforts needed to reach 2030 goals*. https://www.unwomen.org/en/news-stories/feature-story/2023/09/global-gender-equality-in-2023-urgent-efforts-needed-to-reach-2030-goals

Urban, B., & Mapula-e Lehasa, O. (2024). Investing in a social venture to generate social impact or financial return. *Business Perspectives and Research, 12*(4), 633–646. https://doi.org/10.1177/22785337221132614

Uzma, S. H., & Pratihari, S. K. (2019). Financial modelling for business sustainability: A study of business correspondent model of financial inclusion in India. *Vikalpa, 44*(4), 211–231.

van Zanden, J. L. (2023). Examining the relationship of information and communication technology and financial access in Africa. *Journal of Business and Economic Options, 6*(3), 26–36.

Waddock, S. A., & Graves, S. B. (1997). The corporate social performance—Financial performance link. *Strategic Management Journal, 18*(4), 303–319.

White, O., Madgavkar, A., Manyika, J., Mahajan, D., Bughin, J., McCarthy, M., & Spe, O. (2019). *Digital identification: A key to inclusive growth.* McKinsey.

White, O., Madgavkar, A., Townsend, Z., Manyika, J., Olanrewaju, T., Sibanda, T., & Kaufman, S. (2021). *Financial data unbound: The value of open data for individuals and institutions.* McKinsey Global Institute.

Winn, J. K. (2016). Mobile payments and financial inclusion: Kenya, Brazil, and India as case studies. In *Research handbook on electronic commerce law* (pp. 62–88). Edward Elgar Publishing.

Woetzel, J. (2015). *The power of parity: How advancing women's equality can add $12 trillion to global growth.* McKinsey Global Institute.

World Bank. (2014). *Financial inclusion* (Global Financial Development Report). T. W. B. Group. https://openknowledge.worldbank.org/handle/10986/16238

World Bank. (2017). *World development report 2017: Governance and the law.* The World Bank.

World Bank. (2019). *Disruptive technologies in the credit information sharing industry: Developments and implications.* World Bank.

World Bank. (2024). *Financial Inclusion : A key enabler to reducing poverty and boosting prosperity.* The World Bank. https://www.worldbank.org/en/topic/financialinclusion/overview

World Bank, UNESCO, UNICEF, FCDO, USAID, & Foundation, B. a. M. G. (2022). *The state of global learning poverty.*

World Commission on Environment and Development. (1987). *The Brundtland report: 'Our common future'.* Medicine and War.

World Health Organization. (2017). *WHO estimates cost of reaching global health targets by 2030.*

Yang, F., Ye, X., Huang, W., & Zhao, X. (2023, December 01). The impacts on informal financing strategy of small and micro enterprises by interest rate risks and public health emergencies. *International Entrepreneurship and*

Management Journal, 19(4), 1673–1705. https://doi.org/10.1007/s11365-023-00872-3

Yelnikova, Y. V., & Kwilinski, A. (2020). Impact-investing in the healthcare in terms of the new socially responsible state investment policy. *Business Ethics and Leadership, 4*(3), 57-64.

Young, R., & DAI Bruett, T. (2021). 3 How gender-lens investing measures up. *The catalytic effects of DFI investment–gender equality, climate action and the harmonisation of impact standards*, 30.

Yunus, M. (2007). *Banker to the poor*. Penguin Books India.

Zogning, F. (2023, January 02). Financial inclusion, inclusive entrepreneurship, and alternative financing options. *Journal of Small Business & Entrepreneurship, 35*(1), 8–13. https://doi.org/10.1080/08276331.2022.2120345

Index

N. Udohaya, *Impact Investing and Financial Inclusion*, Sustainable Development Goals Series,
https://doi.org/10.1007/978-3-031-87841-1

The manufacturer's authorised representative in the EU is Springer Nature Customer Service Centre GmbH, Europaplatz 3, 69115 Heidelberg, Germany. If you have any concerns regarding our products, please contact ProductSafety@springernature.com

Printed and bound by CPI Group (UK) Ltd, Croydon, CR0 4YY
30/09/2026
02225595-0012